Understanding Material Text Cultures

Materiale Textkulturen

Schriftenreihe des Sonderforschungsbereichs 933

Herausgegeben von
Ludger Lieb

Band 9

Understanding Material Text Cultures

A Multidisciplinary View

Edited by
Markus Hilgert

DE GRUYTER

ISBN 978-3-11-041785-2
e-ISBN (PDF) 978-3-11-041784-5
e-ISBN (EPUB) 978-3-11-042528-4
ISSN 2198-6932

Library of Congress Cataloging-in-Publication Data
A CIP catalog record for this book has been applied for at the Library of Congress.

Bibliografische Information der Deutschen Nationalbibliothek
Die Deutsche Nationalbibliothek verzeichnet diese Publikation in der Deutschen Nationalbibliografie; detaillierte bibliografische Daten sind im Internet über http://dnb.dnb.de abrufbar.

Einbandabbildung: Ausschnitt aus Guido Philipp Schmitt, Ruperto Carola (Ölgemälde, 1886)
Foto: © Atelier Altenkirch, Karlsruhe
Druck und Bindung: Hubert & Co. GmbH & Co. KG, Göttingen
∞ Gedruckt auf säurefreiem Papier
Printed in Germany

www.degruyter.com

Contents

Markus Hilgert

A Multidisciplinary View on Material Text Cultures

Editor's Preface

Generously funded by the Deutsche Forschungsgemeinschaft, the Collaborative Research Center (CRC) 933 "Material Text Cultures: Materiality and Presence of Writing in Nontypographic Societies" was established at Heidelberg University in 2011, in order to promote interdisciplinary research on the material and topological settings of writing in nontypographic societies and the social practices of reception presumably prompted by these settings. Since then, operating within a theoretical framework specifically created for this purpose[1] and using an innovative descriptive vocabulary,[2] researchers from more than 20 disciplines of the Humanities have been able to shed new light on how writing was conceptualized, materialized, and contextualized in societies without widespread means of mass-producing inscribed objects, thereby contributing significantly to our understanding of the material text cultures privileged by these societies.

In order to introduce material text culture research as advanced by the CRC 933 on an international level and to provide young scholars from all over the world with the opportunity to acquaint themselves with the theoretical setting and the research strategies of the research center, the executive board of the CRC 933 decided in 2012 to stage an international competition for 6 research fellowships to be awarded to outstanding young researchers. Applications were invited for innovative, high-risk research proposals pertinent to the CRC 933's overall research scheme. The strategic aim of this decision was threefold. First of all, it was felt that a research center characterized mainly by an interdisciplinary research design should also pay attention to and promote smaller, single-disciplinary projects located outside of, but highly pertinent to the CRC 933, thus adding a multidisciplinary dimension to material text culture research. Second, there existed a curiosity as to the applicability of the theoretical premises and methodology developed and tested by the CRC 933 to research on inscribed artefacts carried out on an international level and in different research environments. Finally, it was hoped that the research made possible by the CRC's research fellowships would contribute to anchoring material text culture research as proposed by the CRC 933 within the tradition and broader context of other research strategies devoted to the material dimension of writing, such as the *filologia materiale*.

1 Hilgert 2010; Hilgert 2014.

2 Ibid.

DOI 10.1515/9783110417845-001,

The present volume comprises 6 highly original studies on material text cultures in different nontypographic societies stretching from the 3rd millennium cuneiform textual record of Ancient Mesopotamia to 20th century Qur'anic boards of northern and central African provenience. Thus, the volume provides a multidisciplinary approach to material text cultures complementary to the interdisciplinary, strongly theory-grounded research scheme of the CRC 933. It illustrates that the questions and aims driving research within the CRC are valid even outside this academic environment and may generate new, exciting research on different subjects and strongly varying evidence.

The first article presented here is an excellent example of the volume's breadth of scope. In her stimulating study "Defining Collectives: Materialising and Recording the Sumerian Workforce in the Third Dynasty of Ur" (p. 5–30), Agnès Garcia-Ventura combines the formal and material analysis of cuneiform administrative records of the late 3rd millennium BCE with "gender studies and, more specifically, feminist epistemologies and postfeminism" (p. 6). Using this innovative approach, Garcia-Ventura proposes "a new way of reading work groups as they are registered in Ur III texts, paying attention to their similarities and differences and focusing less on biological or sexual ties" (p. 6).

Nathan Morello's investigation entitled "A GIŠ on a Tree: Interactions between Images and Inscriptions on Neo-Assyrian Monuments" (p. 31–68) addresses some of the most salient research problems of CRC 933, as it focuses "on the analysis of visual and semantic interactions between images and inscriptions that occur in some Neo-Assyrian monuments when part (i.e. one or more cuneiform signs) of an inscription interplays with the part of the sculpted image (i.e. one element of the image or part of it) that it crosses" (p. 31): "Is it possible to positively identify interactions of this kind? Is it possible to interpret their meaning and to understand their function (or functions)? Were they meant to be a sort of concealed 'game' within the inscription, or were they meant to be seen? And once they were seen, who was the intended recipient, the audience, of such interactions?" (p. 31). Morello is able to define a "complex technique of monument manipulation" (p. 62) that creates "a new level of perception of the work of art, one that is not only visual or literary, but a combination of the two, a combination that produces a new message" (p. 62).

In "From Voice to Papyrus to Wall: *Verschriftung* and *Verschriftlichung* in the Old Kingdom Pyramid Texts" (p. 69–130), Antonio J. Morales deals with the "process of emergence and development of the Pyramid Texts from their oral form to their inscription in the chambers of the late Old Kingdom pyramids of kings and queens" (p. 71). Morales argues in favor of a "re-interpretation of long-standing assumptions on the origin and development of the Pyramid Texts" (p. 119), as he presents evidence for the fact, "that the rituals represented by the Pyramid Texts were already in use by the community before theologians and editors in Heliopolis planned the monumentalization of a king's pyramid with fixed recitations" (p. 71). He furthermore suggests that "the royal corpus incorporated not only mortuary service and temple materials, but also other types of recitations associated with magical practices, guilds' ceremonies, local festivities, and even arcana" (p. 120).

Analysing "Family Cult Foundations in the Hellenistic Age: Family and Sacred Space in a Private Religious Context" (p. 131–202), Sara Campanelli examines five inscriptions from the south-Aegean Doric area with a view to "an aspect that has hardly been touched on in studies on this topic, namely, the physical spaces as an integral part of the foundation system" (p. 133). Campanelli attempts to "look more specifically into how the spaces were meant to be used in material and conceptual terms and to see how they contributed to defining group identity" (p. 133), in order to show "how and to what extent inscriptions can be used as sources for reconstructing the architectural layout of places known only from epigraphic records" (p. 134). As one result of her investigation, Campanelli concludes that "in terms of family self-preservation, a foundation might have been intended to ensure the legitimate transmission of the inheritance under the aegis of the 'ancestral' gods and family heroes. This is all the more likely if one considers that the cult places were themselves an integral part of family assets and were used in most cases as sources of income" (p. 193f.).

Another research area of significant importance for the CRC 933 is covered by Flavia Manservigi's and Melania Mezzetti's study on "The Didyma Inscription: Between Legislation and Palaeography" (p. 203–242). Based on an inscription found in the town of Didyma (Caria) in 1991 and containing a rescript of Justinian I dated to 533 CE, the authors' aim is "to investigate the shift from a majuscule to a minuscule writing system in the Roman world, which started in the 3rd century A.D." (p. 203). Carrying out a scrupulous paleographic analysis, Manservigi and Mezzetti are able to show that "the opening lines of the *gesta praefectoria* of the Didyma inscription can be considered a missing link between the opening script of the reports of proceedings from Egypt and those of the Ravenna papyri" (p. 234). They also contend that "the main function of the inscription, that is, to show a public message to the people, failed, and [that] the writing was the bearer of another function, which was to grant authenticity to a document" (p. 235).

The volume concludes with Anastasia Grib's fascinating investigation into "The Symbolic Repertoire of the Qur'anic Board in Islamic Africa" (p. 243–278). Grib "provides a transcription of the symbolic code of the Qur'anic board based on the study of 124 samples from the Brooklyn Museum, the Gallery of Sam Fogg, the Musée du quai Branly, and other collections" (p. 243). Her aim is to decode "the symbolic language of the MCQ [i.e. Material Culture of the Qur'an] in the local Islamic centres of West and North Africa, where the most striking example of the MCQ is the Qur'anic board *allo*" (p. 244). Among the various results of the study is the observation that "in West and North Africa, one finds a semantic unity between two main types of objects belonging to the Material culture of the Qur'an: the Qur'anic manuscripts and the Qur'anic boards" (p. 277).

The research fellowships of the CRC 933 were awarded with the stipulation that each fellow gives a public lecture and teaches a seminar on their chosen research topic at Heidelberg University. Both lectures and seminars have turned out to be highly stimulating for the CRC's researchers and created new cooperations likely to spark new material text culture research. Therefore, it is the editor's hope that the present volume is not only witness to the dynamic the CRC 933 has created within his field of investigation, but will also serve as a basis for further research on the numerous intriguing questions raised here.

Literature

Hilgert, Markus (2010), "'Text-Anthropologie': Die Erforschung von Materialität und Präsenz des Geschriebenen als hermeneutische Strategie", in: Markus Hilgert (ed.), *Altorientalistik im 21. Jahrhundert. Selbstverständnis, Herausforderungen, Ziele* (Mitteilungen der Deutschen Orient-Gesellschaft 142), Berlin, 85–124.

Hilgert, Markus (2014), "Praxeologisch perspektivierte Artefaktanalysen des Geschriebenen. Zum heuristischen Potential der materialen Textkulturforschung", in: Friederike Elias, Albrecht Franz, Ulrich W. Weiser and Henning Murmann (eds.), *Praxeologie. Beiträge zur interdisziplinären Reichweite praxistheoretischer Ansätze in den Geistes- und Sozialwissenschaften* (Materiale Textkulturen 3), Berlin, 149–164.

Agnès Garcia-Ventura

Defining Collectives: Materialising and Recording the Sumerian Workforce in the Third Dynasty of Ur

Kinship is one of the main aspects that scholars have borne in mind in their attempts to characterise the workforce recorded in administrative texts of the Mesopotamian Third Dynasty of Ur (ca. 2112–2004 BCE). In my view, although some of the hypotheses proposed regarding marital status and filiation can improve our understanding of these texts, others risk distracting our attention from possible complementary or alternative readings.

The twentieth-century scholars who first proposed the translations of Sumerian terms that are currently quoted and used[1] began by considering the biological family and kinship as the main structuring institutions; only rather later did they begin to consider other ways of defining collectives and the workforce. In fact, their first translations may actually have been more a reflection of their own context than of the context they meant to describe. Aware of this possible bias, I propose to reassess some of these terms in an attempt to identify what they highlight, and consequently what they tell us about the ways in which work collectives were built and perceived.

Certain studies of the organisation of work and society in the Mesopotamian Third Dynasty of Ur give the impression that it was arranged on the basis of the heterosexual couple, the typical situation in most of the societies which have carried out research into Assyriology. This means that at different levels—that is, the biological level (for the production of new members of society), and the social or administrative level (for the production and distribution of goods)—the nuclear family is regarded as the foundation. However, extended families might also be focal points for the organisation of labour: great organisations like Mesopotamian temples and palaces could

This article emerged from the Heidelberg Collaborative Research Center 933 "Material Text Cultures. Materiality and Presence of Writing in Non-Typographic Societies". The CRC 933 is financed by the German Research Foundation (DFG). I wish to thank Érica Couto-Ferreira and Lorenzo Verderame for their valuable comments on the first version of this paper. Obviously all remaining errors and omissions are my own responsibility.

1 See the following sections for references—some of them merely quoted, others discussed in more detail. The transliteration of terms follows the Assyriological form: spacing for Sumerian, italicised for Akkadian. Abbreviations follow the usage of the *Reallexikon der Assyriologie und vorderasiatischen Archäologie*. A complete list is also available at http://cdli.ox.ac.uk/wiki/doku.php?id=abbreviations_for_assyriology (last accessed: December 2013). However, to simplify references for readers unfamiliar with the discipline, I include in the footnotes the reading of the abbreviation the first time it is quoted in the paper, together with its correspondence to the author and year as quoted in the final bibliography list.

DOI 10.1515/9783110417845-002,

be considered as extended families, reproducing the relationships of patriarchy and dependency that typify the family defined from the purely biological point of view.

Taking these different levels of social organisation into account, in this paper I intend to show that a careful study of the sources can alert us to the presence of collectives that are larger than the nuclear family and are not based solely on biological ties. In order to do this, I concentrate on a sample of texts dealing with textile production, one of the industries that flourished in southern Mesopotamia at the threshold of the third and second millennia BCE.

The paper is organised in four sections, plus some concluding remarks. First, I offer a short introduction to the context and sources of the Mesopotamian Third Dynasty of Ur. Second, I briefly present the theoretical framework used—that is, gender studies and, more specifically, feminist epistemologies and postfeminism. Third, I discuss how different premises have been used to analyse groups of both males and females, and how these analyses have led to different results, despite the fact that the evidence from the texts from different periods and the evidence we can deduce from archaeological records do not vary greatly. Fourth, I propose a new way of reading work groups as they are registered in Ur III texts, paying attention to their similarities and differences and focusing less on biological or sexual ties.

1 The Third Dynasty of Ur: Historical Context and Sources

The period known as the Third Dynasty of Ur, or Ur III, lasted for roughly 100 years. The exact chronology is still debated, but one of the most widely accepted possibilities is the period between 2112 and 2004 BCE.[2] We know the names of five monarchs who ruled the southern part of ancient Mesopotamia in this period: Ur-Namma (who ruled for 18 years), Šulgi (with the longest reign, 48 years), Amar-Suena (nine years), Šu-Suen (nine years) and Ibbi-Suen (25 years). From this period of approximately 100 years we have a large number of cuneiform texts written in Sumerian, most of them administrative texts. The data vary depending on the study we cite; however, taking BDTNS[3] as a reference, around 120,000 tablets from this period have been unearthed,

---

2 Among the relative chronologies, 2112–2004 BCE is the one that is most widely supported and is quoted by, among others, Marc Van de Mieroop 2004, 282, Jacob Dahl 2007, 2, and Piotr Michalowski 2011, 1. However, Walther Sallaberger proposed some slight differences in his Ur III reference volume (Sallaberger 1999, 123–124) where he suggested the chronology 2111–2003 BCE, and more recently 2110–2003 BCE (Sallaberger 2004, 42).

3 The Online Database of Neo-Sumerian Texts: http://bdtns.filol.csic.es/ (last accessed: December 2013).

of which roughly 80,000 have been published so far.[4] Since this documentation is highly standardised, we can focus on only a few examples in order to broaden our understanding of the whole group.

Usually, the high volume of texts and the information they contain are interpreted as indications of the degree of centralisation and bureaucratisation[5] of the Ur III period. But there are other possibilities: this bureaucracy might have emerged due to institutional mistrust of administrative staff who were breaking the rules for their own personal benefit. If so, bureaucracy would have been a tool for preventing fraud.[6]

In any case, administrative texts are numerous in Ur III. Some of them are related to the textile sector.[7] There are lists showing the names of textile workers organised in teams under a supervisor, sometimes indicating the sum paid for the work. There are also lists of the types of cloth to be sold to the wealthier classes or to be traded abroad. Other administrative texts contain information on the stages of production and include terms related to tasks prior to spinning and weaving, such as sheep shearing and the initial work on the wool. Thus the administrative and economic texts provide both numerical data (prices, wages, the number of workers involved in each task), information on the quality of the cloth, and other data of a social nature, such as the origin of the workforce and the conditions in which they lived and worked.

Finally, we should note that this vast body of written material is not without its limitations as a source for research. One of them is that most clay tablets come from looting and illegal digs; as nothing is known of their archaeological context, a great deal of information is lost. Examples of this are Umma and Puzriš-Dagan, the two main settlements from which most Ur III tablets come, but which have never been systematically excavated.[8]

---

4 Molina 2008, 20.

5 For a reflection on the use of the word "bureaucracy" when referring to Ur III administration, see Civil 1987, 43.

6 Warburton 2005, 174.

7 The main reference on the Ur III textile sector today is the volume edited by Hartmut Waetzoldt in 1972. Subsequent studies have added more information or have tried to summarise and interpret some of the data that Waetzoldt discusses. With a specific focus on the workforce see, among others, Firth 2013; Garcia-Ventura 2012a, 2012b, 2013, 2014a and 2014b; Maekawa 1980, 1989 and 1998; Uchitel 1984 and 2002; Verderame 2008; Verderame/Spada 2013; Waetzoldt 1987, 1988 and 2011; Wright 1996, 1998 and 2008.

8 Zettler 2003, 59–61.

2 Theoretical Framework: Feminist Epistemologies and Postfeminism

The starting point for my analysis of the texts is the debate proposed above all by so-called "feminist epistemologies". The aim of this debate is to question and rethink the process of knowledge production and the choice of topics of study in an attempt to include women in historical discourses, avoiding the preconceptions that have led to androcentric interpretations and conclusions.[9]

In addition, I also start from some debates arising from postfeminism—more specifically, the academic postfeminism that is linked to the principles of post-modernism. I will refer especially to some of the proposals made by Judith Butler.

Of course, the label "postfeminism" groups together diverse, even contradictory schools of thought and trends.[10] Some postfeminist proposals, which are particularly relevant to the topic of this paper, deal with the notion of kinship understood as a cultural entity and not as a natural one. Usually, we tend to define kinship relationships as those established through sexual ties (basically through heterosexual marriage) or biological ties (basically through filiation), and then reduce kinship to them. However, taking into account these proposals and placing the emphasis on friendship and solidarity networks unrelated to sexual and biological ties creates new perspectives of analysis. In this regard the notion of "homosociality"[11] is interesting, referring as it does to same-sex relationships and ties that are not necessarily of a sexual or romantic nature.

The definition of kinship as surpassing biological and sexual ties was an issue raised by thinkers like Gayle S. Rubin and Monique Wittig. In her classic 1975 essay *The Traffic in Women: Notes on the 'Political Economy'*,[12] Rubin analysed the concepts of "family" and "kinship" and discussed some proposals from Marxism (mainly the proposals of Marx and Engels) and from psychoanalysis (mainly from Lacan). Rubin insisted on the concept of kinship defined more as a social entity than as a biological one. For her part, Wittig proposed an idea of heterosexuality not merely as a sexual option but as a political one as well.[13]

---

9 As a classic reference see Alcoff/Potter 1993, including papers by (among others) Helen Longino and Sandra Harding.

10 For a good summary of the diverse trends of postfeminism, see Genz/Bravon 2009 (especially pp. 106–131 for a presentation of "Postmodern (Post)feminism" and "Queer (Post)feminism", chapters 5 and 6). For a summary from a sociological perspective, see Seidman 2008, 235–249.

11 On their potential application to the analysis of ancient Near Eastern written sources, see Lion 2007, 59–64. On their application to archaeological analysis, see Voss 2012.

12 Rubin 1975, 169 and 179–180. See also Rubin 2011, 33–65 for a compilation of her essays, including the one referenced here.

13 Wittig's best-known essay, *The Straight Mind*, was first published in 1980 in the first number of the journal *Feminist Issues* and had previously been read as a lecture in New York in 1978. However, the

More recently, taking these first proposals into account, Judith Butler[14] proposed to define kinship as follows:

> If we understand kinship as a set of practices that institutes relationships of various kinds which negotiate the reproduction of life and the demands of death, then kinship practices will be those that emerge to address fundamental forms of human dependency, which may include birth, child rearing, relations of emotional dependency and support, generational ties, illness, dying, and death (to name a few). Kinship is neither a fully autonomous sphere, proclaimed to be distinct from community and friendship—or the regulations of the state—through some definitional fiat, nor is it 'over' or 'dead'.[15]

Finally, to close this section, I should mention another issue that has been extensively discussed within feminist epistemologies and postfeminism, particularly by Butler herself—the very definition of "woman" as a category of analysis.[16] It is a problematic issue, especially for feminist epistemologies: while questioning certain terms and their uses, they defend and try to deconstruct "women" as a category of analysis.[17] To quote Donna Haraway, in an extract that also alludes to race and class and proposes that all three are culturally constructed:

> There is nothing about being «female» that naturally binds women. There is not even such a state as 'being' female, itself a highly complex category constructed in contested sexual scientific discourses and other social practices. Gender, race, or class-consciousness is an achievement forced on us by the terrible historical experience of the contradictory social realities of patriarchy, colonialism, and capitalism.[18]

At this point, as occurs with "race" as a category of analysis in postcolonial studies, the aim of feminist epistemologies is that the term "woman" should be discussed as a category of analysis in such a manner as to make it wither away. Wittig summarises it thus in her essay *The Straight Mind*:

> 'Man' and 'woman' are political concepts of opposition, and the copula which dialectically unites them is, at the same time, the one which abolishes them. It is the class struggle between women and men which will abolish men and women.[19]

---

classic edition usually quoted is that of 1992, a volume that includes several more essays by Wittig. I quote only the 1992 reference in the final bibliography list and I use it for further citations.

14 Butler 2004, 25–35 and 102–105.

15 Butler 2004, 104.

16 Butler 1990, 3–9.

17 cf. Haraway 1991a, 148: "the identity of 'woman' is both claimed and deconstructed simultaneously".

18 Haraway 1991b, 155.

19 Wittig 1992, 29.

3 Homosociality versus the Harem: Two Patterns for Interpreting Groups

Ur III administrative texts contain two Sumerian words that identify the two main groups of low-ranking workers: $geme_2$ (female workers) and guruš (male workers). Their belonging to a group is usually defined by first considering their family context, understood as being based on biological and sexual ties. However, in this study I propose to analyse it by seeing the group as based on solidarity and kinship networks, which are not totally dependent on biology. In addition, as far as women are concerned, the only collectives considered in the secondary literature (besides the family) are harems. In this section I will discuss both models, focusing on homosociality and solidarity networks to explore the shaping of workers' groups and presenting some criticisms of the specific way in which women's collectives are treated in the secondary literature.

3.1 Homosociality and Solidarity Networks

Textile workers (mainly females) were organised in teams of varying sizes, which shared the same supervisor and performed similar day-to-day tasks. It is clear from the data compiled in the workers' lists that they spent many hours together. Sometimes they were listed together with children, probably their offspring.[20] Therefore, although in these administrative texts their existence is not explicitly stated, it is likely that in this context personal relationships and solidarity networks would have been key elements in these working groups.

One argument in favour of this hypothesis is the fact that female solidarity has supported both productive and reproductive work over the ages.[21] Probably for this reason, it has been feminist thinkers who have stressed the existence of solidarity networks as a characteristic of female work environments. Obviously this does not mean that I defend an essentialist view of female and male work environments—only that certain features of stereotypical femininity and stereotypical masculinity have been constructed, in certain working environments. For example, in the tobacco industry in nineteenth-century Spain, female cigar makers worked in groups, whereas male employees in the same factories performed more individual tasks.[22]

Indeed, throughout history, solidarity networks have traditionally been associated with activities considered pertinent to women's arenas.[23] Stories passed down in

20 For examples of texts, see the following section (number 4).

21 Juliano 1998.

22 Gálvez Muñoz 2000, 237.

23 Juliano 1998, 79–84.

the oral tradition show how certain conceptions of kinship and homosociality operate in women's groups. Today, most of these stories are regarded as oppressive towards women. Nevertheless a closer analysis suggests that, on occasion, they may present opportunities for women: indeed, women have traditionally been the main transmitters of stories, which have sometimes been used to claim a space for freedom built upon solidarity networks.[24] An example is one of the possible interpretations of the *Sleeping Beauty* story: the princess grows up outside the heterosexual nuclear family, in a context in which she is cared for by fairies (all female). This tale is a good example of kinship based not only on biological or sexual ties—an example of a homosocial context.

However, if we go back to ancient Mesopotamia and look now at male workers, the Ur III administrative texts recording only the male workforce (identified as guruš [low rank "male workers"] or $erin_2$ ["male gangs"]) include the same information as those recording only female workers. In other words, the workforces are recorded without any information about the sexual or biological ties among the individuals listed. As in lists of female workers, children are sometimes listed with male workers. As the data and the format of texts are the same, it seems plausible that we will also find examples of solidarity networks and homosociality as well. At a much later date, in eighteenth-century Turin, spinners working in textile production were mainly male: sources show that these men lived outside biologically based family structures, in households composed of fellow workers and neighbours.[25] This can be considered as another example of homosociality.

Another useful example, despite the even greater geographical and chronological distance from the texts of Mesopotamian Ur III, is the analysis of Chinatown in San José, California, at the end of the nineteenth century. The process was characterised by a clear sexual segregation: while the women stayed in China, the men emigrated to the US in order to trade, and created new communities abroad. This situation is not exceptional, as in fact sexual segregation is frequent in migratory processes. Barbara Voss[26] used this example to show how placing the focus on homosociality leads to interesting results. Voss's study shows that we can find evidence of solidarity networks that are usually hidden if the main research question focuses on traditional kinship (i.e., based on sexual and biological ties) and heterosexual relationships. In this regard, Voss notes that when we are asked about the frequency of homosexuality in homosocial contexts we fall into the trap of again imposing political content on the sexual options, as the question is heavily loaded.[27]

---

24 Juliano 1992, 53–54.
25 Carbonell 1997, 52–53.
26 Voss 2012. Cf. Hall 2012.
27 Voss 2012, 187–188.

Moving back to the ancient Near East, it seems logical that solidarity relationships were common not only inside same-sex groups of workers, but among groups containing both males and females. If we focus on status instead of sex or gender in our research questions, then certain situations can be more easily explained. Solidarity networks seem to have worked better among workers of the same or similar status (regardless of their sex) than among workers of the same sex but different rank. A similar panorama is shown by María Rosa Oliver and Eleonora Ravenna in their analysis of certain aspects of Old Babylonian society in the light of some articles in the Code of Hammurabi (reign 1792–1759 BCE, according to the middle chronology).[28] As with the analysis of some Ur III texts, when the starting point is the interest in women in Antiquity and the construction of gender identity, paradoxically, we realise that concentrating on sex and gender alone cannot reflect what is in fact a very complex picture. To achieve our goal we need to combine several factors and apply an intersectional approach. In fact, intersecting gender and class usually produces stimulating results.

All of these reflections suggest that kinship (defined in a broad sense) and homosociality affected males and females of similar status interacting in a variety of arenas. Take, for example, the monastery, an institution that is closer to us as a cultural model and has survived over the centuries. Monastic communities are sexually segregated, and among same-sex groups (i.e., monks and nuns) there arise relationships of solidarity that are not necessarily based on sexual or biological ties. Bridging the gap, probably the females designated *nadītu* in ancient Mesopotamia also lived together in communities, i.e. in communities that were sexually segregated as well.[29] Moreover, even in these sexually segregated contexts there are examples of solidarity networks between male and female communities or individuals. The friendship and collaboration among San Juan de la Cruz and Santa Teresa de Jesús in sixteenth-century Spain is one of the most emblematic examples.

Of course the performance of certain duties—for instance, some of those related to textile work—involved physical proximity and cooperation, which would have facilitated the emergence of these solidarity relationships.[30] It is also interesting to analyse this phenomenon through Butler's thoughts on how subjects are constituted. Butler explains what she terms "stubborn attachment" as the preference we have for being part of a collective (good or bad, better or worse) instead of being part of nothing.[31] Difference feminism also describes this condition as a feeling of lack, and a

---

28 Oliver/Ravenna 2001, especially p. 250.

29 For an overview on *nadītu*, with further references, see Sallaberger/Huber Vuillet 2005, 633–634.

30 Naji 2009.

31 Butler 1997, 31–62.

consequent ongoing search that generates the wish to transcend ourselves through the relationships we establish.[32]

Thus, even assuming that many of the female and male workers registered in Ur III texts were probably members of heteronormative families, it also becomes clear that these families were neither their only resource nor their only daily context. Often, nuclear families survive precisely thanks to solidarity networks that extend beyond their limits. What is more, how a family is defined and formed is dynamic, and changes over a person's lifetime and over history. So we cannot expect a single model to be able to give a full account of the lifetime of a worker in Mesopotamia, as his or her situation may well have changed over time due to age, illness or factors over which they had no control (being taken prisoner in times of war, for example).

3.2 Collectives of Women: The "Harem" under Suspicion

So far, we have seen that we need to add more elements to our research, and not rely solely on gender. But paradoxically, we cannot help but pay special attention to gender because, in my view, it is a key to understanding why the study of groups of females in the ancient Near East focused on "harems" and not on groups of workers. The choice of one group over the other is the result of a process of hypersexualisation of women in traditional historiography.[33] This process places the spotlight on the "harem" (with the sexual connotation that the term implies), while neglecting collectives of work as potential groups in their own right and as scenarios for the creation of solidarity networks. Probably as a consequence, the "harem" has been studied in depth, and the suitability of the use of the term "harem" itself has come in for criticism. Let us now look at some of the main contributions on the "harem", applied to the study of the ancient Near East.

The publication that represents the turning point in the study of the "harem" in the ancient Near East is Nele Ziegler's monograph, published in 1999, entitled *Le harem de Zimri-Lim. La population féminine des Palais d'après les archives royales de Mari*. Ziegler begins by noting some criticisms of the use of the term "harem" applied to the ancient Near East, for instance, the objections raised by Joan Goodnick Westenholz.[34] Moreover, she acknowledges that there are major differences in the conceptions of women's seclusion and the sexual nature attributed to Ottoman harems (the main reference when discussing the term) and the collectives described as "harems" in Mari. However, in the end she accepts the use of the word on the grounds that it is applied to describe a collective of women living together in the same area of a palace,

32 Rivera Garretas 2001, 43.

33 Cf. Assante 2006.

34 Westenholz 1990.

the most private area, and that it has clear connotations of polygamy. She therefore uses "harem" for want of a more appropriate word to refer to

> [...] l'espace habité par les femmes dans le palais royal et plus largement l'ensemble des femmes appartenant à la famille ou au service du roi, qu'elles fussent mères, filles ou épouses du roi ou bien musiciennes, servantes ou gardiennes de portes.[35]

This definition is completed by a list of what is not included in Ziegler's concept of "harem": "Le harem ne comprend donc pas les femmes travaillant dans les ergastules (*nepârum*), tisseuses ou autres".

More recently, Adelina Millet[36] has studied the texts of the Chagar Bazar "harem". The Chagar Bazar and Mari "harems" are contemporary, geographically close, and share certain similarities. Millet quotes Ziegler with regard to the discussion of the suitability of the term, reiterating the differences between the realities of northern Syria at the beginning of the second millennium BCE and the Ottoman empire of the fifteenth and sixteenth centuries. Millet's proposal to define the term "harem" in this context is more inclusive than the previous one: "nous utilisons le terme 'harem' au sens large pour désigner la famille d'un roi ou d'un personnage important et, dans le cas qui nous intéresse, la famille de Sîn-iqišam, dirigeant de Chagar-Bazar".[37] We see, then, that the term is not restricted to women here, but to all those receiving beer allocations, i.e., masculine deities, sons and other characters who can be identified by their proper names as males, alongside females.

Both Ziegler and Millet make convincing criticisms of the use of the term "harem", showing that the evidence from the ancient Near Eastern and the Ottoman era suggest that there are more differences than similarities. However, *faute de mieux*, both defend the use of the word.

Of course all words have connotations, and we are obliged to use them nonetheless. I think, though, that at least we can avoid using the ones that are most heavily loaded. In my view, "harem" is one such word: its definition in the Collins Dictionary is "a group of female animals of the same species that are the mates of a single male". Other definitions in this dictionary and in Webster's always link the term "harem" with women (wives, concubines, servants and so on) even though the "harem" recorded in Chagar Bazar texts also includes men.

Zainab Bahrani[38] contends that the use of the term "harem" is a paradigmatic example of the survival of the concept of "Orientalism" defined by Edward Said in 1978. Following Said, Bahrani observes that scholars use a static concept of the "Orient" as the starting point of their research and as a way to identify some common

35 Ziegler 1999, 8 and 8, footnote 5, respectively for both quotations.

36 Millet Albà 2008.

37 Millet Albà 2008, 239 and 248–249 for a breakdown of who is included in this "harem".

38 Bahrani 2001, 16.

issues shared by both the ancient Near East and the Islamic world. Bahrani also states that the word "harem" is used to identify all the contexts and realities that link women and palaces. So the proposal of avoiding the use of this word aims to show that nothing is static and that the imaginary linked to the Ottoman Empire that we still apply in our research is one created in the nineteenth century, especially as depicted in Western art.[39]

On the other hand, when Ziegler and Millet justify their use of the term, they do it to highlight either polygamy (Ziegler) or the notion of family (Millet). Consequently, they reinforce the definition of kinship as based on sexual and biological ties, leaving aside the potential relevance of other ties like the ones described above, e.g., solidarity networks or friendship. In this regard the Mari letters are especially interesting, as they allow us a glimpse inside the "harem" and of the complex relationships that were created there by hierarchy and inequality—factors that were not restricted to, or totally dependent on, traditional kinship.[40] Millet also reports the participation of males in the so-called "harems" and questions the presence of eunuchs (at least in Mari and Chagar Bazar) at the beginning of the second millennium BCE, as their existence has not been proven. Taking all this evidence into account, we see that the idea of the "harem" as an entity linked exclusively to women fades away.

Finally, other scholars followed Ziegler and Millet in considering that the use of the term "harem" was more problematic than advantageous, and came up with new proposals. Saana Teppo,[41] for instance, proposes "female administrator" to translate the Akkadian *šakintu*, instead of "female manager of the harem". Similarly, Oliver suggests using "house of women" in contexts like Mari, avoiding the use of "harem", which, she says, "se utiliza por la imposición de su uso no porque se asimile al harem islámico ni turco otomano".[42] In taking these decisions, scholars show that kinship ties based on biology or sex are neither "natural" nor the only ones that are useful for interpreting data.

In summary, I think it is important to state explicitly that the use of the word "harem" emphasises certain relationships and renders others invisible. The same happens when we present a group of female or male workers as a work collective: we highlight some links while hiding others. So we should tread carefully when defining groups and collectives, and state explicitly why we privilege certain choices over others. Let us look now at some specific proposals along these lines.

---

39 Graham-Brown 1987.

40 Solvang 2008, 420.

41 Teppo 2007.

42 Oliver 2008, footnotes 10 and 57; see also Oliver 2010, 117, especially footnote 6.

4 Materialising and Recording the Ur III Textile Workforce

In Ur III administrative texts, low-ranking female workers or low-ranking male workers may be recorded together, in the same text, or segregated by sex, i.e., in texts that list exclusively either males or females. When males and females appear segregated by sex, one of the frequent aims in the secondary literature is to look for possible relationships between these male and female workers.[43] These proposals all start from the heterosexual family model, and assess the extent to which the data fit this model. Here I propose to do the opposite, i.e., first analysing how people are grouped in the texts, in an attempt to see whether biological and sexual ties are among the criteria used to group them. I contend that this approach, which does not rely on preconceptions regarding the presence of the heterosexual nuclear family and the sexual relationship between males and females, may shed light on how the scribes who registered the workforce envisaged the collective and personal relationships.

As we are dealing with administrative texts, we have to keep in mind that the primary aim was not to provide information on the family context of the workforce. However, this can be interpreted in two ways. On the one hand, perhaps it was irrelevant to determine the marital status and offspring of the workers. On the other, things that are self-evident are not usually written down, and perhaps there was no need to record kinship relationships based on sexual or biological ties. If we consider this second option, it is impossible for us to determine whether it was evident to them that certain kinship ties did or did not exist.

In any case, I propose that marital status was probably not the main factor in listing the workforce in different groups. Analysing texts listing rations and allocations for the workforce and texts listing workers assigned to specific tasks, it is possible to determine the criteria that were applied to group personnel. Here I highlight four of these criteria: the sexual division of labour,[44] hierarchy, speciality, and workplace.

The first criterion, the sexual division of labour, is evident in the lists segregated by sex. The second criterion (often interacting with the first one) is hierarchy. Accordingly, some lists only include foremen/forewomen and supervisors in various productive sectors; others include only low-ranking workers without separating females and

---

43 See among others, Gelb 1973, 75; Maekawa 1987, 64; Waetzoldt 1988, 41–44; Wright 1996, 89, 91 and 98; Wright 2008, 259 (with previous references).

44 There is some debate concerning the suitability of the term "sexual" in this context. Some scholars propose the use of "gender" instead of "sex". See, among others, proposals like "gendered tasks" or "gendered division of labour" (Asher-Greve 2008, 128–132), or "division of labour by gender" (Harding 1986, 17). Despite the criticisms, here I will use "sexual division of labour" as it is the label that is most commonly used and is easy to understand outside the field of gender studies.

males. The third criterion, speciality, is clear when in the same list we find females and/or males who may or may not share the same status but are listed together simply because they were employed in the same productive sector—for instance, textile production. The fourth and last criterion proposed here is the shared workplace. After examining certain administrative texts that mentioned the workforce involved in textile production, I realised that most texts were governed by one or several of the criteria presented here.[45]

In my view, it is particularly interesting that kinship is rarely made explicit. In fact, there are only occasionally explicit references to offspring, although it is not possible to establish whether they are references to biological offspring.[46] Moreover, even when kinship is made explicit, it is never the main criterion for explaining why a specific group of workers is listed together. Below, I show some texts in transliteration and translation into English, which illustrate how the four criteria here proposed serve to shape collectives. At the end, I also include an example that does not fit in my proposal.

Beginning with the first criterion, there are a number of examples of the use of the sexual division of labour as a criterion for the recording of workers. See for example text 1:[47]

Text 1

obverse	*obverse*
1. 20+2 geme$_2$ u$_4$1-še$_3$	22 female workers for 1 working day
2. ša$_3$ Umma^{ki}	in Umma
3. 2 geme$_2$ u$_4$6-še$_3$	2 female workers for 6 working days
4. ša$_3$ I$_7$-lugal-ka	at the king's channel
5. gir$_3$ Ur-e$_{11}$-e	under the authority of Ur-E'e
6. udu kur-ra ur$_4$-ra	to shear mountain sheep.*

45 The sample I initially used to develop these proposals is a selection of 100 representative texts, in turn selected from a sample of almost 2000. See Garcia-Ventura 2012a, 103–117 and 432 for a description and for some statistics from the first sample (2000 texts) and for the complete list of the second sample (100 texts) respectively.

46 For the state of research on the Sumerian term dumu, usually translated as "son", see Pomponio 2013. Pomponio summarises arguments for and against considering dumu as proof of biological ties among those recorded in Ur III administrative texts. See also Verderame/Spada 2013, 426–427, and Garcia-Ventura 2014b, 305–312.

47 NBC 887 = BPOA 6, 1319 = Sigrist/Ozaki 2009a, t. 1319. The text is from Umma, from the 46th year of Šulgi's reign. Cf. among others with SAT II, 509 (= YBC 376 = Sigrist 2000, t. 509) or BPOA 7, 2108 (= NBC 3259 = Sigrist/Ozaki 2009b, t. 2108), both from Umma, from the same regnal year and listing low-ranking female workers required for seasonal duties.

reverse	*reverse*
7. ki Da-da-ga-ta	From Dadaga
8. kišib ensi$_2$-ka	sealed by the governor.
9. iti ^{d}Dumu-zi	Month: 12
10. mu ki-maš^{ki} ba-hul	Year: 46th of Šulgi's reign
+ seal	*+ seal*

Notes on Text 1: * For the translation "mountain sheep" and their frequent attestation in Umma, see Waetzoldt 1972, 8–9 and Steinkeller 1995, 54.

This text lists female workers using the Sumerian word geme$_2$. It is an example of the texts in which low-ranking female or male workers are required to carry out seasonal duties related to digging channels or to sheep shearing, among other tasks. These workers are usually divided by sex, without their specialisation or age, or any reference to offspring. Many other lists of workers, besides those referring to seasonal duties, follow the same pattern. In these texts the number of workers varies widely. For this reason, any consideration of the idea of the collective and the possible solidarity networks among workers should include the size of the group.

In other words, various scales should be considered when characterising workforce collectives. This is obvious enough in our modern-day employment contexts: groups of people working closely together coexist with other, larger groups. These other groups often comprise members we do not know directly if we work in a big company, but in some way (perhaps only symbolically) we all are part of the same collective, even though we do not share a direct, daily relationship. In my view, this is what we find (among other things), in text Um 2282.[48] In each line we have a number of female workers and the grain allotment they receive. The text includes only low-ranking female workers, geme$_2$ in Sumerian. This sequence is repeated 13 times with similar numbers of female workers: it is an account of the whole year (the third year of Šu-Suen's reign) detailed month by month. Each month, a group of more than 100 female workers is recorded. The fact that this number of workers is larger than the one shown above raises the question of different scales.

Moving on now to the criterion of hierarchy, let us look at text 2:[49]

---

48 Um. 2282 = UTI 3, 2282 = Yildiz/Gomi 1993, t. 2282.

49 U 5086 = UET 9, 38 = Loding 1976, t. 38. The text is from Ur, from the 8th year of Ibbi-Suen's reign.

Text 2

obverse	*obverse*
1. 11 ugula uš-bar	11 foremen/forewomen of the textile workshop
2. 5 sila$_3$-ta	(receive) 5 sila each one
3. 5 sila$_3$ Ur-AB šar$_2$-ra-ab-du	5 sila (for) Ur-abba, *šarrabtû** official.
reverse	*reverse*
4. iti a$_2$-ki-ti	Month: 7
5. mu us$_2$-sa bad$_3$-gal ba-du$_3$ mu us$_2$-sa-bi	Year: 8th of Ibbi-Suen's reign

Notes on Text 2: * I opt not to translate the term, as the duties of these officials are not clear to us. As a guide, I quote the description proposed by Waetzoldt: "a functionary with a scribal education who appears in the documents in the capacity of an inspector. The precise nature of this office remains uncertain" (Waetzoldt 1987, 136).

In this text, 11 foremen or forewomen at the textile workshop receive a payment. Whether they are male or female is not specified, nor are their proper names listed. We do not have any information about the workers under their charge, or their tasks. The absence of these details suggests that the scribes making the record considered them as a collective by then. Indeed the only information given explicitly is their position, which suggests a concern with hierarchy.

For the third criterion—the speciality or productive sector—text 3[50] is a good example:

Text 3

obverse	*obverse*
1. 600 geme$_2$ uš-bar 1 sila$_3$ ninda-ta	600 female weavers, 1 sila of bread (for each one)
2. 12 ^{lu_2}azlag$_2$ 1 sila$_3$ ninda-ta	12 fullers, 1 sila of bread (for each one)
3. 8 ugula uš-bar 1 sila$_3$ ninda-ta	8 foremen/forewomen at the textile workshop, 1 sila of bread (each one)
4. 8 ra$_2$-gaba 1 sila$_3$ ninda-ta	8 messengers, 1 sila of bread (each one)
reverse	*reverse*
5. geme$_2$ uš-bar ^{lu_2}azlag$_2^{!}$	female weavers and fullers
6. ugula uš-bar u$_3$ ra$_2$-gaba	foremen/forewomen at the textile workshop (and) messengers
7. igi-kar$_2$-de$_3$ gen-na	going to pass inspection
8. Gir$_2$-su^{ki}-ta Gu$_2$-ab-ba^{ki}-še$_3$	from Girsu to Guabba
9. šu ba-ab-ti	received.
10. gir$_3$ Lu$_2$-kal-la	Under the authority of Lukalla,

50 AO 27476 = DAS 255 = Lafont 1985, t. 255. The text is from Girsu, from the 1st year of Šu-Suen's reign.

11. gir$_3$ Inim-^{d}Ba-u$_2$- i$_3$-dab$_5$	under the authority of Inim-Ba'u
12. u$_3$ Lu$_2$-kiri$_3$-zal dumu na-mu	and Lu-Kirizal, son of Namu (all them) received.
13. iti gu$_4$-ra$_2$-izi-´mu$_2$-mu$_2$`	Month: 2
14. mu ^{d}Šu-^{d}Suen ´lugal`	Year: 1st of Šu-Suen's reign

In lines 5 and 6, female weavers and fullers, foremen/forewomen of the textile workshop, and messengers are mentioned. Therefore, there is neither a sexual division of labour nor a hierarchical criterion, as all of them have to pass the inspection. The heterogeneous group includes male and female workers of different ranks. But in this case, unlike others, the speciality is stated: with the exception of the messengers, all the others are explicitly linked to textile production.

An example of how some of the criteria mentioned to date (sexual division of labour, hierarchy, speciality or productive sector) might be combined is text 4,[51] where sexual division of labour and productive sector are presented together.

Text 4

obverse	*obverse*
1. 1 tug$_2$ Ša$_3$-igi-na engar	1 garment for Ša-igina, the farmer
2. 1 tug$_2$ Ur-^{d}Utu dumu bar-ra	1 garment for Ur-Utu, son of Barra
--- *(1 line blank)*	--- *(1 line blank)*
3. ugula Šeš-kal-la dumu da-da	foreman Šeškalla, son of Dada
4. 2 tug$_2$ ka-guru$_7$	2 garments for the granary supervisor
5. 2 tug$_2$ Ur-^{d}Nin-tu ugula uš-bar	2 garments for Ur-Nintu, foreman at the textile workshop
6. 1 tug$_2$ Šeš-kal-la dumu tir-gu	1 garment Šeškalla, son of Tirgu
--- *(1 line blank)*	--- *(1 line blank)*
7. lu$_2$ didli-me	they are for one (they are not part of a team)
8. 1 tug$_2$ Ur-Gu$_2$-eden-na engar`	1 garment for Ur-Guedena, the farmer
9. 1 tug$_2$ A-kal-la ša$_3$-gu$_4$	1 garment for A(ya)kalla, the oxherd
reverse	*reverse*
--- *(1 line blank)*	--- *(1 line blank)*
10. ugula Lugal-nesag-e	foreman: Lugal-nesag'e
==== *(blank space)*	==== *(blank space)*
11. tug$_2$ mu-kuX iti min-eš$_3$ (=DU)	garments were delivered the 7th month
12. mu ^{d}Šu-^{d}Suen lugal	year: 1st of Šu-Suen's reign

This text describes cloth allocations for several males (there are no females) sharing a professional sector in a broad sense, as all of them are related to animal husbandry and agriculture in some way. The text mentions two farmers, a granary supervisor, a foreman at the textile workshop and an oxherd. Someone called Ur-Utu receives allo-

51 Crozer 79 = Rochester, 106 = Sigrist 1991, t. 106. The text is from Umma, from the 1st year of Šu-Suen's reign.

cations as the son or apprentice of Barra (dumu), although the occupations of these two characters are not stated. The same applies to Šeškalla, the son or apprentice of Tirgu. Two foremen are also mentioned as superiors (from the hierarchical point of view) of the farmers and the oxherd; they are recorded as distributing payment rather than receiving it. All the workers in their charge receive one garment, while the supervisor and the foreman of the textile workshop, apparently not listed under the authority of others, receive two. Here, then, the number of garments received plus the presence or absence of a superior are elements that reveal differences in rank but, despite these differences, they are listed together. So in this case the criterion used to group them seems to have been not hierarchy, but professional sector and sex.

Moving on now to the fourth and last criterion, the workforce's shared workplace, we find good examples in a number of texts listing working groups and their foreman or supervisor. When a duty is mentioned in these texts, I suggest that the whole group would have worked together in the same place, carrying out the same task. Some of the texts mentioned above evidence this, but below I will show two slightly different examples, texts 5 and 6:[52]

Text 5

obverse	*obverse*
1. 300 sa gi	300 reed bundles*
2. gu-nigin$_2$-ba 16 sa-ta	there are 16 bundles in each bale**
3. ša$_3$-gu$_4$-˹ke$_4$˺ ga$_6$-ga$_2$	carried by the oxherd
4. 1200 s[a-gi]	1200 reed bundles
5. gu-[nigin$_2$-ba 16 sa-ta]	there are 16 bundles in each bale
6. ša$_3$ [giš-gi]	from the reed bed
reverse	*reverse*
7. geme$_2$ [uš-bar ga$_6$-ga$_2$]	carried by the female weavers
8. ga$_2$-nun e$_2$-lu[gal]-/ka ku$_4$-r[a]	to the royal storehouse
9. ugula Ur-lugal	foreman: Ur-lugal
10. Kišib A-kal-la	sealed by A(ya)kalla
11. mu bad$_3$ mar-/tu ba-du$_3$	year: 4th of Šu-Suen's reign
seal	*seal*
1. A-kal-l[a]	A(ya)kalla
2. dub-s[ar]	scribe
3. dumu Lu$_2$-sa$_6$-[ga]	son of Lu-saga

Notes on Text 5: * On the sorts of reeds and their use in Ur III texts from Umma, see Waetzoldt 1992. On the collection and management of this type of reed as a duty of foresters in Umma, see Steinkeller 1987, especially 92–93. ** For the transliteration gu-nigin$_2$ instead of gu-kilib (Sumerian terms for "bale") in texts from Umma, see Heimpel 2003 (with previous references).

52 Text 5: MM 381 = AuOrS 11, t. 187 = Molina 1996, t. 187 // Text 6: MM 853 = AuOrS 11, t. 611 = Molina 1996, t. 611. Both are from Umma, from the 4th year of Šu-Suen's reign.

Text 6

obverse	*obverse*
1. ´1800+60`[(+x) sa gi]	1860 reed bundles
2. gu-nigin$_2$-ba 10[+6$^?$ sa-ta]	there are 16 bundles in each bale
3. ša$_3$-gu$_4$-k[e$_4$ g]a$_6$-[ga$_2$]	carried by the oxherd
4. 1200 sa gi	1200 reed bundles
5. gu-nigin$_2$-ba 16 sa-ta	there are 16 bundles in each bale
6. ša$_3$ giš-gi	from the reed bed
reverse	*reverse*
7. geme$_2$ uš-bar ga$_6$-/ga$_2$	carried by the female weavers
8. ga$_2$-nun E$_2$-lugal-/ka ku$_4$-ra	to the royal storehouse
9. ugula Ab-ba-sig$_5$	foreman: Abbasig
10. kišib A-kal-l[a]	sealed by A(ya)kalla
11. [m]u bad$_3$ [mar-tu ba-du$_3$]	year: 4th of Šu-Suen's reign
seal	*seal*
1. [A-a-]-kal-[la]	A(ya)kalla
2. dub-sar	scribe
3. dumu Lu$_2$-sa$_6$-[ga]	son of Lu-saga

These texts count reed bundles and bales taken by the oxherd and the female weavers to the royal storehouse. We have no information about the number of female weavers involved in this operation. Nevertheless, we do have other data: we know that both texts refer to the same year (the fourth year of Šu-Suen's reign), that both were sealed by A(ya)kalla, and that the foreman supervised the work being performed. Taking all this into account, here I propose that there are probably two different foremen controlling two different gangs of workers, and so the members of each group would have shared the same place or places where the transport tasks were carried out. In addition, the two groups would sometimes have coincided. Another possibility is that the two texts actually mention only one group, and that it was the same group controlled by two different foremen at two different points in time. In my view, it is difficult to determine which option is more plausible, as we lack other potentially helpful data such as month names.

Other examples of this fourth criterion, the shared workplace, are texts that list female weavers and millers together. In some cases it is not specified how many were weavers and how many were millers; they are considered as a group, and consequently were not listed separately. In text 7,[53] for instance, female weavers and female millers are referred to as a collective receiving allocations of oats, fat and bread:

53 Text 7: NBC 476 = BPOA 6, 1072 = Sigrist/Ozaki 2009a, t. 1072. The text is from Umma, from the 1st year of Amar-Suena's reign.

Text 7

obverse	*obverse*
1. 0.3.0 ninda-gen	3 *barig* of regular quality bread
2. 0.0.1 nig_2-ar_3-ra sig_5	10 silas of good quality oats
3. 0.0.1 3 $sila_3$ i_3-$šah_2$	13 silas of fat
4. [...] $geme_2$ uš-bar $geme_2$ kinkin-na	[...] for the female weavers and the female millers
5. [...]-$še_3$ de_6-a u_3	[...] (some of them) going to [...] and
6. [...]-uru-sag tuš-a	[...] (the others) settled at [...]-irisag
reverse	*reverse*
7. gir_3 Ur-e_{11}-e	under the authority of Ur-E'e
8. mu ^{d}Amar-^{d}Suen lugal	year: 1st of Amar-Suena's reign

These cases of female weavers and millers show how several criteria coexisted in shaping an idea of the collective, e.g. the sexual division of labour, hierarchy and work-place—but, in this case, not speciality. The example is especially interesting because it highlights a well-known feature often mentioned in Ur III workforce studies, i.e., that female weavers and female millers sometimes worked together and might even exchange duties. Lorenzo Verderame has compiled several texts and references to this phenomenon, and has shown that the same people might supervise both sectors.[54]

However, this relationship between female weavers and millers has led to some misunderstandings in the translation and presentation of texts. Although their tasks may be interchanged in the records, if they are explicitly mentioned as female weavers and female millers and not only as female workers, this should be reflected in the translation. An example is the case of two texts published recently in transliteration by Marcel Sigrist and Tohru Ozaki.[55] Both mention garment allocations for female millers ($geme_2$ kinkin), but both texts are presented in the catalogues of the two volumes as "garments for weaver women".

To complete the examples for this proposed classification, let us look at a particular case that does not fit into my model—the exception that proves the rule. Text 8[56] lists several proper names with their garment allocations.[57] The last lines of the text specify specialties of these preceding proper names, and specialties of some other individuals who also received garment allocations. Below are a transliteration and

54 Verderame 2008, 114, footnote 25; Verderame/Spada 2013, 439–441.

55 NBC 637 = BPOA 6, 1204 = Sigrist/Ozaki 2009a, t. 1204. The text is from Umma, from the 9th year of Šu-Suen's reign / NCBT 1315 = BPOA 7, 2614 = Sigrist/Ozaki 2009b, t. 2614. The text is from Umma, from the 8th year of Amar-Suena's reign.

56 Text 8: YBC 13419 = BPOA 6, 21 = Sigrist/Ozaki 2009a, t. 21. The text is from Umma, from the 2nd year of Šu-Suen's reign.

57 Each line, from line 1 to line 21, reproduces the structure "1 garment for PNx".

translation into English of the last lines of the text in which these occupations are explicitly stated:

Text 8

reverse	*reverse*
[...]	[...]
22. lu$_2$-tir-me ugula Ur-e$_2$-maš	They are foresters;* foreman: Ur-emaš;
23. 7 nar-munus	7 (garments) for the female musicians
24. 1 Šeš-kal-la šu-ku$_6$	1 (garment) for Šeškalla, the fisherman
25. 1 Ze$_2$-ze$_2$-ga šu-ku$_6$	1 (garment) for Zezega, the fisherman
26. 1 Lugal-sig$_5$ šu-ku$_6$	1 (garment) for Lugal-sig, the fisherman
27. 1 [... šu-k]u$_6$ ugula Šeš-pad$_3$-da	1 (garment) for [NP?], the fisherman, foreman Šeš-pada
28. mu-kuX mu ma$_2$ ^{d}En-ki ba-ab-du$_8$ (=DU)	Delivery. Year: 2nd of Šu-Suen's reign

Notes on Text 8: * Referring to all workers listed in previous lines (lines 1 to 21).

This text records garment allocations for foresters (previously mentioned individually with their names), for female musicians (referred to as a collective, without listing their personal names), and for fishermen, again listed in detail along with their personal names. In my view, the data detailed in the text do not indicate why they were listed together, and were considered in the same register for the distribution of the garment allocation.

Consequently, none of the criteria mentioned here apply to text 8. Sexual division of labour does not apply, because males (foremen and fishermen) and females (musicians) are explicitly mentioned together. Equally, neither speciality nor shared workplace serves as a criterion: it is difficult to imagine a context in which foresters, fishermen and musicians would share a workplace. Only one of our criteria remains: hierarchy or status. Indeed, all of these occupations probably shared the same status, but as far as I know there is no clear evidence to support this claim. An additional (fifth) criterion to be considered would be a geographical one, so listing together workers with a shared origin or having been stationed in the same place during periods of compulsory service to the state, like the bala.[58] Whether this criterion would work for this and other texts that do not fit the previous proposal is a possible path for future research.

In any case, in this example, it is difficult to defend the idea of the collective as we have done in this section. What it shows, I think, is that the proposal works for most texts, though not for all. For this reason, my proposal may be useful for explaining who is listed together in accounts including relatively small numbers of workers, but

58 This geographical criterion has been suggested by Lorenzo Verderame (personal communication).

not for general accounts in which the idea of the collective is blurred by the diversity of the people listed together and/or for the size of the group considered. So the proposed model appears to have been valid for both the scribes writing the texts and the workers who appeared in them.

5 Some Concluding Remarks

One frequent research question when analysing Ur III lists of workers is whether the men and women registered had a "family life", that is, whether they were married, whether they had offspring, and who they lived and shared their lives with when they were not at work. Here I have outlined two main objections to this approach. First, asking this question presumes that the nuclear family, understood as it is today in our Western societies, was also a structural feature of ancient Mesopotamia several millennia ago. Second, the question has been applied in different ways to male workers and to female workers, thus generating different results.

Regarding the first objection, I have suggested considering kinship in a broad sense following the proposals made in gender studies—among others, those of Judith Butler. With this broad definition, I do not propose to abandon the idea of the heterosexual couple as one of the axes of ancient societies, but rather to highlight the greater complexity of the concept of kinship. In this regard, then, kinship is understood as including all the practices that establish relationships with the aim of managing production, reproduction, the maintenance of life, death, and physical and emotional dependency. From this perspective, whether or not these practices were carried out by people with blood ties loses some of its relevance and, in my view, bearing this definition in mind when analysing Ur III administrative texts will help to enrich the resulting picture. Rather than (or in some cases in addition to) widows, single women, abandoned sons and daughters and happily married men, we find solidarity networks and working groups that provide the context for kinship ties in a broad sense.

In brief, if we move the focus of the research from biological and sexual ties to how collectives were shaped, new possible readings appear. Shifting the emphasis from marital status or filiation to non-family contexts allows us to see that the people who registered the workforce applied certain criteria in order to group workers, criteria which were not related to biological kinship. In other words, here I have suggested that collectives and groups of workers are valid categories of analysis that help us to gain a better understanding of how the workforce was organised.

Moving to the second objection (the differential treatment of men and women in the lists in secondary literature) I discussed the use of the term "harem". Collectives of men are usually described on the basis of sharing a professional relationship. This is clear from the Sumerian term erin$_2$, usually translated as "male gangs", a translation that highlights the idea of men working together. Groups of women, however, are usually explained via the model of the "harem", i.e., emphasising sexual ties and kinship as based on blood ties. This differential treatment reinforces the links "men: professional world" and "women: sex & body", and the ideal of the function of the male (working outside the home, concentrating on productive work) in contrast to that of the female (working at home, concentrating on reproductive work).[59]

Needless to say, this model is wholly inapplicable to the Mesopotamia of the end of the third millennium BCE, where most of the population worked only part-time for great organisations like temples or palaces. Therefore, most of the workforce would not have been registered. To quote Marc Van de Mieroop: "The common worker in the state sector of the Ur III period was thus only part-time employed by the state, worked with other family members, and spent a substantial amount of time engaged in a world inaccessible to us".[60] While agreeing and acknowledging that most issues are "inaccessible to us", I would contend that applying explicit theoretical approaches like the ones proposed by feminist epistemologies would help us to gain a better and closer understanding of the realities which we might, to a greater or lesser extent, find to be "accessible".

Bibliography

Alcoff, Linda/Potter, Elizabeth (1993), *Feminist Epistemologies*, New York.

Asher-Greve, Julia M. (2008), "Images of Men, Gender Regimes, and Social Stratification in the Late Uruk Period", in: Diane Bolger (ed.), *Gender Through Time in the Ancient Near East*, Lanham/New York/Toronto/Plymouth, 119–171.

Assante, Julia (2006), "Undressing the Nude: Problems in Analyzing Nudity in Ancient Art, with an Old Babylonian Case Study", in: Silvia Schroer (ed.), *Images and Gender. Contributions to the Hermeneutics of Reading Ancient Art*, Fribourg/Göttingen, 177–207.

Bahrani, Zainab (2001), *Women of Babylon. Gender and Representation in Mesopotamia*, London.

Butler, Judith (1990), *Gender Trouble. Feminism and the Subversion of Identity*, London/New York.

Butler, Judith (1997), *The Psychic Life of Power. Theories in Subjection*, Stanford.

Butler, Judith (2004), *Undoing Gender*, London/New York.

Carbonell i Esteller, Montserrat (1997), *Sobreviure a Barcelona, dones, pobresa i assistència al segle XVIII*, Vic.

Civil, Miguel (1987), "Ur III Bureaucracy: Quantitative Aspects", in: McGuire Gibson and Robert D. Biggs (eds.), *The Organization of Power. Aspects of Bureaucracy in the Ancient Near East*, Chicago, 43–53.

Dahl, Jacob (2007), *The Ruling Family of Ur III Umma. A Prosopographical Analysis of an Elite Family in Southern Iraq 4000 Years Ago*, Leiden.

Firth, Richard (2013), "Considering the Finishing of Textiles Based on Neo-Sumerian Inscriptions from Girsu", in: Marie-Louise Nosch, Henriette Koefoed and Eva Andersson Strand (eds.),

---

59 Garcia-Ventura 2014b.

60 Van de Mieroop 1999, 92.

Textile Production and Consumption in the Ancient Near East. Archaeology, Epigraphy and Iconography, Oxford/Oakville, 140–151.
Gálvez Muñoz, Lina (2000), *Compañía arrendataria de tabacos 1887–1945. Cambio tecnológico y empleo femenino*, Madrid.
Garcia-Ventura, Agnès (2012a), *El trabajo y la producción textil en la Tercera Dinastía de Ur*, PhD Thesis, University Pompeu Fabra.
Garcia-Ventura, Agnès (2012b), "Tejedor@s en la Tercera Dinastía de Ur: nuevas perspectivas de estudio", in: *Claroscuro (Revista del Centro de Estudios sobre diversidad cultural, de la Universidad Nacional de Rosario)* 11, 13–31.
Garcia-Ventura, Agnès (2013), "Trabajar en tiempos de guerra en Mesopotamia", in: Jordi Vidal and Borja Antela-Bernárdez (eds.), *Más allá de la batalla. La violencia contra la población en el Mundo Antiguo*, Zaragoza, 1–25.
Garcia-Ventura, Agnès (2014a), "Ur III Biopolitics. Reflections on the Relationship between War and Work Force Management", in: Davide Nadali and Jordi Vidal (eds.), *The Other Face of the Battle. The Impact of War on Civilians in the Ancient Near East* (Alter Orient und Altes Testament 413), 7–23.
Garcia-Ventura, Agnès (2014b), "Mano de obra y relaciones de parentesco en Mesopotamia: madres trabajadoras versus hombres 'ganadores de pan'", in: *Arenal. Revista de historia de las mujeres* 21 (2), 297–316.
Gelb, Ignace J. (1973), "Prisoners of War in Early Mesopotamia", in: *Journal of Near Eastern Studies* 32, 70–98.
Genz, Stéphanie/Brabon, Benjamin A. (2009), *Postfeminism. Cultural Texts and Theories*, Edinburgh.
Graham-Brown, Sarah (1987), *Images of Women*, London.
Hall, Martin (2012), "Sexuality and Materiality. The Challenge of Method", in: Barbara L. Voss and Eleanor Conlin Casella (eds.), *The Archaeology of Colonialism. Intimate Encounters and Sexual Effects*, Cambridge, 323–340.
Haraway, Donna (1991a), "'Gender' for a Marxist Dictionary: The Sexual Politics of a Word", in: Donna Haraway (ed.), *Simians, Cyborgs, and Women. The Reinvention of Nature*, London/New York, 127–148.
Haraway, Donna (1991b), "A Cyborg Manifesto: Science, Technology, and Socialist-Feminism in the Late Twentieth Century", in: Donna Haraway (ed.), *Simians, Cyborgs, and Women. The Reinvention of Nature*, London/New York, 149–181.
Harding, Sandra (1986), *The Science Question in Feminism*, Ithaca/London.
Heimpel, Wolfgang (2003), "gu-nigin$_2$, 'bale'", in: Cuneiform Digital Library Notes 3, 1 http://cdli.ucla.edu/pubs/cdln/php/index.php (last accessed: 29.09.2015).
Juliano, Dolores (1992), *El juego de las astucias. Mujer y construcción de modelos sociales alternativos*, Madrid.
Juliano, Dolores (1998), *Las que saben. Subculturas de mujeres*, Madrid.
Lafont, Bertrand (1985), *Documents administratifs sumériens provenant du site de Tello et conservés au Musée du Louvre*, Paris.
Lion, Brigitte (2007), "La notion de genre en assyriologie", in: Violaine Sébillotte Cuchet and Nathalie Ernoult (eds.), *Problèmes du genre en Grèce ancienne*, Paris, 51–64.
Loding, Darlene (1976), *Economic Texts from the Third Dynasty*, London/Philadelphia.
Maekawa, Kazuya (1980), "Female Weavers and Their Children in Lagaš -Pre-sargonic and Ur III", in: *Acta Sumerologica (ASJ)* 2, 81–125.
Maekawa, Kazuya (1987), "Collective Labor Service in Girsu-Lagaš: The Pre-sargonic and Ur III Periods", in: Marvin A. Powell (ed.), *Labor in the Ancient Near East*, New Haven, 49–71.
Maekawa, Kazuya (1989), "Rations, Wages and Economic Trends in the Ur III Period", in: *Altorientalische Forschungen* 16, 42–50.

Maekawa, Kazuya (1998), "Ur III Girsu Records of Labor Forces in the British Museum (1)", in: *Acta Sumerologica (ASJ)* 20, 63–110.
Michalowski, Piotr (2011), *The Correspondence of the Kings of Ur. An Epistolary History of an Ancient Mesopotamian Kingdom*, Winona Lake.
Millet Albà, Adelina (2008), "Le harem de Sîn-iqišam", in: Önhan Tunca and Abdul-Massih Baghdo (eds.), *Chagar Bazar (Syrie) III. Les trouvailles épigraphiques et sigillographiques du chantier I (2000–2002)*, Louvain/Paris/Dudley (MA), 239–279.
Molina, Manuel (1996), *Tablillas administrativas neosumerias de la Abadía de Montserrat (Barcelona). Transliteraciones e índices con un apéndice de Hartmut Waetzoldt: Siegelliste*, Sabadell.
Molina, Manuel (2008), "The Corpus of Neo-Sumerian Tablets: An Overview", in: Steven J. Garfinkle and J. Cale Johnson (eds.), *The Growth of an Early State in Mesopotamia. Studies in Ur III Administration*, Madrid, 19–53.
Naji, Myriem (2009), "Gender and Materiality in the Making: The Manufacture of Sirwan Femininities Through Weaving in Southern Morocco", in: *Journal of Material Culture* 14 (1), 43–73.
Oliver, María Rosa (2008), "Entre lechos, alianzas y alta política: las mujeres como botín de guerra durante el reinado de Zimrî-Lîm de Mari", in: *Claroscuro (Revista del Centro de Estudios sobre diversidad cultural, de la Universidad Nacional de Rosario)* 7, 11–34.
Oliver, María Rosa (2010), "La perspectiva de género en el análisis de las relaciones entre centros alternativos de poder en el antiguo reino de Mari (reinado de Zimrî-Lîm) a través de la correspondencia femenina", in: *Rivista degli Studi Orientali* 83 (1–4), 115–132.
Oliver, María Rosa/Ravenna, Eleonora (2001), "Identidades, desigualdades y solidaridades femeninas en la época Hammurabiana", in: *Claroscuro (Revista del Centro de Estudios sobre diversidad cultural, de la Universidad Nacional de Rosario)* 1 (1), 235–251.
Pomponio, Franco (2013), "The Ur III Administration: Workers, Messengers, and Sons", in: Steven Garfinkle and Manuel Molina (eds.), *From the 21st Century B.C. to the 21st Century A.D.* Proceedings of the International Conference on Sumerian Studies Madrid, 22–24 July 2010, Winona Lake, 221–232.
Rivera Garretas, María Milagros (2001), *Mujeres en relación. Feminismo 1970–2000*, Barcelona.
Rubin, Gayle S. (1975), "The Traffic in Women. Notes on the 'Political Economy'", in: Rayna R. Reiter (ed.), *Toward an Anthropology of Women*, New York/London, 157–210.
Rubin, Gayle S. (2011), *Deviations. A Gayle Rubin Reader*, Durham/London.
Sallaberger, Walther (2004), "Relative Chronologie von der späten Frühdynastischen bis zur altbabylonischen Zeit", in: Jan-Walke Meyer and Walter Sommerfeld (eds.), *2000 v. Chr. Politische, Wirtschaftliche und Kulturelle Entwicklung im Zeichen einer Jahrtausendwende*, Berlin, 15–43.
Sallaberger, Walther/Huber Vuillet, Fabienne (2005), "Priester A. I", in: *Reallexikon der Assyriologie und vorderasiatischen Archäologie*, vol. 10 (7/8), 617–640.
Sallaberger, Walther/Westenholz, Aage (1999), *Mesopotamien. Akkade-Zeit und Ur III-Zeit*, Göttingen.
Seidman, Steven (2008), *Contested Knowledge. Social Theory Today*, Oxford.
Sigrist, Marcel (1991), *Documents from Tablet Collections in Rochester*, New York/Bethesda.
Sigrist, Marcel (2000), *Texts from the Yale Babylonian Collections*, vol.1 (Sumerian Archival Texts 2), Bethesda.
Sigrist, Marcel/Ozaki, Tohru (2009a), *Neo-Sumerian Administrative Tablets from the Yale Babylonian Collection*, vol. 1, Madrid
Sigrist, Marcel/Ozaki, Tohru (2009b), *Neo-Sumerian Administrative Tablets from the Yale Babylonian Collection*, vol. 2, Madrid.
Solvang, Elna K. (2008), "Classifying Women: The 'Harem' and What it Does and Doesn't Tell Us about Women", in: Robert D. Biggs, Jennie Myers and Martha T. Roth (eds.), *Proceedings of the*

51st Rencontre Assyriologique Internationale Held at The Oriental Institute of The University of Chicago, July 18–22, 2005, Chicago, 415–420.
Steinkeller, Piotr (1987), "The Foresters of Umma: Toward a Definition of Ur III Labor", in: Marvin A. Powell (ed.), *Labor in the Ancient Near East*, New Haven, 73–115.
Steinkeller, Piotr (1995), "Sheep and Goat Terminology in Ur III Sources from Drehem", in: J. Nicholas Postgate and Marvin A. Powell (eds.), *Domestic Animals of Mesopotamia*, vol.1 (Bulletin of Sumerian Agriculture), Cambridge, 49–70.
Teppo, Saana (2007), "The Role and the Duties of the Neo-Assyrian *šakintu* in the Light of Archival Evidence", in: *State Archives of Assyria Bulletin* 16, 257–271.
Uchitel, Alexander (1984), "Women at Work: Pylos and Knossos, Lagaš and Ur", in: *Historia. Zeitschrift für Alte Geschichte* 33, 257–282.
Uchitel, Alexander (2002), "Women at Work: Weavers of Lagaš and Spinners of San Luis Gonzaga", in: Simo Parpola and Robert M. Whiting (eds.), *Sex and Gender in the Ancient Near East*, Helsinki, 621–631.
Van de Mieroop, Marc (1999), *Cuneiform Texts and the Writing of History*, London/New York.
Van de Mieroop, Marc (2004), *A History of the Ancient Near East, ca. 3000–323 BC*, London.
Verderame, Lorenzo (2008), "Il controllo dei manufatti tessili a Umma", in: Massimo Perna and Francesco Pomponio (eds.), *The Management of Agricultural Land and the Production of Textiles in the Mycenaean and Near Eastern Economies*, Paris, 111–133.
Verderame, Lorenzo/Spada, Gabriella (2013), "Ikalla, Scribe of (Wool) Textiles and Linen", in: Steven Garfinkle and Manuel Molina (eds.), *From the 21st Century B.C. to the 21st Century A.D.* Proceedings of the International Conference on Sumerian Studies Madrid 22–24 July 2010, Winona Lake, 425–444.
Voss, Barbara L. (2012), "The Scale of the Intimate. Imperial Policies and Sexual Practices in San Francisco", in: Barbara L. Voss and Eleanor Conlin Casella (eds.), *The Archaeology of Colonialism. Intimate Encounters and Sexual Effects*, Cambridge, 173–192.
Waetzoldt, Hartmut (1972), *Untersuchungen zur neusumerischen Textilindustrie*, Rome.
Waetzoldt, Hartmut (1987), "Compensation of Craft Workers and Officials in the Ur III Period", in: Marvin A. Powell (ed.), *Labor in the Ancient Near East*, New Haven, 117–141.
Waetzoldt, Hartmut (1988), "Die Situation der Frauen und Kinder anhand ihrer Einkommensverhältnisse zur Zeit der III. Dynastie von Ur", in: *Altorientalische Forschungen* 15, 30–44.
Waetzoldt, Hartmut (1992), "'Rohr' und dessen Verwendungsweisen anhand der Neusumerischen Texte aus Umma", in: J. Nicholas Postgate and Marvin A. Powell (eds.), *Trees and Timber in Mesopotamia* (Bulletin of Sumerian Agriculture), Cambridge, 125–146.
Waetzoldt, Hartmut (2011), "Die Textilproduktion von Garšana", in: David I. Owen (ed.), *Garšana Studies*, Bethesda, 405–454.
Warburton, David A. (2005), "Working", in: Daniel C. Snell (ed.), *A Companion to the Ancient Near East*, London, 169–182.
Westenholz, Joan Goodnick (1990), "Towards a New Conceptualization of the Female Role in Mesopotamian Society", in: *Journal of the American Oriental Society* 110, 510–521.
Wittig, Monique (1992), *The Straight Mind and Other Essays*, Boston.
Wright, Rita P. (1996), "Technology, Gender, and Class: Worlds of Difference in Ur III Mesopotamia", in: Rita P. Wright (ed.), *Gender and Archaeology*, Philadelphia, 79–110.
Wright, Rita P. (1998), "Crafting Social Identity in Ur III Southern Mesopotamia", in: Cathy Lynne Costin and Rita P. Wright (eds.), *Craft and Social Identity*, Virginia, 57–69.
Wright, Rita P. (2008), "Gendered Relations and the Ur III Dynasty: Kinship, Property, and Labor", in: Diane Bolger (ed.), *Gender Through Time in the Ancient Near East*, Lanham/New York/Toronto/Plymouth, 247–279.
Yildiz, Fatma/Gomi, Tohru (1993), "Die Umma Texte aus den Archäologischen Museen zu Istanbul", vol. 3: *Nr. 1601–2300*, Bethesda.

Zettler, Richard L. (2003), “Archaeology and the Problem of Textual Evidence for the Third Dynasty of Ur”, in: *Bulletin of the Canadian Society for Mesopotamian Studies* 38, 49–62.
Ziegler, Nele (1999), *Le harem de Zimri-Lim. La population féminine des Palais d’après les archives royales de Mari*, Paris.

Nathan Morello

A GIŠ on a Tree: Interactions between Images and Inscriptions on Neo-Assyrian Monuments

1 Introduction

The study of the relationships between images and writing in the Ancient Near East is a broad field, which includes different approaches of historical and philological analysis. In this article, I would like to focus on an aspect of such relationships that has been only partially analysed in previous studies, and which I find particularly interesting in the sphere of a comparative study on materiality and presence of writing in non-typographic societies, such as the one offered by the present occasion.[1]

I will focus on the analysis of visual and semantic interactions between images and inscriptions that occur in some Neo-Assyrian monuments when part (i.e. one or more cuneiform signs) of an inscription interplays with the part of the sculpted image (i.e. one element of the image or part of it) that it crosses. More specifically, the purpose of this article is to try to answer the following questions: is it possible to positively identify interactions of this kind? Is it possible to interpret their meaning and to understand their function (or functions)? Were they meant to be a sort of concealed "game" within the inscription, or were they meant to be seen? And once they were seen, who was the intended recipient, the audience, of such interactions? As will become evident in the course of the article, my belief is that there are no unequivocal answers to these questions.

To develop my analysis I will concentrate primarily on four examples coming from the decorative record of Aššurnasirpal II's North-West Palace (henceforth NWP) in the Assyrian capital Nimrud/Kalhu (in modern north-eastern Iraq).[2] The examples come

This article emerged from the Heidelberg Collaborative Research Center 933 "Material Text Cultures. Materiality and Presence of Writing in Non-Typographic Societies". The CRC 933 is financed by the German Research Foundation (DFG). I would like to thank particularly Prof. Markus Hilgert, for allowing me, through this project, to deepen my analysis of the subject presented in this article. I am grateful to Prof. F. Mario Fales for his kind remarks and his correction of the English text, and to Cinzia Pappi for her remarks and stimulating comments during the writing of this article. I would also like to thank Prof. Stefan M. Maul, Kamran V. Zand, Alexander Edmonds, Christian W. Hess and Simone Bonzano for discussing my research with me and for their suggestions. Any mistakes or errors are the sole responsibility of the author.

1 For a theoretical approach to the subject of text as artefact and the many questions that such a perspective raises, see Hilgert 2010. For the materiality and visual aspects of Assyrian monumental inscriptions see Russell 1999, with previous bibliography.

2 The palace was excavated by Layard (1845–47, 1849–51), Rassam (1853–54), Mallowan (1949–53,

DOI 10.1515/9783110417845-003,

from three bas-reliefs on orthostats attached to the walls of three different rooms of the palace: specifically, one on relief No. 23 from throne room B, one on No. 4 from room H, and two on No. 3 from room G.[3]

Besides being the first and one of the best-preserved cases in Assyrian architecture in which bas-reliefs were used to decorate almost all of its public spaces, the NWP is the only Assyrian building in which more than half of its reliefs are crossed in their central part by an inscription of ca. 20 lines. All of the inscriptions are more or less exact copies of the same text, repeated hundreds of times throughout the palace, the so-called Standard Inscription (SI).[4]

I will first try to analyse how to define an image-inscription interaction. When, in other words, can we almost certainly identify an intentional interaction (as opposed to a mere coincidence), and what are its main visual and, possibly, semantic characteristics. Namely, the de-contextualization of one or more signs of the inscription, the de-contextualization of the corresponding part of the image, the direct or indirect relation between the two of them, and the possible meaning and/or message inferred by the visual-literary "game" resulting from their interplay.

Secondly, I will present four examples of interactions coming from the NWP with some discussion of previous interpretations.

Then, I will try to show how the interactions on the NWP's orthostats should be considered as part of a broader compositional project that was behind at least some of the most important reliefs of the palace, a project that involved the SI and the ways in which it was engraved on the palace's slabs.

I will then try to analyse whether these interactions were meant to be visible to the public or not. I will argue that the nature of both Ancient Near East figurative

1956), by the Iraqi Directorate General of Antiquities (1959–60, 1969–74) and by a joint expedition between the IDGA and a Polish mission directed by J. Meuszyński (1974–75), see Postgate/Reade 1980, 304–307, and Oates/Oates 2001, with previous bibliography. For a study that combines NWP's reliefs and inscriptions, see Russell 1999. The reconstruction of the decorative program of the NWP, with the exact position of each of the surviving sculpted orthostats, was carried out first by J. Meuszyński (1981) and then by S. M. Paley and R. P. Sobolewsky (1987). See also Russell 1998.

3 The first two examples have already been noted by Brian Brown in his article on the NWP's reliefs and cultic architecture (Brown 2010), but while I agree with the majority of his conclusions, I believe that something can be added to his analysis. The interactions from slab G-3 were noted by me during a recent visit to the British Museum conducted for the purpose of this research.

4 RIMA 2 A.0.101.23. The SI contains all the topics of a typical royal inscription: the king's name and genealogy, some of his epithets, mentions of his military conquests (in geographical, rather than chronological, order), some other epithets and a description of the king's works at Kalhu with a focus on the construction of the NWP. A. K. Grayson counted 406 specimens of the inscription (RIMA 2, 274). We shall use here the same distinction between "text" and "inscription" used by Russell (1999, 7): "A 'text' is a verbal composition, while an 'inscription' is the physical result of replicating all or part of such a text in a durable medium." However, like Russell, we will make an exception for the Standard Inscription's text, for conventional reasons.

art and cuneiform writing suggests that they were meant to be seen, though how explicitly they were to be understood depended on the context, and on the intended recipient of the interactions. With the aim of thoroughly analysing the question of the intended public, I will make use of a comparison with another Assyrian monument, the stele erected by King Esarhaddon (680–669 BC) at Sam'al/Zincirli (Turkey). This stele has already been analysed by B. N. Porter in a study on Assyrian propaganda, in which she noticed some examples of image-inscription interactions that will be useful for our purpose.

Finally, I will try to sum up some conclusions, with a reflection on the methodology that I think we should follow when treating image-inscription interplay and its possible meanings and/or functions in relation to other research questions, especially those focused on the functions of the reliefs in their architectural context.

2 Identifying Image-Inscription Interactions[5]

The main problem, when seeking to define an interaction between part of an inscription and the image that it crosses, is the possibility that the interaction occurred casually, and the observer's wish to identify it confers meaning where there is none. The very nature of multiple reading, which characterizes cuneiform writing and which (as we will see in the examples below) often makes an interplay between image and inscription possible through specific word games, can in other cases be misleading and lead the observer to an erroneous identification. Put more simply, the presence of a word "in the right place" is not a sufficient substantiation for the existence of the relative interaction with the image behind it.

5 These kinds of image-inscription interactions can only be effectively investigated with a precise knowledge of where the signs of the inscription are positioned on the monument. Modern editions of royal inscriptions, though providing in-depth information on the philological and literary aspects of the inscriptions' texts, do not allow such an analysis, and it is only by constantly comparing them with photos retrieved from other sources, or with the direct examination of the originals, that it is possible to perform this kind of analysis. Some older studies represent a good example of a more useful edition for our purpose. For instance, in C. F. Lehmann's copy of Aššurbanipal's Stele S^3, in *Šamaššumukîn, König von Babylonien 668–648 v. Chr.* (1892), the exact position of each sign is precisely indicated, making it easier to analyse any possible interactions between image and inscription. If we compare Lehmann's study with that of Grayson—which is currently the most widely used (and rightly so) edition of Assyrian royal inscriptions, but doesn't offer copies of the inscriptions—it is clear that the interactions described here will inevitably be overlooked. Now, it is obvious that an edition like Lehmann's of, for example, the more than four hundred versions of the SI from Aššurnasirpal II's NWP would be nearly impossible, especially if we consider how very few are the clear cases that we will define here as image-inscription interactions. Nevertheless, it is my belief that a digital project, in which high-definition photos of the monuments could be compared with the transliterations, the translations and the positions of their inscriptions would prove to be of great benefit.

We shall take as a negative example fig. 1, where it is possible to see the arm of a protective two-winged genie (*apkallu*) sculpted on slab S-a-2 (upper, right).[6] At line 5 of the relief, the signs that form the word "arms" (*i-da-at*) are easily visible right on the genie's arm. These signs are taken from the phrase "When Aššur, my Lord, who called me by name (and) made my sovereignty supreme, placed his merciless weapon in my lordly *arms*".[7] However, we may note in the first place the presence, on the same genie's arm, of the sign *-na* from the preposition *ana*, which precedes the word for "arms", and secondly we see the absence of any additional expedient, such as an empty space before or after the word, aimed at highlighting it. As a matter of fact, we lack any possible evidence that the word was deliberately put in that particular position.

Therefore, it is necessary to meet specific conditions in order to determine a visual interaction between image and inscription. Essentially, these conditions are two: the sign or group of signs have to be isolated or partly separated from the rest of the inscription by leaving empty space before and/or after it (but also above and/or below it); and the image element that interplays with the sign or group of signs has to encircle (and so isolate) it/them. Sometimes, the two conditions may occur together, acting as a further confirmation of the existence of the interaction.

Once the actual existence of the interaction is proven, the possible kinds of interplay between image and cuneiform sign/s will be worth considering. As may be seen in the table at the end of the article (Table 1, on p. 64–65), where all of the examples considered here have been summarized, images and inscriptions can interact in different ways:

a. *Possible de-contextualization of the sign or group of signs from the inscription to which they belong.* a.1) The sign or even the group of signs can belong to a split word, i.e. only part of the word will be included in the interaction. a.2) Even when the word is entirely included in the interaction, the fact of being part of this interaction puts it in a new context, which can be (and usually is) different from that of the inscription's text. a.3) Because of the nature of cuneiform writing, one sign can be read in different ways depending on its syllabic or logographic values. A different reading of the sign is then possible but will necessarily depend, first, on its being isolated from the rest of the inscription and, second, on a sort of "indication" given by the image with which the sign interacts. The first example that we will see below of the "GIŠ on the tree" meets all of these conditions.
b. *Possible de-contextualization of the image.* As part of the interaction, the image, or rather, the element of the image that plays with the inscription, can be extrapolated from its original context. For instance, in the interaction example noted by Porter

6 Paley/Sobolewsky 1987, 47, pl. 3. See also Stearns 1961, pl. 19.

7 RIMA 2, A.0.101.23, 5–6: *e-nu-ma aš-šur* EN *na-bu-ú* MU-*ia mu-šar-bu-ú* MAN-*ti-a* GIŠ.TUKUL-*šú la pa-da-a a-na* <u>*i-da-at*</u> EN-*ti-a lu-ú it-muḫ*. I have preferred to keep here the transliterated form, for the sake of clarity, in dealing with the single cuneiform signs.

Fig. 1: Relief S-a-2 from North-West Palace. In the frame the signs (a-)*na i-da-at*, "(in)to (my lordly) arms" are visible © Metropolitan Museum of Art, New York (elaboration by the author).

in Esarhaddon's stele (see below), the king's sculpted feet "lie" upon a written text that describes them in the act of trampling on the rulers that did not surrender to Esarhaddon.

c. *The existence of a direct or indirect relation between image and sign/s.* The interacting sign/s and image can have a *direct* relation, i.e. a relation in which the sign or word identifies the visual element on which it is written. The example described below from slab H-4, on which the name of king "Aššurnasirpal" is encircled by the hand of a sculpted king, would be, if the two were the same man, a perfect example of a direct relation. Elsewhere, when more complex reasoning is necessary to guess the relation between image and sign/s, we will have an *indirect* relation. The degree and quality of knowledge necessary to understand this indirect relation will then define the typology of the interaction and, to a certain extent, its intended audience.

The analysis of the above-mentioned characteristics allows us to study the possible meaning(s) and function(s) an interaction might have. In the following examples, we will see how in certain cases an interpretation of such meanings and functions was attempted. In other cases, such an interpretation was not possible, a fact that had the consequence of partly undermining previous theories.

3 A GIŠ on a Tree

Slab B-23 is one of the most studied reliefs from the NWP. It is located on the eastern wall of throne room B, behind the throne dais, and it shows two kings, traditionally identified as two portraits of Aššurnasirpal II, facing each other with pointing fingers in the typical gesture of greeting.[8] Behind both kings, in a symmetrical pattern, two human-headed *apkallu*s are sculpted. Each carries a ritual bucket in its left hand, while with its right it points a pine-cone-like object towards the back of the king's neck. Both bucket and pine cone object are closely connected to the image of a stylised tree, which stands between the two kings as the axis of the entire scene: the so-called Assyrian Sacred Tree, whose image is constantly reproduced (about one hundred times) all over the palace's rooms.[9] Above the tree floats a winged disc, and within it, a god greets the king on the right side of the panel, showing him the divine ring (fig. 2).

In this particular relief, the SI is written with evident care: the signs are clear, well-separated, and not indiscriminately carved on every part of the image they

8 The interpretation that the two figures represent the same king has been undermined by recent studies, see Brentjes 1994 and Brown 2010.

9 See Brentjes 1994 and Richardson 1999–2001. Bucket and pine-cone object should most probably be identified with the Akkadian terms *banduddû* and *mullilu*, see Porter 2003, 27.

Fig. 2: Relief B-23 from North-West Palace (taken from Budge 1914, pl. XI).

cross. The AST, in particular, is left almost completely untouched. This is not the only element of the relief to be free from signs (the same holds for the fringes of the characters' dresses, as well as the hands of the king and of the *apkallu* on the left side, and almost the entirety of the king's sceptre on the right side). This is, however, the only relief in the NWP in which the AST is left completely bare, with not a single element of it inscribed by cuneiform signs,[10] except for one sign right in the middle of the tree trunk: a GIŠ, the Sumerian logogram for "tree" (fig. 2a).

As part of the inscription, the sign gets its syllabic value /*is*/, as the last sign of the participle *mukabbis* in the sentence NÍTA *dan-nu mu-kab-bi-is* GÚ *a-a-bi-šú* "strong

10 There are other ASTs that are partially left uncrossed by the inscription, but none of them is completely untouched by it. I have found four of them, see Meuszyński 1981, pl. 12, H-30, H-32; pl. 15, L-21 and pl. 16, N-2. Slab B-13 is an almost exact copy of B-23. Like B-23, it was set in a very important position, directly in front of the entrance D that led into the throne room from the external courtyard. The only noteworthy difference between the two slabs is in the iconography of the god in the floating winged disc, who faces left (= east) instead of right (= south; note that both gods' gestures point at the doorway that leads from throne room B to room F, see Brown 2010, 28). Moreover, the floating god greets the king with his open right hand (instead of the divine ring), while keeping in his left an undrawn bow. Unfortunately, there are no pictures of slab B-13 with high enough resolution to check how its inscription fits on the tree. See Meuszyńsky 1981, pl. 2 for a line drawing of the slab, and Reade *Iraq* 27 (1965), pl. XXVIII for a photo of the upper-left corner of the slab, where the heads of one king and one *apkallu* are visible. The only complete (but too small) picture available of B-13 is the one in Meuszyńsky 1974, 59.

Fig. 2a: Relief B-23 from North-West Palace, detail of the GIŠ sign (photo by the author).

Fig. 2b: Relief B-23 from North-West Palace, detail. Arrows indicate two examples of uninscribed space left for the sake of the tree's integrity. In the black circle the sign GIŠ is visible. White circles: 1) sign GÚ, slightly deformed to fit into the space between branches; the same happens with 2) sign UGU, and 4) sign GAL. Sign Ú (3) was too long to fit into the spaces between the branches, and was therefore engraved in the first possible space before the end of the tree (photo and elaboration by the author).

male who treads upon the necks of his foes".[11] The fact that the sign has been intentionally highlighted by the slab's artisans is apparent for three reasons:

a. A clear equivalence of GIŠ with "tree". "Tree" is the first meaning of GIŠ as a logogram, and the same sign is used as a determinative for names of trees.[12]
b. In no other cases from the NWP (as far as I am aware) is a sign left as "isolated" on a relief as this one. As already mentioned, save for our GIŠ, not a single element of the tree is touched by the inscription. Looking closely at the slab, it is clear that even the space between the branches was filled only with signs, whose shape would not invade the area of the sculpted tree. There would have been, for example, not enough room to fit the Ú at line 12 until the last empty space before the end of the tree. Looking at certain signs, such as the GÚ of line 3 (the sign just after our GIŠ), the UGU (*/muḫ/*, from the sentence *a-na i-da-at* EN-*ti-a lu-ú it-muḫ*[13]) at line 5 and the GAL (from the sentence É.GAL TÚG.MEŠ[14]) at line 15, one can even see how they were partially deformed so as not to touch (or at least, in the case of GAL, to touch only very lightly) the tree's branches (fig. 2b).
c. Among the variants of the Standard Inscription copies noted by Le Gac, one is the use of the sign SI instead of GIŠ.[15] Needless to say (however forced this might be), no interaction between trees and signs are to be spotted in the corresponding reliefs.

In this example, then, the double reading of the sign reflects a kind of interaction in which the presence of a peculiar image (the tree) forces the reader to extrapolate a sign from its textual context (*/is/*) and to read it alone as a logogram (GIŠ), in direct connection with the image itself.

Brown's interpretation of the meaning of this interaction starts from the Assyrian Sacred Tree. Different scholars[16] have argued that the sacred tree—which very much resembles a palm tree (a feature that could put it in a close relation with kingship, see below)—served as an index that referred back to several characteristics of Assyrian kingship.[17] On the one hand, however, recent studies have convincingly undermined the supposed equivalence between the AST and a palm tree,[18] and on the other, it has been argued that the AST was probably not an actual tree, but a wooden and metal

11 Aššurnasirpal II, RIMA 2, A.0.101.23, 3–4.
12 Labat/Malbran-Labat 1988, 136–137[1].
13 "He put in my lordly arms", RIMA 2, A.0.101.23, 6.
14 "Palace of boxwood", RIMA 2, A.0.101.23, 18.
15 Le Gac 1907, 154.
16 Winter 2000, Richardson 1999–2001.
17 Brown 2010, 31–33.
18 Giovino 2007; Seidl/Sallaberger 2005–2006.

object that *represented* a tree.[19] Consequently, Brown does not concentrate directly on the meaning of the sacred tree to explain the equivalence between sign and image. Instead, he considers the equivalence between the sign GIŠ and the name of the legendary king, prototype and index of the very institution of kingship: Gilgamesh of Uruk. As a matter of fact, as shown by Parpola, the simple addition of a pre-determinative for "god" (AN = ^{d}) to the sign is enough to allow us to read the equivalence ^{d}GIŠ = Gilgamesh.[20] Brown, then, suggests a double symbolism for this sign on the AST: the "prototypical warrior and builder king serving as a symbol of the larger institution of kingship".[21]

Before passing to the next example of interaction, I would add to Brown's conclusions only one consideration, because I believe his reasoning too complex, although I don't rule it out altogether. I would not reject a simpler equivalence of GIŠ = AST, even with the necessary prudence in doing so. Very briefly, the innumerable studies on the AST (one of the most popular subjects in the history of Assyriology) have now—thanks to Giovino's *The Assyrian Sacred Tree: A History of Interpretations* (2007)—reached a stage where we can posit a close interrelation between the sacred tree and the secrecy of kingship. The interpretations of what is or is not an AST, in any case, allow us to place it within a rather restricted symbolic area. It certainly does represent a tree, even if not a palm tree, nor a real tree. However, even if they are not directly linked to one another, the AST and the palm tree share a proximity to the king's sphere, since for example, as Richardson[22] notes, the palm and the king (Sennacherib) can be defined with the same Sumerograms GIŠ.GIŠIMMAR, to say nothing of the many metaphorical relations between a tree (particularly the palm tree) and its seeds, and the king's offspring, which often occur in royal inscriptions. All together, these different theories do not put any relevant distance between "tree" (GIŠ) and kingship. Hence, I consider Brown's equivalence with the "divine tree" Gilgamesh plausible, but too indirect to be the primary one. I would say that the GIŠ on the tree is not the direct symbol of Gilgamesh, even though it shares with Gilgamesh (and one of the possible renderings of his name) the institution of kingship, in the same way that the AST is *not* a palm-tree, but shares with it the symbolism of kingship (as in Sennacherib's title). In a very similar way, I will admit that the AST was *not* an actual tree, but I do not see any reason, after so many convincing studies on this subject, to reject the idea that those who ritually saw, used, sculpted and finally recognized an AST on a palace relief would not have called it a "tree". In the same way, for thousands of years since the end of cuneiform culture, people in the western world have used the tree to illustrate genealogy—though at first the sole prerogative of kings and nobles, in today's

19 Giovino 2007, 177–196.
20 Parpola 1998, 324.
21 Brown 2010, 33.
22 Richardson 1999-2001, 162.

internet world, anybody can have his own digital "genealogy-tree", even though we know it has never been an actual tree.

4 A Name on a Hand (Aššurnasirpal)

Slab H-4 from the NWP shows a standing king performing rituals, and behind him, a two-winged human-headed *apkallu*. Brian Brown noted that on the king's left hand is inscribed the name of Aššurnasirpal ([1]*aš-šur*-PAP-A). It will be worth noting, also for the sake of what we will argue below, exactly how this is presented (fig. 3).

Fig. 3: Relief H-4 from North-West Palace © Daderot, CC BY-SA 3.0, via Wikimedia Commons (courtesy of the Brooklyn Museum, New York).

The Assyrian king (traditionally identified with Aššurnasirpal himself) is shown facing left (south), performing a libation ritual. With his right hand he raises a drinking bowl, while he keeps his left arm horizontally at waist-level, showing the back of the hand, which holds a bow from its upper end (the bow being held in rest position with the lower end on the ground, "opened" externally). The king's figure is sculpted very close to the left edge of the slab, and so it happens that the lower part of the king's left hand cuts across the sign of the very first part of the SI's first line, "Palace of Aššurnasirpal", consequently encircling Aššurnasirpal's name. No space is left before or after the hand; the inscription continues normally. It should also be noted that the very lowest part of the hand is crossed by the upper part of some of the signs from the lower line (fig. 3a). Therefore, even if it is true that the king's name is clearly visible, highlighted by the presence of the crossing hand, I believe that it cannot be argued with absolute confidence that the scribe decided to highlight the king's name deliberately, by means of this particular image-inscription interaction.

Fig. 3a: Relief H-4 from North-West Palace, detail. In the frame the name Aššurnasirpal (^{1}*aš-šur*-PAP-A) is visible © Daderot, CC BY-SA 3.0, via Wikimedia Commons (courtesy of the Brooklyn Museum, New York).

Nevertheless, we will admit that the king's name clearly catches the eye, and it will be worth discussing its possible meaning(s). Brown's interpretation is directly linked to the main subject of his article, i.e. the interpretation of the reliefs and their relation to the functions of the palace's rooms in the sphere of the ancestors' cult.

Brown rightly underscores the fact that this is the only relief in which Aššurnasirpal's name is interacting with an image in the NWP's reliefs. He also considers this detail while describing the difference, in particular, between NWP rooms G and H. As Brown notes, there is a strong similarity between all the kings depicted in room G, while in room H the kings are different from one another.[23] Consequently, while the kings in room G can all be identified as Aššurnasirpal II himself (an identification that is also due to the functions of the cult that Brown posits for the room), those depicted in room H are different kings, probably ancestors. The name on the hand, then, identifies the *only* king that actually *is* Aššurnasirpal II or his eponymous Aššurnasirpal I, an identification that can become a substitution once Aššurnasirpal II dies.[24]

Notwithstanding the value of Brown's interpretation of the NWP's architectural functions, we will try to show in the next paragraph how attributing actual direct functions to image-inscription interactions can be misleading and risks undermining part of the general theory.

5 A Name on a Throne (Shalmaneser)

Slab 3 from room G is the central part of a three-slab relief showing a banquet theme. The king is depicted seated on a backless throne, holding a drinking bowl with his right hand, while his left hand rests on his legs. Left of the king (i.e. in front of him) stand a eunuch and, behind him, a winged human-headed *apkallu*. Similarly, to the right of the king (behind him) are two eunuchs and an *apkallu* with the same features. The scene is flanked by two other slabs bearing one large AST each. The inscription runs on a band that crosses the figures' bodies from under their waists to the upper half of their legs (the lower half in the case of the seated king). Each slab contains a complete copy of the SI (fig. 4).

Looking closer at G-3's inscription, one can see how many parts of the image have been left uninscribed in the area around and including the backless throne (fig. 4a). Considering the inscription as a horizontal band that crosses the relief from left to right, we see how it invades the left part of the king's figure, covering his bent legs up to the fringes of his clothing. Then, following the line of the fringes it stops, leaving intact the rest of the king's vest, his backside, the back drape of his clothing, the throne's cushion and its ornamental ram-heads. On the throne, the inscription runs along its side and legs. The space between the throne's legs is filled with an undulating motif that represents textile fringes coming down from the throne's side. This part is not inscribed, so that in this area the inscription is carved only on the throne's

23 Brown 2010, 10–12.
24 Brown 2010, 11, 38.

legs. To the right of the king and his throne, the inscription starts again, continuing uninterrupted to the end of the slab. There is one more distinctive trait: the last line, which consists of only eight signs, is carved after the right leg of the throne instead of starting from the left edge of the slab.

Fig. 4: Relief G-3 from North-West Palace © The Trustees of the British Museum.

Fig. 4a shows how the scribe roughly followed the curved line of the king's clothing, creating a smooth aesthetic effect, but was not as careful in the lower part of the inscription, where the lines are at throne-level. In comparing those spaces with the lines carved on the throne, we get the impression that he deliberately left the exact space that he needed to fit the signs he wanted. In this relief, we find two examples of image-inscription interactions, one on the throne's side and one on the throne's legs.

The side of the throne can be divided in two parts. An upper part corresponds to the seat, on which the cushion is laid. In its real appearance, this part was probably made of wood, while to its front and back metallic ram-heads were attached. All of the

Fig. 4a: Relief G-3 from North-West Palace, detail. The arrows indicate spaces intentionally left empty. In the frame: the last eight signs of the Standard Inscription, starting at the right of the throne instead of at the left edge of the slab © The Trustees of the British Museum (elaboration by the author).

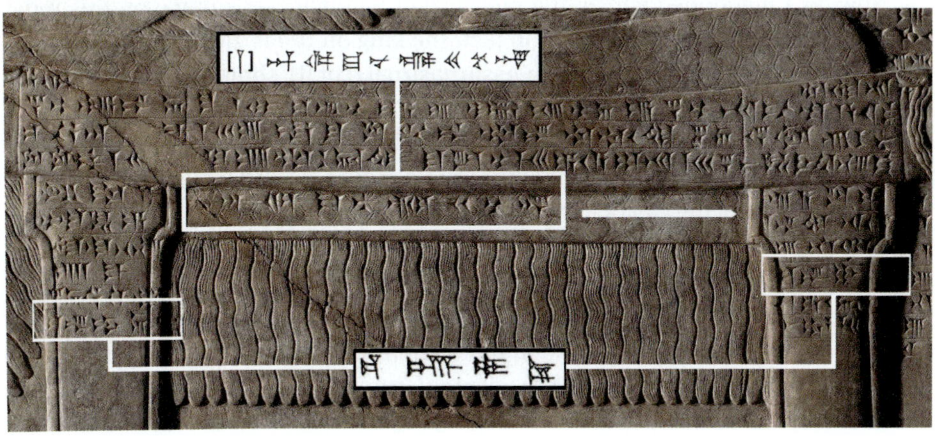

Fig. 4b: Relief G-3 from North-West Palace, detail. The arrow indicates the empty space left after the line, in the "wooden" band of the throne. The black frames are: (upper) the signs for the name of Shalmaneser ([1]d*šùl-ma-nu*-MAŠ) and for his epithet, "King of Assyria" (MAN KUR *aš-šur*); (lower) two interactions between the throne's legs and the signs for *meskannu* wood (GIŠ *mes-kan-ní*) © The Trustees of the British Museum (elaboration by the author).

"wooden" parts of the throne's seat are inscribed, i.e. three complete lines of inscription are carved, while where the outline of the seat bends to receive the cushion, only the lateral parts of the seat are inscribed, so as not to touch the "fabric". This upper part lies on the throne's legs (which show six inscribed lines each). Seat and legs together frame a lower part of the throne's side, formed by a strip of "wood" and the band of "textile" fringes hanging from it.

The "wooden" lower side of the throne is engraved with only nine signs (fig. 4b), with a long empty space between them and the throne's right leg (the only part of the relief with some empty space on the right side of the inscription). If compared with the line above, this space looks large enough to have enclosed a further 6–7 signs. Furthermore, the space below the signs could have been filled with a further line, which starts on the throne's left leg but continues only on the right one. The thus isolated nine signs form the name of Shalmaneser and his epithet, "King of Assyria" (= [1]d*šùl-ma-nu*-SAG MAN KUR *aš-šur*[25]). As part of the inscription, the king's name refers to Shalmaneser I (1274–1245 BC), who is said to have been the founder of the city of Kalhu, rebuilt by Aššurnasirpal. Therefore, the name is highlighted in two ways: by encircling it in the frame of the throne's lower side, and by leaving some empty space to the right and below it. Here, there is no different reading of the signs when following the inscription or when looking at them alone. For its part, however, the role of the image is unclear. Is the name referring to the king seated on the throne? Does it refer to the throne itself, as a sort of property label? As far as I have been able to make out, this is the only other king's name highlighted on the NWP's reliefs, besides that of Aššurnasirpal on H-4.

Now, as I mentioned above, the presence of Shalmaneser's name on this relief could undermine Brown's theory. If, in fact, the king on H-4 is given a name because he is the only one to be identified with Aššurnasirpal (I or II), shouldn't the king seated on G-3—the very room (and right on its "banquet" relief) that, as Brown convincingly explains, was decorated with only Aššurnasirpal's representations—be the only one not to be identified as Aššurnasirpal, but as Shalmaneser? And which Shalmaneser? The first, founder of the city of Kalhu, or the third, Aššurnasirpal's son and legitimate successor? My inclination tends to neither of these. This king is not Shalmaneser (I or III), nor can we have any certainty that the one on H-4 is or is not Aššurnasirpal. I find it similarly difficult to presume a functional relationship between bas-relief, room functions, and image-inscription interaction as the one suggested for H-4.

Rather, I would look for an ideological interpretation of these interactions that more broadly stressed the celebration of kingship, similar to what happens with the example of the "GIŠ on a tree". Besides deities and kings, the throne may belong to members of the royal family, members of the assembly, judges, and spirits of the dead.[26] Furthermore, the throne often has a metaphoric meaning of firmly establishing kingship (*kussī šuršudu*).[27] As suggested by C. Pappi, (and also keeping in mind

25 RIMA 2, A.0.101.23: 15. The DIŠ used as determinative ([1]) before the king's name was probably lost during the restoration of the relief.

26 Salonen 1963, 65.

27 Pappi 2013, 635. See CAD R, 189. Cf., for example, the 7th-century letter from the exorcist Urad-Gu-

the role of a king's ancestors in the NWP, as underlined by Brown in his analysis), this could be the ideological meaning behind the presence of the name Shalmaneser—the king who is celebrated as the prime founder of Nimrud in the inscription—on the throne of Aššurnasirpal II, his natural heir.[28]

6 Two Wooden Legs

The second interaction to note on slab G-3 is, admittedly, less evident than the one mentioned above, but it is rather interestingly repeated twice in the same way. On the right leg of the throne, fourth line from the bottom, and on left leg, second line from the bottom, we find the signs for the word "*meskannu* wood" (^{giš}*mes-kan-ni*) from the final description of the construction of the NWP: "I founded therein a palace of cedar, cypress, *daprānu* juniper, boxwood, *meskannu* wood, therebinth, and tamarisk as my royal residence and for my lordly leisure for eternity [...] I hung doors of cedars, cypress, *daprānu* juniper (and) *meskannu* wood in its doorways" (fig. 4b).[29]

The *meskannu/musukkannu* (also *mesukannu*, *meskannu*, *mismakannu*, *usukannu*, Sum. mes.má.kan.na, original meaning "mes tree of Makkan/Magan", possibly to be identified with the *Dalbergia sissoo*[30]) is a type of tree known to have been imported from Magan (presumably modern Oman). By the first millennium BC, the *musukkannu* tree also grew in the Near East, as we have attestations of Assyrian kings planting it (Aššurnasirpal II and Sennacherib), while Tiglath-pileser III (722–705) found a plantation of it in Babylonia.[31] In the royal inscriptions, *musukkannu* wood is defined as *iṣṣu dārû* "lasting wood",[32] and is known to have been used for the decoration of buildings and furniture of value, especially legs and bands.[33] It is particularly noteworthy that the *musukkannu* is known as one of the valuable woods used for thrones.[34]

la SAA 10, 294: 10, "May they (the gods) keep the foundation of the throne of your kingship as firm as a mountain rock forever (*išdi kussī šarrūtika kīma šipik šadî lišaršidu ana ūmē ṣâti*)."

28 Pappi, personal communication.

29 RIMA 2, A.0.101.23, 18–21.

30 Gershevitch 1957, 317ff.

31 Postgate 1992, 183.

32 See entry *musukkannu* on the CAD M II, esp. p. 238. See also for example the inscriptions from Tukulti-Ninurta II (890–884): RIMA 2, A.0.100.5, 70–73a "from Ilī-Ibni, governor of the land of Suḫu (I received as tribute): three talents of silver, 20 minas of gold, an ivory couch, three ivory chests, 18 tin bars, 40 furniture legs of *meskannu* wood, a bed of *meskannu* wood, six dishes of *meskannu* wood, a bronze bathtub, linen garments, garments with multi-coloured trim, purple wood, oxen, sheep, bread, (and) beer."

33 Salonen 1963, 221.

34 Pappi 2013, 633. See also, from the Mari archives, the throne in *musukkannum* wood that Zimrī-Lîm

Except for the sign *-ni* that "falls" out of the right leg, both words are precisely fitted in the "wooden" parts of the image and I believe that this position, with no inscribed space before and after the signs (except for the right side of the right leg), leaves room for the possibility that they were set there deliberately. If so, we would have here an interaction in which the word is kept (almost) in its entirety on the image, keeps its reading and is partly extrapolated from its original context, even though, when Aššurnasirpal claims to have built a palace of *mesakanni* wood, he might also mean its valuable furniture like the ritual throne. Moreover, if my interpretation of the aforementioned interaction between the throne and the king's name proves to be correct, a relation could be seen between the everlasting character of the *meskannu* tree and the ideological motif of *kussī šuršudu*, providing an interesting example of two interactions conceived around the same image, both aimed at conveying a similar message.

7 Image-Inscription Interactions in the Frame of the NWP Decorative Project

The examples shown above need to be analysed in the broader framework of the construction of Kalhu and its palace. In particular, studies from the past forty years have permitted more detailed interpretations of the role of inscriptions (especially the SI) in the decorative project of the NWP.

Besides their textual characteristics (narrative, linguistic, etc.), monumental inscriptions have a well-known symbolic and propagandistic meaning in their visual impact, fully aside from the ability of the viewer to read them. In its very essence, the inscription stands as a representation of the king's knowledge, and consequently, of his power and legitimacy.[35] The visual effect provided by monumental inscriptions is particularly evident in the NWP, where each relief-slab had its copy of the SI, which is visually connected to the ones before and after it, giving the impression of a continuous text. This inscription formed a horizontal band about 1.5 meters high that either crossed the images or had a register of its own, dividing the reliefs in two separate scenes.[36] A closer look at each slab clearly shows that, for the sake of this visual

donates to Nagatmiš, king of Tigunani (north of Šubat-Enlil), to reward him for his victory against the king of Ašlakka (ARM 30, 424–426).

35 Russel 1999, 47: "the power of these inscriptions must reside not primarily in their content but rather in the very fact of their existence".

36 Most of the separated slabs had images of a narrative and historical type, such as battle scenes and royal hunts of lion and bull. These slabs were posted in the throne room of the palace (B) and in the rooms to the west of the inner court Y (WG, WH, WK / BB). Other slabs carried images that covered their entire surface and bore one of three different subjects: 1) the king performing different kinds of

effect, when a slab did not have enough space to contain the entire text, the inscription was truncated, often mid-sentence. This procedure has been shown to have been carried out in accordance with some criterion.

With the studies of Paley (1976), de Filippi (1977), Sobolewski (1982), Reade (1985) and Russell (1999), it has been possible to identify and analyse two versions of the SI. These versions, Type A and B, differ in few aspects, sometimes of possible chronological nature, as in the two ways in which the text defines the northernmost limits of Aššurnasirpal's conquests: "to the interior of the land Nirib" (*adi māt Nirib ša bītāni*) in Type A, and "to the land of Urartu" (*adi māt Urarṭi*) in Type B. Type B is thus dated to a later period, after the eighteenth year of Aššurnasirpal's reign, when he conducted his first campaign against Urartu.[37]

Following a reconstruction proposed by Reade (1985) and elaborated by Russell (1999), Type A, the earlier version, was carved on all or most of the unsculpted slabs in a period that extends from the king's ninth year (875 BC) to the campaign in Urartu. In this period, the sculpted slabs were still under construction. There are no known examples of Type A on sculpted reliefs, nor any cases of truncated inscriptions on unsculpted slabs. When the slab was too narrow to contain the entire inscription, another text with similar literary motifs, the so-called Palace Wall Foundation Text,[38] was used. Somehow, after the Urartian campaign, and by the time the sculpted reliefs were ready to be inscribed, Type B had supplanted Type A and any other text. All sculpted slabs, in fact, are inscribed with Type B, no matter what size they are. When a slab is too narrow, the text is left truncated, often in mid-sentence, and no Wall Foundation Text, nor any other, is used to fill narrow spaces.[39]

This reconstruction, besides exhaustively explaining the distribution of the texts in the palace's inscriptions, suggests a period of time long enough for a group of royal scribes, entrusted with the composition of the relief inscriptions, to define in detail a work of great complexity and refinement, a work that could have included the opportunity to insert variations in the obsessive repetition of the same inscription, without corrupting the original text.

Sometimes, certain elements of the image were left untouched, usually the most elaborate ones (wings, fringes of clothing, bowstrings), which could be adversely affected by the addition of cuneiform signs, and on which, moreover, the signs would

---

rituals (throne room B and rooms C, F, G, H, S); 2) protective deities with human or bird heads (present in all the sculptured areas of the palace); and 3) foreign people presenting tribute to the king (on the throne room façade D, ED, E).

37 For other differences between Types A and B, see Russell 1999, 30–41.

38 Russell 1999, 24–28.

39 Russell 1999, 38.

have been less readable.[40] If we compare slab B-23 (the one with our GIŠ) and B-26,[41] for instance, we see that the writing is placed indiscriminately on many parts of the latter (e.g. including wings and fringes), and is therefore far less legible than on the former (fig. 5).[42]

We will imagine these scribes planning the composition of the most important reliefs of the palace. It is also possible to imagine that they would have painted the SI's signs on the sculpted slab to see the possible outcome of the whole work before ordering the stone-cutters to start working.[43]

On the other hand, the rest of the orthostats were inscribed without particular care or distinctive traits, and in some cases they were even left truncated at mid-sentence if space on the slab was insufficient. Yet, even when truncating occurred, a possible criterion is visible, related to the sculptured subject. As Russell notes, on the narrow slab 11 in room G (1.12 m), where a king is carved, the SI is included in its entirety, with a visible overcrowding of signs. Close to G-11, slab G-9 is decorated with a simple courtier and is only 7 cm narrower, but has a truncated text.[44]

The creation of the palace reliefs appears to have been a highly complex compositional project, of which the image-inscription interactions we are dealing with here were but one feature.

8 Visibility and the Public

If the complex compositional project behind the interactions, and more generally behind the whole effort of the decorative plan of the NWP, is now clearer, a question remains: how visible were these interactions meant to be? Were they meant to be concealed in the complex figurative decoration, or were they meant to be seen clearly?

The concealed hypothesis seems to be supported by two substantial elements. The first is the overall difficulty in understanding some of these interactions, as even the most evident "GIŠ on the tree" needs a well-trained mind to understand it in its deepest meaning. The second feature is their inaccessibility, their location in the rooms of the NWP exposing them to a very limited group of people.[45] Furthermore,

---

40 See for example Stearns 1961 (with the corresponding locating numbers from Meuszyński 1981 in brackets), pl. 2 (H-33), 3 (H-4), 5 (G-11), 9 (L-17), 10 (G-18), 11 (L-35), 16 (N-5), 17 (L-10), 18 (L-8).
41 Sterns 1961, pl. 51. For the identification of this slab's original position, see Meuszyński 1981, pl. 1.
42 Note that in many cases, in which different objects are left untouched by the inscription, no specific need for a clear reading can be presumed. Ritual buckets, parts of furniture, and hands are all elements that would not be affected by carving, nor would they render the inscription illegible.
43 For an example of painted cuneiform signs, see Reade 1986, 217.
44 Russell 1999, 40. See also for example Slab H33 (AfO Beich. 15, pl. 2), 1.04 m wide, showing a king and a complete SI, compared to slab B-26 (fig. 5), 1.31 m wide, showing an *apkallu* and a truncated SI.
45 Fales 2009, 253.

Fig. 5: Relief B-26 from North-West Palace. If compared with B-23 this slab looks as if it had been inscribed with less care. No part of the sculpted image was left uninscribed. This slab also carries a truncated Standard Inscription (which ends with the signs for *ek-ṣu-te a-pi-ir* – RIMA 2, A.0.101.23: 13), although it is 1.31 m wide, almost 20 cm more than G-11 (1.12 m), where the SI is complete (Russell 1999, 40) (taken from Budge 1914, pl. X).

sometimes their very positioning in the architectural arrangement made them even less visible. One should not forget that even the "GIŠ on the tree" was placed behind the throne dais, therefore behind the throne itself, leaving very few occasions for possible contemplation.

Nonetheless, one should take into account how Ancient Near East figurative art and cuneiform writing—the two intimately related forms of art involved in image-inscription interactions—not only show a predisposition for approaches of multiple reading. They are also conceived following arrangements that include the possibility of focusing the attention of the observer on one particular element of the (literary or figurative) work by graphically highlighting it.[46]

Ancient Near East figurative art appears, at least to our eyes, naturally suitable for double or multiple readings and multiple significance. It is evident at first glance that the ritual scenes sculpted on the orthostats of the NWP do not represent simple celebrations of single historically identifiable moments. On the contrary, they bear a number of different messages relating to many aspects of the cult and institution of kingship. Furthermore, such works of art are not conceived only to be contemplated. In fact, they are meant to involve their observer in a periodical (and in some ways perpetual) ritual.[47] The presence of image elements characterized by multiple readings (the AST is a clear example) gives the bas-reliefs not only a quality of polysemy but also one of performativity, i.e. the possibility of doing things, or performing actions with and through them.[48]

---

46 As many scholars have pointed out, it is not possible to study figurative art without textual analysis, especially when they are two parts of the same work of art, i.e. the monument with its image(s) and inscription(s). In the introduction to their brief study on the codes of visual representations in the reliefs of the Assyrian king Sennacherib (705–681), Carlo Lippolis and Marco Benetti argue that only by considering the centrality of the bounds between visual art and cuneiform writing can we hope to coherently interpret Mesopotamian art. The two authors do not refer simply to the message of royal propaganda shared by the two forms of art, but especially to the way of conceiving visual representation as a written enunciation (Benetti/Lippolis 2011, 79). Their study develops a less recent idea proposed by Irene Winter, who, in her article on the royal rhetoric in Neo-Assyrian reliefs (Winter 1981), summarizes the main (physical, ideological, historical, etc.) relationships between Assyrian royal inscriptions and reliefs, starting from the case of Aššurnasirpal II's NWP. In particular, she indicates that writing and sculpted reliefs share, on the one hand, "a similar syntax of action and explicit articulation of action to consequence", and, on the other, "the same grammatical and compositional play on the ambiguity between the subject of the action / the king in the first and third persons" (Winter 1981, 21). Furthermore, in a more recent article, she underlines the close relationship between the structural organizations of visual narrative in the reliefs and the very nature of the Akkadian language (Winter 1997, 362).

47 Nadali 2013.

48 The ritualistic property of figurative art is not limited to reliefs of mythological themes. For similar characteristics in narrative reliefs see Nadali 2001–2003. The subject of performativity of the Ancient Near East monuments is in many ways close to the subject of the agency they were to perform on the society in which they were conceived. See Marian Feldman's article on the Hammurabi codex-stela

This active role of reliefs is also made possible through their graphic arrangement, which allows for different points of view for the same work of art, following a multi-centric principle. As we know, Assyrian reliefs were not conceived as static graphic representations of a given subject that reproduce reality following the principles of modern (western) perspective. On the contrary, in Mesopotamian reliefs, space is conceived and represented as non-perspective and multi-dimensional.[49] This characteristic is particularly evident in the reliefs of narrative subjects, where different scenes are gathered on a single sculpted slab. Here, the different sizes of the elements do not depend on the fixed perspective of Renaissance conception (e.g. the closer objects appear larger). On the contrary, the principle that rules the size of the depicted elements is grounded in the spatial, psychological and temporal relationships between them.[50] In the narrative orthostats from the NWP's throne room, for instance, we see how the size of the soldiers depends on their status, the role they play in the narrative, and their relation to a particular part of the scene. Looking at slab B-4 (lower register), we see that the assault on the fortified city can be divided into different sub-scenes placed around the city, depending on the size of the characters and their relation to time and space (fig. 6). Close to the right edge of the slab, two officials, one archer and one shield bearer, have the same height as the city and represent the main narrative theme: the overwhelming attack conducted by the Assyrian army against the enemy. For the sake of dynamism, the enemies, shown at almost half the height of the two soldiers on the right, are proportionally too big if compared with the walls they are standing on. In the centre of the relief, two Assyrian soldiers are almost the same size

(Feldman 2010), where she analyses the consequences of the introduction of perspective in spatial rendering of the sculpted scene, and the work on agency marked and agency ascribed in Mesopotamian monuments by Irene Winter (Winter 2007). Winter's article, in which she reviews Alfred Gell's *Art and Agency. An Anthropological Theory* (1998), also deals with the agency of language in monumental inscriptions, but focuses only on linguistic aspects and semantic relationships with the respective monuments' images. I believe that it would be worth adding to her conclusions a study on the active role of monumental inscriptions that included both their linguistic characteristics *and* their visual features. Such a study would start from their physical incorporation onto the monuments—a subject that Winter mentions in her work on the conception of royal image in Assyrian ideology (Winter 1997, 74)—and would include the kind of image-inscription interactions we are dealing with here, in the light of social theory.

49 In their article, Benetti and Lippolis describe the non-perspective nature of Assyrian reliefs. Differently than in western art, where space is conceived following the principles of perspective, in the Assyrian reliefs there is no fixed point of view that defines the shape of objects, deforming them according to their position relative to the viewer. On the contrary, the objects are depicted "simultaneously", showing the relationships between them in space and time. As recipients of the work of art, we can appreciate the represented objects as very close—with minute, clearly visible details—and very far, as single parts of the whole (Benetti/Lippolis 2011, 100). For a critical analysis of western perception of non-western art (especially Ancient Near Eastern), according to the principles of perspective, see among others Bahrani 2003, 73–95 (but cf. Fales 2009), and Feldman 2010, 151–159.

50 Benetti/Lippolis 2011, 86.

as the enemies with whom they are directly interacting, as they try to pull a soldier down from the walls with a pair of chains (fig. 6a). Still of another size are four Assyrian soldiers, one pair on top of a battering ram (shown as big as the city walls) and another pair intent on crushing the city walls with maces. The result of such multi-dimensional representation is that we are able to "zoom in" on the different parts of the battle scene, so as to better appreciate the narrated events.

Fig. 6: Relief B-4 lower band, from North-West Palace © The Trustees of the British Museum.

Quoting the neurologist R. L. Gregory, Lippolis and Benetti explain this way of conceiving space in the Assyrian reliefs as a consequence of the very nature of visual perception, which is not passive and objective (as though seen through a modern camera). In fact, visual perception is the result of a decisional process based on limited sensory evidence, for which the brain dynamically creates an interpretation of what the eyes perceive.[51] Moreover, the eye is naturally attracted to objects that are in different ways (size, position, etc.) more evident than others.

Cuneiform writing is also well known for its characteristic of multiple reading. The single cuneiform sign has two main readings, one (inherited from its Sumerian origins) as logogram, i.e. one sign = one (or more different) word(s), and one as syllabogram, i.e. one sign = one (or more different) phoneme(s). The example of the "GIŠ on the tree" shows this double nature well. In Mesopotamian culture, as has been explained by such scholars as Bottéro, the relation between the cuneiform sign and the signified realities is extremely deep, and it is bound to the notion that every element or phenomenon of reality (including writing) can be interpreted as a message

51 Benetti/Lippolis 2011, 83.

Fig. 6a: Relief B-4 lower band, from North-West Palace, detail © The Trustees of the British Museum.

from the divine world.[52] Under certain circumstances, the fact that two (or more) realities can be identified by the same sign (polysemy) or that two (or more) signs meaning different words can be read in the same way (homophony) can create substantial relationships. Furthermore, such relationships can sometimes—in specific texts of high literary importance—be actively modified for specific purposes. Similar word games, then, are not conceived on a merely linguistic basis.[53] They are deeply rooted

52 Bottéro 1987, passim.

53 See, for example, a case from Esarhaddon's inscriptions (RINAP 4, 104 ii 2–8), where the god Marduk himself plays with cuneiform signs—in this particular case, fractions. After having punished Babylon and its inhabitants by designating a 70-year period of destruction (i.e. Sennacherib's devas-

in the very idea of knowledge as conceived in Mesopotamia. As Glassner summarizes, "analogy was woven into the very nature of language. We are struck by the amazing deftness with which the Mesopotamians plied the resources of their languages to express the relationships between words and things. It is here that we touch on the core of their thought, for the analogical relationships were not only consecrated by words, they were founded by them".[54]

The hermeneutical importance of writing is also based on the fact that words and signs have, as figurative art, a quality of performativity. It should be noted here that, already in the Sumero-Akkadian royal inscriptions of Old Babylonian times, the pronounced and/or written word represented not only a descriptive means, but in fact the necessary condition to create a communicative relation with the gods. As Seminara points out, the king's (building or military) undertakings needed to be perceived by the gods through their senses, and writing had a central role in such ritual performance.[55] In a recent article, I used this point of view as a basis to analyse the notably large number of references to sensorial perceptions (especially sight, hearing, and smell) used in the descriptions of landscape from Sargon II's *Letter to Aššur*. In the text, the Assyrian king (722–705 BC) writes directly to the main god of the Assyrian pantheon, giving an account of his victorious military campaign against Urartu in 714 BC.[56] My opinion in that article is that such an abundance of references to sensorial perceptions in a text of high literary and cultic value has the performative function of facilitating the opening of a communicative channel between king and god during the reading of the letter.[57]

The performativity of writing, or more simply its feature of "doing" different things at one time due to the possibility of multiple reading, is also visible in examples in which a particular sign seems to be used specifically to attract the eye of the observer. We should note, for instance, Irene Winter's textual analysis of one of Gudea's royal inscriptions (22nd century BC). In Cylinder A, where the Sumerian king celebrates the building of a new temple for the god Ningirsu, Winter points out that after the end of the construction works, the temple is described with the words "it stood to be marvelled at". Here, a central role is taken by the Sumerian logogram U_6, "marvellous", which is composed by two other signs, IGI, the logogram for "eye", and

---

tation of the city in 689 BC) on the "Tablet of Destinies", the mighty god reconsiders his decision and reverses the order of the two wedges that form the number 70. In this way 70 years become 11, and Esarhaddon thereby finds the divine authorization to rebuild the ruined city without dishonouring his father. As Glassner has pointed out, "none of those operations was gratuitous; they were the very basis of Mesopotamian intellectual production." (Glassner 1995, 1819).

54 Glassner 1995, 1818. See also Maul 1999 for the use of orthography and etymology as a hermeneutical method of Babylonian scholars, and more recently Selz 2013.

55 Seminara 2004, 537.

56 Thureau-Dangin 1912.

57 Morello 2013.

É, the logogram for "house/temple". It would be difficult to deny that any sufficiently trained cuneiform reader would notice (and be attracted by) the use of such a composite logogram, which uses the very objects involved in the action of admiring: the temple and the public's eyes.[58]

In considering such characteristics of Assyrian figurative art and cuneiform writing, we will find it less difficult to understand the nature of the interactions between image and inscription. In this light, it becomes more plausible to suppose the existence of compositional projects, behind the reliefs from the NWP, that combine the possibility of putting certain words or signs in evidence, focusing (zooming) attention on a particular spot, and creating possible relations between those words or signs and the part of the image they cut across.

On the other hand, we should not overestimate this phenomenon and conclude that all written sources coming from the Mesopotamian world were bearers of this kind of polysemic double meaning. In fact, as Fales recently warned us, we will not forget that such "games" of multiple significance are confined to very few examples gathered from particular kinds of literary texts.[59]

At the same time, we will also agree with Fales on another aspect. When such examples of multiple readings occur, they seem to be the outcome of the presence, in the king's entourage, of scribes and scholars expert in the technical competences involved in the relations between the divine and human world—what Pongratz-Leisten has defined as *Herrschaftswissen*.[60] Thanks to the correspondence of Esarhaddon and Aššurbanipal,[61] we know these relations to have been particularly intense during the 7th century, but we can assume they were also at least partly extant in previous periods. Those belonging to this restricted circle, namely the king himself, his intellectual entourage and a few others, were the primary recipients, the intended public, of these complex literary and artistic operations.[62]

---

58 Winter 2007, 55–56.

59 Fales 2009, 253: "the vast majority of Assyro-Babylonian written documents, i.e. those which for more than three millennia were concerned with administration, law, epistolary communication, international relations, and historiography, do not exhibit this extraordinary 'game' of multiple and multidimensional significance—and fortunately so, otherwise we would have never been able to piece together the skeleton of Mesopotamian history as we have painstakingly done for the last 160 years! In other words, these thousands of texts 'speak straight' to us almost 100% of the time—with terms, syntax and especially logic which we, some two thousand years after the last cuneiform tablet was written, have fully come to follow and understand, despite the innumerable linguistic difficulties and contextual obscurities involved."

60 Pongratz-Leisten 1999.

61 See esp. SAA 13.

62 The more refined and complex example of this kind of sapiential dialogue, conceived for the inner court only, is the one that we could also define here as the strongest example of interplay between figurative art and writing, in which the images are themselves words: the so-called "astroglyphs", small depictions of animals, anthropomorphic elements, objects of everyday life and astral symbols that,

Hence, our conclusion on the image-inscription interactions coming from the NWP's reliefs will be that these complex visual-literary interplays, even if partly concealed by their position in the palace's spaces, were meant to be seen. Their ideal recipient would have belonged to one of the three categories indicated by Fales in his chart of "possible types of audiences for Assyrian bas-reliefs": the inner court (scribes, scholars, and other in-groups), gods, and the king himself. Consequently, the respective functions of the interactions were those of political-ideological internal propaganda (preaching to the converted), feedback for divine support (through the celebration of kingship and its bond with the divine world), and self-gratification of the king, in addition to his intent to preserve the memory of himself for future generations, especially princes.[63]

All this said, we shall take into account how very similar image-inscription interactions prove to have been used in a diametrically different context, for a different kind of recipient and with a different purpose: on a public stele, displayed outdoors and in a frontier territory. The stele was erected by Esarhaddon in the city of Sam'al/Zincirli Höyük (in the Anti-Taurus Mountains of modern southern Turkey), two centuries after the foundation of the NWP (fig. 7). The evidence provided by this monument, as analysed by B.N. Porter, shows a different use of image-inscription interactions, aimed at a much more direct, and in a way simpler, propagandistic message. The interactions briefly described below were meant for two possible classes of people: Assyrian officials based in frontier regions, and possibly, if we imagine the occurrence of a public reading of the stele, perhaps with the aid of a translator, a local non-Assyrian public.

The propagandistic functions of the Sam'al stele have been already extensively analysed by Porter in her article;[64] here we will only describe the image-inscriptions as they appear on the monument.

The stele is 3.46 meters high and it was erected for public display in the gate leading to the citadel. It shows the king raising an emblem of royal power in one hand, while the other holds the sceptre and two ropes with which he tames two small captives, who are kneeling before him: they are the rebellious king Abdi-Milkutti of Sidon (defeated in 677 BC)[65] and the crown prince of Egypt, captured during the Egyptian campaign described in the stele's inscription.

---

through the filter of the aforementioned sapiential knowledge (especially astronomy), can be "read" or "translated" as names and well-attested epithets of the kings who commissioned them (see esp. Finkel/Reade 1996, Scurlock 1997 and Roaf/Zgoll 2001).

63 Fales 2009, 281, Chart 1.

64 Porter 2003, 75–77.

65 This ruler has also been identified with Ba'al of Tyre (Pettinato 1975).

In the Sam'al stele, we see three different kinds of interactions:

a. the sign *-ia*, suffix of the 1st person singular possessive pronoun "mine", repeated at the end of each of the first ten lines of the inscription, and highlighted by the space left to isolate it from the rest of the signs (fig. 7a);[66]

b. the ligature *aš-šur* right in the middle of Abdi-Milkutti's beard, isolated by means of empty space left between the last four signs of line 13 (fig. 7a);

c. the positioning of the sentence "all the non-submissive, the kings who would not bow to him, like swamp reed he cut down and trampled at his feet",[67] in the area of the stele under the king's feet (fig. 7b).

Fig. 7: Stele of Esarhaddon (Sam'al/Zincirli) (photo by the author).

66 RINAP 4, 98: 1–10.

67 RINAP 4, 98: 32–33: *kullat lā mā<gi>rēšu malkī lā kanšūtišu kīma qan api / uḫaṣṣiṣma ušakbisa šēpuššu.*

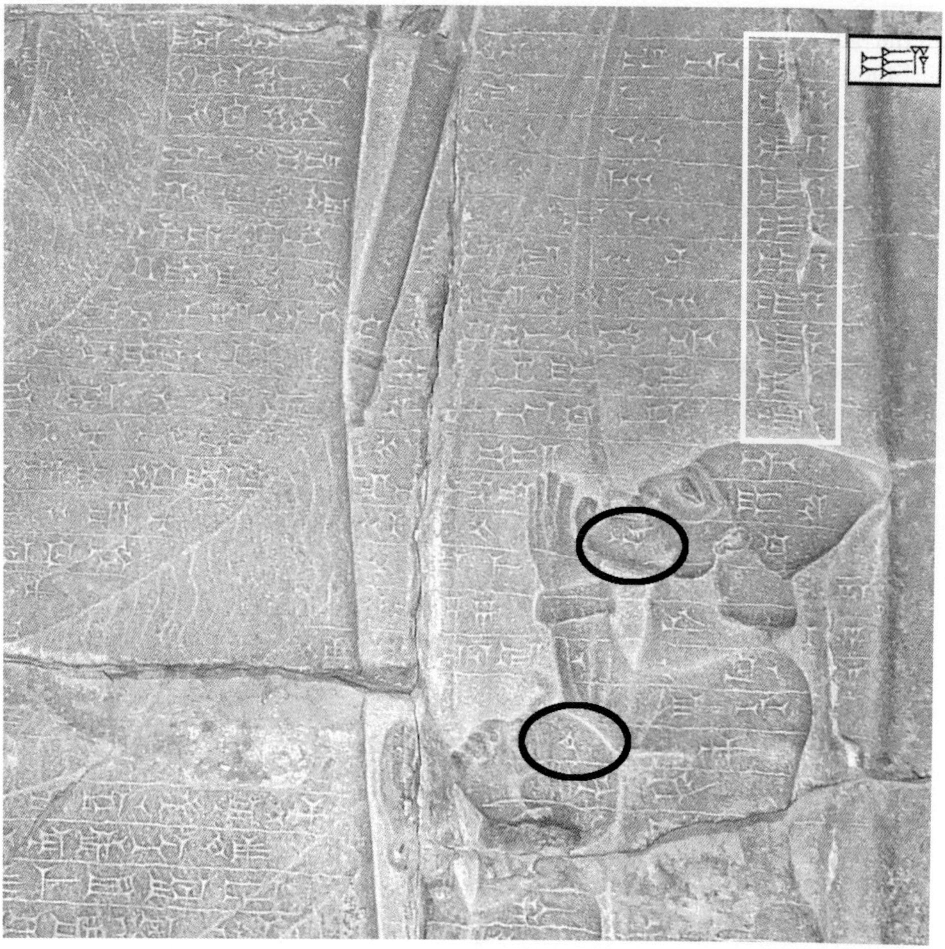

Fig. 7a: Stele of Esarhaddon (Sam'al/Zincirli), detail. In the white frame: repetition of possessive pronoun *–ia*. In the black circles: (upper) the ligature *aš-šur* on captive's beard; (lower) sign ŠÁR (< AN.ŠÁR = Aššur) on captive's head (photo by the author).

Fig. 7b: Stele of Esarhaddon (Sam'al/Zincirli), detail. Inscription engraved under the king's feet (photo by the author).

In the first case (a), the lines are visibly and excessively spaced in the inscription area, so that the pronoun "mine" seems to stand apart from the other signs at the very end of each line. Here, the "interaction" is created by the absence of any image: the signs are isolated in the void (filled only by the ruling lines and by the signs themselves) that constitutes the background of the sculptured scene.

In absence of a particular image, it is nevertheless possible to imagine an indirect relation between the sign ("mine") and the stele itself, as well as the place where the stele was displayed, the inhabitants of that city, and the two rulers at the king's knees.

In the case of the word "Aššur" on Abdi-Milkutti's beard (b), Porter points out that "the spacing of the signs in this line [i.e. 13] and the line before make it clear that this placement was deliberate [...]. The ligature *aš-šur* was carefully isolated on the beard, separated from its pre- and post- determinatives [i.e. KUR, "land" and KI, "earth", both determinatives for toponyms] by space for a total of about three signs."[68] Porter suggests interpreting this "Aššur" on the beard as an "ironic"[69] property label, and she compares it with other objects taken as booty and similarly inscribed with a line of cuneiform labelling them as property of the king, certainly a plausible hypothesis. In this way, the ruler's portrait is enriched with a further detail that asserts his tamed condition.[70] In contrast to the case of the GIŠ on the AST, here the word "Aššur", even if separated from its determinatives, keeps the same reading in the inscription and when extracted from its textual context and applied to the image of the captive king. The relation between image and signs is indirect because it needs "historical" knowledge to support it, suggested by the image.

In the third case (c), the interaction between image and inscription is not due to their intersection, but to their juxtaposition. Yet the carved feet "trample" on the sentence, as the king tramples on his foes. Here, the feet are extrapolated from their iconographic context to take part in a game with the words deliberately positioned below them.

9 Conclusions

The examples of interactions between images and inscriptions that we have analysed in these pages are all characterized by extraordinary techniques of elaboration—

68 Porter 2003, n. 182.

69 Personally here, I do not see any irony nor mockery. Rather I take it as a plain and "objective" (with all its propagandistic consequences) declaration of property.

70 There may also be a similar interaction, although this is admittedly more speculative, on the Egyptian captive's head. Here too, the scribe left more than enough space between the last four signs of the line. In fact, there would have been enough space to fill the line with even more signs than in line 13. The only sign inscribed on the prince's head is a ŠÁR, part of the word AN.ŠÁR, again " (god) Aššur".

through continuous acts of de-contextualization and re-contextualization—of the elements that coexist in the same monument. In the form of an inscription, writing acquires materiality, and it can be physically "moved", until it finds its place on the image sculpted behind it. Still, the complexity of this operation resides in the fact that inscription and image keep their normal reading, even when a new one is added, so that we end up having two (or more) texts for one inscription. This happens because of the possibility of multi-significance created by isolating particular words, but also by the different possible readings of the single sign. The image as well, or rather its part, is extrapolated from its original context to interact with the signs. We are able to see the creation of a new level of perception of the work of art, one that is not only visual or literary, but a combination of the two, a combination that produces a new message.

This complex technique of monument manipulation can then be applied to completely different contexts, which require different messages meant for different recipients. The stele of Esarhaddon at Sam'al has a clear propagandistic purpose, which is surely performed not only by the interactions between images and inscription. As Porter has pointed out, the analysis of this monument—and Til Barsip's possible comparison with monuments with the same iconographic theme—has revealed a complex project, partly based on information acquired from local tradition and political conventions, which is conceived especially for its audience. A certain kind of text, certain ways to depict the characters (their clothing and their proportions), and a certain place for the monument display, are all elements carefully manipulated for the desired purpose.

The case of the reliefs from the NWP is also the result of a complex project, but within a very different (inner) context. Here, we may observe a similarly careful choice in the reliefs' subjects, their arrangement within the architectural spaces of the palace, and their engraving with many different versions of the same text. However, the analysis of the interactions in the palace of Aššurnasirpal II has shown how, in contrast with the analysis of Esarhaddon's stele, we keep missing the functional relations between interactions, subjects of the reliefs and architectural spaces.

I would conclude with a very brief reflection, based on this lack of certainty. Especially in the case of the NWP, as we have mentioned in previous paragraphs, I believe we should reduce our attempts to interpret functional relations between interactions, reliefs and architectural spaces, and certainly we should avoid unambiguous solutions that could have the unfortunate outcome of limiting our research. In other words, as we understand that putting a GIŠ on an AST, even if it clearly creates a relation between the two, does not imply an equivalence between AST and "tree" (i.e. does not represent a sufficient datum for the debate on the subject of the nature of the AST), we will also avoid oversimplified identifications between names and kings. The name of Shalmaneser on the throne in G-3, or that of Aššurnasirpal on the king's hand on H-4, does not represent sufficient evidence to identify either the name of the depicted king nor the function of the room. As a matter of fact, what we have seen of

Esarhaddon's stele shows that, even in that case, the "scope" of the interactions does not seem to be that of creating simple tags to indicate who is who on the monuments.

Rather, by leaving the task of decoding, case by case, the reasons behind the realization of these "games" of image and signs somewhat open, I would propose to take into account the apparently simplistic function of opportunity.

The reconstruction that we have seen of the phases of realization of the NWP's reliefs has clarified the fact that the decorated orthostats have been inscribed only with the Type B SI, and that this operation happened only after they had been sculpted in their entirety. The scribe or group of scribes entrusted with the realization of the complex project of composition had the opportunity to work with ready-sculpted reliefs and a rather brief fixed text. On one hand, the example of the GIŠ on the AST—on the very relief that would have stood behind the king's throne—appears quite similar to Esarhaddon's stele, in terms of how the general message of the relief given by the image "rules" the creation of its interaction with the inscription. On the other hand, in the reliefs of slabs H-4 and G-3, the depicted scenes seem to create a graphic basis on which it was *possible* to elaborate interactions, e.g. the king's throne and its legs, and the king's hand positioned at the right height to be where the SI would have begun. In these last two examples, the realization of interactions between images and inscriptions appears to have been the result of "simple" opportunity. As a way of conceiving and perceiving writing in which there is an interplay between elements originally not conceived to interact with one another, this produces variations on the main theme—the celebration of kingship—that are worth imagining, realizing and communicating, in a permanent dialogue performed in two directions: from the king to mankind, and from the king to the gods.

Table 1: Image-inscription interactions; visual interactions, image and inscription de-contextualization and direct/indirect relationships between sign/s and (part of the) image.

<table>
<tr><th>Monument</th><th>Image element</th><th>Inscription element</th></tr>
<tr><td rowspan="2">NWP B-23</td><td>Assyrian Sacred Tree</td><td>GIŠ (A.0.101.23: 3, mu-kab-bi-<u>is</u>)</td></tr>
<tr><td colspan="2"><u>Visual interaction:</u> The sign is the only one on the whole tree.</td></tr>
<tr><td rowspan="2">NWP H-4</td><td>King's hand</td><td>^{1}aš-šur-PAP-A (A.0.101.23: 1)</td></tr>
<tr><td colspan="2"><u>Visual interaction:</u> The hand's line encircles the name.
No space left (note: part of signs from lower line are visible)</td></tr>
<tr><td rowspan="2">NWP G-3</td><td>Throne's side</td><td>$^{[1]d}$šùl-ma-nu-SAG MAN KUR aš-šur
(RIMA 2, A.0.101.23: 15)</td></tr>
<tr><td colspan="2"><u>Visual interaction:</u> the words are isolated on the throne's side.
Space left: R. of the words (ca. 8 signs' space); under the words on the throne's side (1 line space); under the words, out of the "wooden" part of the throne's side (4 lines space).</td></tr>
<tr><td rowspan="2">NWP G-3</td><td>Throne's legs</td><td>L. leg: GIŠ.mes-kan-ni
(A.0.101.23: 21 / G-3 l. 2nd from bottom)
R. leg: GIŠ.mes-kan-ni
(A.0.101.23: 18 / G-3 l. 4th from bottom)</td></tr>
<tr><td colspan="2"><u>Visual interaction:</u> inside the legs (exc. –ni on R leg)
Space left: L. leg.: no sign on L and R; R. leg: no sign on L, remaining inscription on R.</td></tr>
<tr><td rowspan="6">Esarh.'s Stele of Sam'al</td><td>Empty stele's background</td><td>-ia suffix pronoun 1st p. sing. "mine"
1+9 times at the end of the first 10 lines
(RINAP 4, 98: 1-10).</td></tr>
<tr><td colspan="2"><u>Visual interaction:</u> The -ia is isolated on an empty background.
Space left: about 1 sign each line (exc. 1st).</td></tr>
<tr><td>Abdi-Milkutti's beard</td><td>aš-šur (RINAP 4, 98: 13, KUR.aš-šur.KI)</td></tr>
<tr><td colspan="2"><u>Visual interaction:</u> the ligature is on the beard.
Space left: L. of the ligature: ca. 2-3 signs space; R. of the ligature: ca. 3 sign space.</td></tr>
<tr><td>King's feet</td><td>kul-lat la ma-<gi>-re-r-šú mal-ki la man-šú GIM GI a-pi // ú-ḫa-ṣi-ma ú-šak-bi-sa še-pu-uš-šú
"All the disobedient and unsubmissive rulers, like a reed in the swamp // he broke down and trampled (them) under his foot" (RINAP 4, 98: 32-33).</td></tr>
<tr><td colspan="2"><u>Visual interaction:</u> The king's feet "lie" on the sentence.
No space left.</td></tr>
</table>

Image de-context.	Inscription de-context.	Direct/indirect relationship
No	Word-split: yes Diff. reading Diff. meaning Diff. context: yes	<u>Indirect</u> → "philological relation" requires knowledge to appreciate: 1. different reading; 2. semantic relation (GIŠ = tree); 3. symbolic relation (GIŠ [= Gilgameš] = Kingship = AST)
No	Word-split: no Same reading Same meaning Diff. context: ?	Identity? Yes → <u>direct;</u> No → <u>indirect</u> (room's function (?), kingship)
No	Word-split: no Same reading Same meaning Diff. context: ?	Identity? Yes → <u>direct;</u> No → <u>indirect</u> (room's function (?), kingship), possibly *kussī šuršudu* "establishing firmly the throne (= kingship)"
No	Word-split: no Same reading Same meaning Diff. context: yes	<u>Direct</u> → same semantic realm between *meskannu* wood (used for furniture, e.g. thrones) and the wooden legs of the throne. *and* <u>Indirect</u> → everlasting character of the *meskannu* tree → relation with *kussī šuršudu*?
No/Yes	Word-split: yes Same reading Same meaning Diff. context: yes	<u>Indirect</u> → No interaction with an image's element. Relation with the medium, with the place, with the characters etc.
Yes	Word-split: yes Same reading Same meaning Diff. context: yes	<u>Indirect</u> → "historical relation" requires knowledge of Abdi-Milkutti's identity and destiny.
Yes	Word-split: no Same reading Same meaning Diff. context: no	<u>Indirect</u> → different action "performed" by the feet

Abbreviations

CAD	The Assyrian Dictionary of the University of Chicago.
RIMA 2	Grayson 1991.
RINAP 4	Leichty 2011.
SAA	State Archives of Assyria, Helsinki.
SAA 10	Parpola 1993.
SAA 13	Cole/Machinist 1998.

Bibliography

Bahrani, Zainab (2003), *The Graven Image. Representation in Babylonia and Assyria* (Archaeology, Culture and Society), Philadelphia.

Barnett, Richard D./Lorenzini, Amleto (1975), *Assyrian Sculpture in the British Museum*, Toronto.

Benetti, Marco/Lippolis, Carlo (2011), "Osservazioni sui codici di rappresentazione visiva nei rilievi di Sennacherib", in: Carlo Lippolis (ed.), *The Sennacherib Wall Reliefs at Nineveh* (Monografie di Mesopotamia 15), Florence, 79–111.

Bottéro, Jean (1987), *Mésopotamie. L'écriture, la raison et les dieux* (Bibliothèque des Histoires), Paris.

Brentjes, Burchard (1994), "Selbstverherrlichung oder Legitimitätsanspruch? Gedanken zu dem Thronrelief von Nimrud-Kalah", in: *Altorientalische Forschungen* 21, 50–64.

Brown, Brian (2010), "Kingship and Ancestral Cult in the Northwest Palace at Nimrud", in: *Journal of Ancient Near Eastern Religions* 10 (1), 1– 53.

Budge, E. A. Wallis (1914), *Assyrian Sculptures in the British Museum*, London.

Cole, Steven W./Machinist, Peter (1998), *Letters from Assyrian and Babylonian Priests to Kings Esarhaddon and Assurbanipal* (State Archives of Assyria 13), Helsinki.

Fales, Frederik M. (2009), "Art, Performativity, Mimesis, Narrative, Ideology, and Audience. Reflections on Assyrian Palace Reliefs in the Light of Recent Studies", in: *KASKAL* 9, 237–295.

Feldmann, Marian (2010), "Object Agency? Spatial Perspective, Social Relations, and the Stele of Hammurabi", in: Sharon R. Steadman and Jennifer C. Ross (eds.), *Agency and Identity in the Ancient Near East: New Paths Forwards*, London, 148–165.

Filippi, W. De (1977), "The Royal Inscriptions of Assur-Nāṣir-Apli II (883–859 B.C.). A Study of the Chronology of the Calah Inscriptions Together with an Edition of Two of These Texts", in: *Assur* 1 (7), 123–169.

Finkel, Irving L./Reade, Julian E. (1996), "Assyrian Hieroglyphs", in: *Zeitschrift für Assyriologie und Vorderasiatische Archäologie* 86, 244–280.

Gershevitch, Ilya (1957), "Sisoo at Susa", in: *Bulletin of the School of Oriental and African Studies* 19 (2), 317–320.

Giovino, Mariana (2007), *The Assyrian Sacred Tree. A History of Interpretations* (Orbis Biblicus et Orientalis 230), Göttingen.

Glassner, Jean-Jacques (1995), "The Use of Knowledge in Ancient Mesopotamia", in: Jack M. Sasson (ed.), *Civilizations of the Ancient Near East*, vol. 3, 1815–1823

Grayson, Albert K. (1991), *Assyrian Rulers of the Early First Millennium BC*, vol. 1 (Royal Inscriptions of Mesopotamia, Assyrian Periods 2), Toronto.

Hilgert, Markus (2010), "'Text-Anthropologie'. Die Erforschung von Materialität und Präsenz des Geschriebenen als hermeneutische Strategie", in: Markus Hilgert (ed.), *Altorientalistik im 21.*

Jahrhundert. Selbstverständnis, Herausforderungen, Ziele (Mitteilungen der Deutschen Orient-Gesellschaft 142), Berlin, 85–124.

Labat, René/Malbran-Labat, Florence (1988[6]), *Manuel d'épigraphie akkadienne. Signes, syllabaire, idéogramme*, Paris.

Le Gac, Yves (1907), *Les Inscriptions d'Assur-naṣir-aplu III, Roi d'Assyrie (885–860 av. J.-C.)*, Paris.

Leichty, Erle (2011), *The Royal Inscriptions of Esarhaddon, King of Assyria (680–669 BC)* (The Royal Inscriptions of the Neo-Assyrian Period 4), Winona Lake.

Maul, Stefan M. (1999), "Das Wort im Worte. Orthographie und Etymologie als hermeneutische Verfahren babylonischer Gelehrter", in: Glenn W. Most (ed.), *Commentaries/Kommentare, Aporemata. Kritische Studien zur Philologiegeschichte*, vol. 4, Göttingen, 1–18.

Meuszyński, Janusz (1974), "The Throne-Room of Assur-nasir-apli II (Room B in the North-West Palace at Nimrud)", in: *Zeitschrift für Assyriologie und Vorderasiatische Archäologie* 64, 51–73.

Meuszyński, Janusz (1981), *Die Rekonstruktion der Reliefdarstellungen und ihrer Anordnung im Nordwestpalast von Kalḫu (Nimrūd). Räume: B.C.D.E.F.G.H.L.N.P* (Baghdader Forschungen 2), Mainz.

Morello, Nathan (2013), "Percezioni sensoriali del paesaggio montano nell'ottava campagna di Sargon II, 714 a.C.", in: Stefano Magnani (ed.), *Le aree montane come frontiere. Spazi d'interazione e connettività*, Rome, 441–462.

Nadali, Davide (2001–2003), "Guerra e morte. L'annullamento del nemico nella condizione del vinto", in: *Scienze dell'Antichità* 11, 51–70.

Nadali, Davide (2013), "When Ritual Meets Art. Rituals in the Visual Arts Versus the Visual Arts in Rituals: the Case of Ancient Mesopotamia", in: Claus Ambos and Lorenzo Verderame (eds.), *Approaching Rituals in Ancient Cultures. Questioni di Rito: rituali come fonti di conoscenza delle religioni e delle concezioni del mondo nelle culture antiche*. Proceedings of the Conference Held in Rome, November 28–30, Pisa/Rome, 209–226.

Oates, David/Oates, Joan (2001), *Nimrud. An Assyrian Imperial City Revealed*, London.

Paley, Samuel M. (1976), *King of the World. Ashur-nasir-pal II of Assyria, 883–859 B.C.*, Brooklyn.

Paley, Samuel M./Sobolewski, Richard P. (1987), *The Reconstruction of the Relief Representations and their Positions in the Northwest-Palace at Kalḫu (Nimrūd)*, vol. 2: Rooms: *I.S.T.Z, west-wing* (Baghdader Forschungen 10), Mainz.

Pappi, Cinzia (2013), "Thron A. I.", in: *Reallexikon der Assyriologie und Vorderasiatischen Archäologie* 13, 633–635.

Parpola, Simo (1993), *Letters from Assyrian and Babylonian Scholars* (State Archives of Assyria 10), Helsinki.

Parpola, Simo (1998), "The Esoteric Meaning of the Name of Gilgamesh", in: Jiri Prosecky (ed.), *Intellectual Life in the Ancient Near East. Papers Presented at the 43rd Rencontre assyriologique internationale, Prague, July 1–5 1996*, Prague, 315–329.

Pettinato, Giovanni (1975), "I rapporti politici di Tiro con la Siria alla luce del 'Trattato tra Asarhaddon e Baal'", in: *Rivista di Studi Fenici 3*, 145–160.

Pongratz-Leisten, Beate (1999), *Herrschaftswissen in Mesopotamien. Formen der Kommunikation zwischen Gott und König im 2. und 1. Jahrtausend v. Chr.* (State Archives of Assyria Studies 10), Helsinki.

Porter, Barbara N. (2003), *Trees, Kings, and Politics. Studies in Assyrian Iconography* (Orbis Biblicus et Orientalis 197), Göttingen.

Postgate, John N. (1992), "Trees and Timber in the Assyrian Texts", in: *Bulletin on Sumerian Agriculture* 6, 177–192.

Postgate, John N./Reade, Julian E. (1980), "Kalhu", in: *Reallexikon der Assyriologie und Vorderasiatische Archäologie* 5, 303–323.

Reade, Julian (1985), "Text and Sculptures from the North-West Palace, Nimrud", in: *Iraq* 47, 203–214.

Reade, Julian (1986), “Archaeology and the Kuyunjik Archives”, in: Klaas R. Veenhof (ed.), *Cuneiform Archives and Libraries. Papers presented at the 30th Rencontre Assyriologique Internationale Leiden, 4–8 July 1983* (Uitgaven van het Nederlands Historisch-Archaeologisch Instituut te Istanbul 57), Leiden, 213–222.

Richardson, Seth (1999–2001), “An Assyrian Garden of Ancestors. Room I, Northwest Palace, Kalhu”, in: *State Archives of Assyria Bulletin* 13, 145–214.

Roaf, Michael/Zgoll, Annette (2001), “Assyrian Astroglyphs. Lord Aberdeen’s Black Stone and the Prism of Esarhaddon”, in: *Zeitschrift für Assyriologie und Vorderasiatische Archäologie* 91 (2), 264–295.

Russell, John M. (1998), “The Program of the Palace of Assurnasirpal II at Nimrud. Issues in the Research and Presentation of Assyrian Art”, in: *American Journal of Archaeology* 102 (4), 655–715.

Russell, John M. (1999), *The Writing on the Wall. Studies in the Architectural Context of Late Assyrian Palace Inscriptions* (Mesopotamian Civilizations 9), Winona Lake.

Salonen, Armas (1963), *Die Möbel des Alten Mesopotamien nach sumerisch-akkadischen Quellen. Eine lexikalische und kulturgeschichtliche Untersuchung* (Suomalaisen Tiedeakatemian toimituksia, Serie B 127), Helsinki.

Scurlock, JoAnn (1997), “Assyrian Hieroglyphs Enhanced”, in: *Nouvelles Assyriologiques Brèves et Utilitaires* 92, 85–86.

Seidl, Ursula/Sallaberger, Walther (2005–2006), “Der ‘Heilige Baum’”, in: *Archiv für Orient-forschung* 51, 54–74.

Selz, Gebhard J. (2013), “Text. Textual Bilingualism, and the Evolution of Mesopotamian Hermeneutics”, in: Michaela Bauks, Wayne Horowitz and Armin Lange (eds.), *Between Text and Text. The Hermeneutics of Intertextuality in Ancient Cultures and Their Afterlife in Medieval and Modern Times* (Journal of Ancient Judaism, Supplements 6), Göttingen, 47–65.

Seminara, Stefano (2004), *Le iscrizioni reali sumero-accadiche di età paleobabilonese. Un’analisi tipologica e storico-letteraria* (Accademia Nazionale dei Lincei, Classe di Scienze Morali, Storiche e Filologiche, Serie 9, vol. 18, fasc. 3), Rome.

Stearns, John B. (1961), *Reliefs from the Palace of Aššurnaṣirpal II* (Archiv für Orientforschung, Beihefte 15), Graz.

Sobolewski, Richard P. (1982), “Beitrag zur theoretischen Rekonstrution der Architektur des Nordwest-Palastes in Nimrud (Kalhu)”, in: D. Papenfuss and V. M. Strocka (eds.), *In Palast und Hütte. Beiträge zum Bauen und Wohnen im Altertum von Archäologen, Vor- und Frühgeschichtlern. Tagungsbeiträge eines Symposiums der Alexander von Humboldt-Stiftung Bonn-Bad Godesberg, 25.–30. November 1979, Berlin*, Mainz, 237–250.

Thureau-Dangin, François (1912), *Une relation de la huitième campagne de Sargon (714 av. J.-C.)*, (Textes Cunéiformes du Louvre 3), Paris.

Winter, Irene (1981), “Royal Rhetoric and the Development of Historical Narrative in Neo-Assyrian Reliefs”, in: *Studies in Visual Communication* 7 (2), 2–38.

Winter, Irene (1997), “Art in Empire. The Royal Image and the Visual Dimensions of Assyrian Ideology”, in: Simo Parpola and Robert Whiting (eds.), *Assyria 1995. Proceedings of the 10th Anniversary Symposium of the Neo-Assyrian Text Corpus Project Helsinki, 7–11 September 1995*, Helsinki, 359–381.

Winter, Irene (2000), “Le Palais imaginaire: Scale and Meaning in the Iconography of Neo-Assyrian Cylinder Seals”, in: Christoph Uehlinger (ed.), *Images and Media: Sources for the Cultural History of the Near East and the Eastern Mediterranean (1st Millennium BCE)* (Orbis Biblicus et Orientalis 175), Fribourg, 51–88.

Winter, Irene (2007), “Agency Marked/Agency Ascribed. The Affective Object in Ancient Mesopotamia”, in: Robin Osborne and Jeremy Tanner (eds.), *Art’s Agency and Art History* (New Interventions in Art History), Malden/Oxford, 42–69.

Antonio J. Morales

From Voice to Papyrus to Wall: *Verschriftung* and *Verschriftlichung* in the Old Kingdom Pyramid Texts

In memoriam Harold M. Hays

1 Introduction

Since the initial discovery of the corpus of Pyramid Texts by Gaston Maspero in 1880,[1] Egyptologists have been attracted by the large collection of texts that were inscribed on the walls of the crypts in some Old Kingdom pyramids, first at the end of the Fifth Dynasty in the pyramid of Wenis, and then later in the monuments of the kings of the Sixth Dynasty buried in Saqqara. For some debatable reasons, the corpus of Pyramid Texts was also used for some queens of the Sixth Dynasty, royal wives and mothers, from the reign of Pepi I onwards, as well as later for (at least) two other kings of the First Intermediate Period: Ibi and the Herakleopolitan Wahkhare Khety.

This phase of development of the corpus began around 2345 BCE and expanded for over 160 years until 2184 BCE.[2] For this period, scholars have mostly emphasized the pivotal role of the monumentalization of the corpus and the configuration of mortuary literature in ancient Egypt. However, this phase does not represent the initial

This article emerged from the Heidelberg Collaborative Research Center 933 "Material Text Cultures. Materiality and Presence of Writing in Non-Typographic Societies". The CRC 933 is financed by the German Research Foundation (DFG). I am indebted to this research project and to Professors Markus Hilgert, Ludger Lieb, and Joachim Friedrich Quack for supporting my nomination and facilitating my work as a research fellow at the University of Heidelberg in 2012–2013. This study also owes much to my current research on the transmission of religious knowledge and practices under the auspices of the SFB 980 "Episteme in Bewegung: Wissenstransfer von der Alten Welt bis in die Frühe Neuzeit" and its research division Altägyptische Philologie at the Freie Universität Berlin. I am grateful to Jochem Kahl, Bernard Mathieu, and Ludwig Morenz for comments on a draft of this paper, bibliographic references, and discussion on the oral and literary mechanisms for the transmission of religious knowledge, as well as to the audience at the University of Heidelberg for commenting on issues arising from this material.

1 On the discovery of the Pyramid Texts, see Verner 2001, 39–41, and Ridley 1983, 74–80. For the dispute over the real discoverer of the Pyramid Texts, cf. Gestermann 2005, 10, n. 29 and Sledzianowski 1976, 5. In this study, the rubric PT refers to the *Spruch* numerical designation, Pyr. §§ to the sectional assignments of Sethe 1908–1910; and sPT to texts numbered by the MafS (Leclant/Berger-El Naggar/Pierre-Croisiau 2001). Superscript sigla designate the pyramid sources for the texts: W= Wenis; T= Teti; P= Pepi I; M= Merenre I; and N= Pepi II.

2 Following Lloyd 2010, xxxiv.

DOI 10.1515/9783110417845-004,

stage of this long-lasting tradition.[3] Before the reign of Wenis, recitations and performances of the same nature certainly co-existed in the domains of orality and the more restricted realm of writing.[4] In fact, these materials constituted the backbone of the Pyramid Text assemblages composed in the Fifth and Sixth Dynasties, and that factor accounts for the variation and multiplicity in organization, typology, and function of the corpus components (e.g. various categories, themes, groups, sequences). Consequently, one should identify a stage of the process before monumentalization in which similar materials were used and transmitted orally or in constrained written form.[5] Such a phase entails the emergence of the corpus and its initial development in writing until its commitment on the walls of the pyramids, and involved: i) recitation and performance of rituals; ii) *Verschriftung* or entextualization, i.e. fixation in writing of these materials; iii) editorial selection and recension; and iv) *Verschriftlichung* or textualization.[6]

One of the most interesting aspects for the study of this earlier development of the corpus is that such a process of configuration left traces in the resulting compositions used for kings and queens of the late Old Kingdom. In other words, one can identify some formal features of the corpus—as observed in the pyramid assemblages—that do not resonate with the contrived nature of each composition and the mortuary setting in which they appear. Contrarily, these features reveal that each composition incorporated texts of different origin, nature and function, whose existence can be proved in diverse settings before their use in the inner chambers of the pyramids. In addition, there is ample evidence for ritual practices and knowledge among the elite in the earlier part of this period that bridges the gap between the use of these recitations (i.e. in the oral and performative arena) and their monumentalization in stone (i.e. pyramid inscription).

In classical examinations of the corpus, the Pyramid Texts inscribed on the walls of the royal monuments were taken as intentionally composed mortuary literature, a body of texts with the purpose of protecting the king's body, facilitating his transfig-

3 Altenmüller 1972, 278.

4 See Baines 2004, 17–18, who points out that oral performances accompanied rituals and, when written down for the first time, around the late Second or Third Dynasty in his opinion, they were alluded to in the form of fictionalized statements such as speeches of gods. In the same vein, Strudwick 2005, 1; Mathieu 2004, 253; Allen 2001, 97; Mathieu 1996, 289; Altenmüller 1984, 20.

5 In a personal communication (24.02.2014), Bernard Mathieu suggested three strata in the redactional history of these texts: I. strate archaïque (Dynastie 0–IIIe dynastie); II. strate héliopolitaine pré-osirienne (IIIe–IVe dynasties); and III. strate hélipolitaine post-osirienne (V^{e}–VIe dynasties).

6 These initiatives prompted the configuration of the earliest stage of mortuary literature tradition, in general, and the construction of a miscellaneous corpus for the kings in particular. Additionally, three other traits of adaptation materialized during the process: decontextualization, recontextualization, and monumentalization. These mechanisms will be discussed below.

uration into an Akh, and ensuring his well-being in the netherworld.[7] Later, further research took the interpretation of the corpus to such divergent avenues as a faithful reflection of the funerary rites for the king (*Auferstehungsritual*),[8] or personal magical equipment for the afterlife (*Grabinventar*).[9] In recent years, scholars have advanced categorically beyond these postulations in the knowledge and comprehension of the Pyramid Texts corpus. As a matter of fact, it is now believed that these texts were not initially composed for the underground chambers of the late Old Kingdom pyramids, a setting that came to incorporate the corpus only secondarily.[10]

As the original setting of the Pyramid Texts was not the tomb, it is conceivable that one might identify particular textual features of the grammar, syntax and even the epigraphic aspect of the inscriptions that betrayed their original *Sitze im Leben*. Because of the vagaries of redactional history, however, the feasibility of associating these diagnostic features with written records before the late Fifth Dynasty is limited. Notwithstanding this flaw, the repertoire of peculiarities in the Pyramid Texts relates some constituents of the corpus with recitations and performances found in the context of sacerdotal mortuary rites and personal magical practices. Thus, it is possible to have an extensive discussion on the original settings of the Pyramid Texts as well as on the process that brought sacerdotal voices and ceremonies, personal recitations, and magical incantations into writing, first in the form of operative scripts in papyri, and later as monumental inscriptions in the pyramids of kings and queens in the late Old Kingdom.

In this paper, I shall address the process of emergence and development of the Pyramid Texts from their oral form to their inscription in the chambers of the late Old Kingdom pyramids of kings and queens. In addition, I identify and comment on some of the original settings of these recitations outside the pyramids. The confirmation that some texts are found in non-royal contexts pre-dating the inscription of the pyramid of Wenis does not support the idea of a gradual diffusion of this material to non-royal people, and indicates that the rituals represented by the Pyramid Texts were already in use by the community before theologians and editors in Heliopolis[11] planned the monumentalization of a king's pyramid with fixed recitations.

---

7 Hornung 1999, 5; Barta 1981, 71.

8 Altenmüller 1974; Altenmüller 1972; Spiegel 1971; Spiegel 1956; Spiegel 1953; Ricke 1950, 123–124; and Schott 1950, 153–154. Cf. the comparative analysis of the textual function and motifs, and the emphasis on the *Sitze im Leben* of the texts observed in Roeder 1993, 81–82, n. 8.

9 Barta 1981, 66–69 (cf. Hays' interpretation on the ramifications of Barta's misinterpretation of the corpus as a kind of "treatise of beliefs", in Hays 2012, 9–10).

10 See Hays 2012, 80, who points out that "prior to the innovation of inscriptional decoration in the tombs, established sets of texts must have already existed within the body of literature from which the Pyramid Texts were drawn [...] Pyramid Texts were drawn from existing, external rituals and collections of rites which had not been entirely canonized".

11 As observed in Mathieu 1998, 71–78; Bickel, 1997, 113–122; Willems 1996, 254 (n. 1408); and Bickel

2 *Verschriftung* and *Verschriftlichung*

In modelling the interim forms that generated the Pyramid Texts, the general procedures by which the texts were transmitted will be hypothesized in accordance with theoretical universal observations in the fields of orality and performance studies. In applying these theories, I have drawn on the studies of the linguist Wulf Oesterreicher,[12] anthropologists Richard Baumann and Charles Briggs,[13] and the recent work on the subject of the oral-compositional form of the Pyramid Texts by the Egyptologist Chris Reintges.[14] Having made this claim, I aimed at sketching out the contours of an extensive program of textual adaptation in media and material that resulted in the advent of the mortuary literature tradition in the Fifth and Sixth Dynasties with visible roots in the early Old Kingdom.

The emergence of ancient Egyptian mortuary literature in the third millennium BCE is the history of the adaptation of recitational materials to the eventualities of media and materiality. Upon a gradual development "as a layering of processes",[15] the transformation of the oral discourse into writing began with the use of papyri for transcribing the guidelines of ritual performances,[16] and culminated with the concealment of sacerdotal voices and deeds in the sealed-off crypt of Wenis. Thus, the genuine innovation of the Memphite priests, the *real landmark in the use of mortuary literature*, was the recontextualization of discursive and scriptural materials into the architecture of the pyramid of Wenis at the end of the Fifth Dynasty, *not the invention* of a corpus. As a result, the (pyramid) texts inscribed on the walls of the king's underground chambers no longer belonged to the realm of verbal art, nor functioned as operative instructions for the priests performing rituals.[17] Now, these inscriptions represented a new systematization of ritual practices—a novel ritual syntax—that perpetuated in stone the ultimate goal of the mortuary ceremonies: the transfiguration of the deceased into a successful spirit in the afterlife, an Akh.[18]

1994, 285–298, the influence of the Heliopolitan priesthood on the composition and control of materials is manifest.

12 Mainly Oesterreicher 2005, 1998, 1993.

13 Bauman/Briggs 1990.

14 Reintges 2011.

15 Baines 2004, 16.

16 For a general examination of the nature and interaction of orality and writing in ancient Egypt, see Morenz 1996, 20–43; Eyre 1993, esp. 117–118; and Eyre/Baines 1989, 103–114.

17 Hays 2006, 298: "After being inscribed in the tomb, the role of a text was necessarily different. Sealed off from the eyes of any living priest, it only represented the rite [...] Whereas the original efficacy of the text as ritual script included the vocalic dimension of its words being uttered by a priest, after its inscription upon walls the efficacy of the text inhered to the hieroglyphs alone, independently of any human voice or effort".

18 Mathieu proposes a stimulating approach that examines the deictic referential frame of the language in the Pyramid Texts (see Mathieu forthcoming). According to his analysis, the use of proximal

It is evident that the process of commitment of ritual and magical recitations into *scriptio continua* or *soluta*[19] in the Pyramid Texts was subjected to several stages of adaptation, detachability, and recentering. In mapping the dimensions of transformation, one can identify a series of transformational procedures, mainly entextualization and textualization, accompanied by decontextualization, recontextualization, and monumentalization. While the first two mechanisms of discourse transformation superseded each other, the latter three procedures were the result of transforming the discourse and separating it from their original cultural and religious contexts of use (i.e. decentering the text).

The initial adaptation of the recitations into the realm of writing, that is, the transfer from *phonique* to *graphique*, is a process known as *Verschriftung* or entextualization:[20] "the process of rendering discourse extractable, of making a stretch of linguistic production into a unit—a text—that can be lifted out of its interactional setting".[21] By means of this process, the oral materials used above-ground in cultic and ritual activities were cemented in writing, first in manuscripts (papyrus scrolls), which could be stored in temple libraries and used as ritual guidelines by the sacerdotal class. At this point, these materials were still considered to be supports for the ceremonies performed in the original settings,[22] although in scriptural form, decontextualized from the oral domain and recontextualized in the realm of

deixis in reference to the deceased (*Wnjs pn*), the absence of distal deictics (*N pf*), and the interconnection of this type of deixis with the pyramid space (*mḥr pn ḥwt-nṯr tn*), the deceased's corpse (*sˁḥ pn*), and the monumentalized inscriptions reflect the role of the Pyramid Texts as an "élément du mobilier funéraire" or "objet". My thanks are due to the author for allowing me to read his manuscript before publication and for his observations on the invalid application of modern distinctions between text and object to this particular phenomenon in antiquity.

19 Oesterreicher 2005, 203.

20 See Oesterreicher 2005, 202–203; and Oesterreicher 1993, 271–284. For the Pyramid Texts particularly, see Reintges 2011, 25–28. The process has been termed "mise par écrit" in French and "graficación" in Spanish (see Oesterreicher 1993, 272, ns. 9–10). Cf. Assmann 1988, 26–27, who encapsulates this notion under the phenomenon of *Schrifterfindung*.

21 Bauman/Briggs 1990, 73 (in the quotation, italics are the authors' emphasis). For an example of the application of the entextualization phenomenon to explicit Pyramid Texts materials, see Hays 2012, 90–92 ("The Entextualization of Group A"), and 198–203 for a comprehensive explanation of the entextualization in the Pyramid Texts. As Greg Urban argues, a given instance of a discourse is unique by virtue of its formal properties; the "transduction" or carrying over of parts of these formal attributes—"metadiscursive markers"—can help to discern the social and religious implications that, in the case of the transmission of rituals and their texts such as the Pyramid Texts corpus, were prominently preserved by a community (Urban 1996, 21).

22 See, for example, the diverse layers of meaning that the scene of a lector priest presenting a papyrus scroll or reciting from it to the deceased could imply, in Manuelian 1996, 561–588. Cf. the comments about the reading of texts from papyri, not from walls, in Hays 2012, 91. For the general idea that many texts of a ritual nature in Egypt contain elements that point to an oral setting, see O'Rourke 2001, 407–410; and Walle 1965, 122–124.

literacy.[23] Otherwise, no traces of monumentalization had yet occurred. Therefore, at that point the process of entextualization altered the media of recitational materials that still reflected the activities for which they were first used.[24] However, a second stage of entextualization followed the fixation of recitational materials in papyrus scrolls (transcripts) and came to transfer these texts onto the walls of the pyramid of Wenis (inscripts),[25] divorced from human practice or scriptural *aide-mémoire*. On that account, the second stage of the process of *Verschriftung* certainly conveyed a new juncture of decontextualization, recontextualization, and monumentalization.

The practice of collecting recitations, writing them down in papyri, and transferring these scriptural materials to the walls of a pyramid is considered an intellectual undertaking[26] that the Heliopolitan sacerdotal class of the Fifth Dynasty initiated for the benefit of the king's afterlife. Thus, religious authority in the environs of the royal court, temples and repositories of Memphis[27] must have approved the constitution of the corpus.[28] This decision implied recording oral recitations, collecting scrolls, copying, editing and even transferring them to Memphis if they were selected and collected from other parts of the country.[29] In addition, to understand the initial history of the Pyramid Texts it is relevant to highlight that this initiative shaped the corpus with a particular structure (i.e. symmetric structure), which would experience gradual modifications—first slightly, in the Sixth Dynasty, and later, more dramatically, in the

23 Both aspects are part of the same process: Bauman/Briggs 1990, 75.

24 I agree with Reintges that the fixation in writing occurs "to sustain the immediacy of the performance and secure it for future recall" (Reintges 2011, 20). As observed in the related studies on early Mesopotamian poetry, the emergence of poetic texts was the result of premeditated oral composition (Teffeteller 2007, 69).

25 For the transference of the oral discourse into the written record as a process that involves the conversion of oral compositions into transcripts that are later converted into (in)scripts (i.e. textual monumentalizations), which is implied in the conversion of "I-transcriptions" to "he-inscriptions", see Svenbro 1998, 48–52. For a definition of the terms "transcript" and "(in)script" in the process of entextualization and textualization, see Nagy 1998, 79. Note that one must not take the relationship of the two contexts, oral and written, as a translation of verbal transmitted knowledge into texts. For the latter idea and its development, see Assmann 1999, esp. 10.

26 See Mercer 1956, 5, who refers to this phenomenon as both "recension" and "redaction".

27 For the suggestion that the significance of the solar cult and the priesthood of Heliopolis, features strongly associated with the inception of the Pyramid Texts, might have materialized as early as the early Dynastic Period, see Cervelló-Autuori 2011, 1125–1149; Kahl 2007, 2–3; and the brief history of the raising of the solar cult in Quirke 2001, 115–128, although the author notes that in the formative phase of Egyptian art and writing, ra did not refer to a deity but to the solar entity (p. 22).

28 See Bauman/Briggs 1990, 76–77: "To decontextualize and recontextualize a text is thus an act of control [...] legitimacy is one of being accorded the authority to appropriate a text such that your recentering of it counts as legitimate".

29 Baines 2004, 21. For the notion of *Transportabilität* and its relationship with the transmission of religious texts, mainly in scriptural form, see Kahl 1996, 11.

First Intermediate Period—resulting in a varied collection of structures in the Middle Kingdom (i.e. doctrinal structures).[30]

Furthermore, the recitational materials transferred from the scriptural domain to the monument of the king experienced an ensuing modification known as *Verschriftlichung* or textualization.[31] By means of this process, the entextualized oral discourse acquired attributes of the literary format and detached itself from the discursive style.[32] This shift in the style of the Pyramid Texts can be observed in the modification of the original performance structure from interlocutive to delocutive, that is, from the active presence of the first-person pronoun for the ritualist or beneficiary—typical in the speech acts—to the consideration of the addressee as a passive recipient in the second or third person.[33] Although the process of entextualization might be considered immediate, the phenomenon of textualization seems to have been characterized by its gradualness,[34] and involved the conscious editing of the discursive style by scribes in libraries and repositories.[35]

The discussion just now culminated with the contribution of the scribal and religious class from Memphis to the execution of these two processes of transformation of the corpus, its fixation in writing (*Verschriftung*) and style adjustment (*Verschriftlichung*). As categorical as this definition of the two adaptative procedures might be, it by no means embodies the entire process of the royal Pyramid Texts' inception. In fact, the decisive factor that led up to the configuration of the corpus as witnessed in the pyramid of Wenis was the monumentalization of the texts, a process that—parallel to the theological architectural vertex achieved at this time[36]—transformed the repertoire of recitations kept in scriptural form into carved, monumentalized inscriptions.[37] Such texts no longer belonged to the domain of the scriptural media, but to

30 For a summary of the development of the corpus during the Sixth Dynasty and its gradual transformation from a symmetric structure to doctrinal structures, see Morales 2013, 43–44 (n. 95), and 94–104.

31 See Oesterreicher 2005, 202–204; and Oesterreicher 1993, 271–284. For the Egyptian materials specifically, see Reintges 2011, 28–31; and Assmann 2000, 81–82. Cf. also Assmann 1988, 26–27, who uses the term *Schriftverwendung* in alluding to the phenomenon of appropriation of the discourse in a new setting and the transformation of the original discourse.

32 Oesterreicher 1998, 12; Oesterreicher 1993, 273–275.

33 Reintges 2011, 28.

34 See Bakker 1999, 29–47.

35 See the implications of the modifications from poetic-compositional style to literary character in Foley 2004, 101–120; Honko 2000, 3–54; and Nagy 1998, 79.

36 More precisely, the architectural model imitated in the later royal complexes is achieved in the reign of Wenis's predecessor, Djedkare Izezi: see Hays 2012, 79; and Billing 2011, 58.

37 See Hays 2012, 11–13, 257–262; Billing 2011, 58; Hays 2009b, 218–219; and Assmann 2005, 238–241. See also Zumthor 1983, 58, who, in relation to the transfer of settings (decentering) from the domain of the *vocale* to the *textuel*, emphasizes the secondary nature of the new setting: "[l]e monument se constitue à un autre niveau, mettons "poétique", défini par une structure seconde, intentionelle et

the domain of epigraphy; therefore, they were recontextualized (or recentered)[38] by concealing them in a new setting, the mortuary chambers of the king.[39] The whole transfer from voice to papyrus to wall reached a pinnacle with the encapsulation of the effective ritual performances in the crypt in which the king's corpse and spirit resided.

Now, the monumentalization of the recitations extant in scriptural media (papyrus scrolls) might have demanded the composition of one or several master copies conveying the final assemblage of texts that would be used by the scribes working at the monument. Presumably, this procedure would have allowed the production of master documents in the temple repository, while minor modifications could have been executed onsite. For instance, one can observe the modification caused by remodelling the allusion to the beneficiary of recitations as written in operative papyri[40]—a roll with nmemonic observations which the ritualist used to address *anyone*, *this Osiris so-and-so* or *this king*—to the explicit reference to the particular king for whom the texts had been selected, copied and, by this method, customized.[41] Another feature of the Pyramid Texts that ratifies what has been suggested on the basis of operative papyri is the actual epigraphic correction of non-explicit references in order to carve the king's name in the pyramid.[42] As Bernard Mathieu has pointed out, some

résultant d'un travail, réorganisant des éléments déjà organisés en structures primaires".

38 Bauman/Briggs 1990, 75.

39 Oesterreicher 2005, 210–211 also defines this process as *autonomisation* of the corpus.

40 Unfortunately, the only instances available date to the Middle Kingdom: see pap. MafS T2147, a remarkable example of Middle Kingdom operative papyri found by the MafS in the complex of Pepi I at Saqqara. Observe that the published verso and recto sides of T2147 are rather two fragments of different sheets that adhered to each other. The verso of the papyrus shows the use of the non-explicit reference *Wsjr mn pn*, and therefore betrays the operative nature—as an *aide-mémoire*—of the document. See Berger-El Naggar 2004, 85–90, fig. 1 (recto with PT217, Pyr. §§155d–159: cols. 4, 6, 13, 14, 18), fig. 2 (verso with PT690, Pyr. §§2096c–2101b: cols. 1, 3, 14). See also the use of *mn* "whoever" in Gardiner 1955, 11, cols. 18, 19 (pl. II), 91 (pl. V), and frg. a (pl. VI); with remarks in Eyre 2013, 46. A related phenomenon is the nature of some Pyramid Texts in which the role of a non-royal deceased—rather than a royal one—seems to fit better with the contents of the recitations: see Smith 2009, 6–7 (referring to examples PT456, PT467, PT486, and PT571).

41 As only the owner of the corpus could benefit from the efficacy of the perpetuated rites, it was essential to personalize the texts by including the name of the king. One of the features corroborating the scriptural origin of the recitations is the attestation of terms such as "whoever" (*mn*) and "king" (*nsw*) in the Pyramid Texts, which were for the most part substituted by the name of the king: see n. 40 above, and Mathieu 1996a, 290–291.

42 See, for instance, Mathieu 1996a, 290, figs. 2–3: PT23 (Pyr. §16a)[W]: *Wsjr jṯ n=k msḏḏw nsw nbw mdw m rn=f ḏw* has been modified to *Wsjr jṯ n=k msḏḏw Wnjs nbw mdw m rn=f ḏw* "Osiris, take for yourself all those who hate the king/Wenis and anyone who speaks bad of his name". In such cases, the scribe skipped the non-explicit reference, retaining the vague address to the beneficiary as extant in the original recitation on papyrus, and leaving out any explicit reference to the king: *n(j) kw mn nṯr pw* "O whoever, you belong to that god" (PT215, Pyr. §147a)[W].

corrections were executed by the scribe in the process of transferring the texts from papyri to the walls, although one should also expect a final reading to rectify minor mistakes.[43]

A great gap in our understanding of the construction of the Pyramid Texts could be bridged through attention to the Old Kingdom (temple) repositories or libraries in which the operative papyri and master copies for the pyramid assemblages were produced.[44] In principle, the operative papyri could be stored in these institutions;[45] otherwise, they would have been carried by specialized ritualists such as the lector priests.[46] The nature of the transmission of ritual knowledge, texts and iconography during the Old Kingdom defines the functional and cultural-binding role of these repositories associated with the temples of Memphis and its cemeteries. In Jan Assmann's opinion, the normal means of circulation were oral, while writing became significant only in those fields of communication that required the use of artificial storage, such as religious ceremonies.[47] Unfortunately, no traces of Old Kingdom operative papyri with religious compositions have been retrieved.

---

43 See Mathieu 1996a, 292–293 (n. 18), and 311. Cf. the use of the term *zẖ-qdt* for "sketch, draft" in the tomb of Senedjemib Inti as an example of master copies being used for the carving of texts and the practice of final reading in tombs. The topic is discussed in further detail below.

44 The role of manuscript collections, which certainly existed and were of primary importance in the transmission and production of texts, was also critical in the transference of recitations from the domain of speech to the context of monumentalization in the pyramid of Wenis. See Baines 2004, 26–30; and Müller 1984, 244–246. Concerning the early existence of the "house of life" (*pr-ꜥnḫ*), where texts were composed, and the "repositories" (*pr-mḏꜣt*), where they were stored, some kind of institutional bureau must have existed in the Old Kingdom: see remarks in Eyre 2013, 311; Zinn 2011, 183–184 (n. 9); Burkard 1980, 85; and Kessler 1984, 929.

45 It is unclear whether the system for dividing the textual units in some pyramid assemblages, namely the usage of *ḥwt* to indicate the conclusion of a text or "stanza", could have been used as a method for organizing papyrus sheets and storing them in any repository. For the use of *ḥwt* in the Pyramid Texts, see Eyre 2013, 49–50; Blackman 1938, 64–66; and Grapow 1936, 35–36.

46 Although the earliest attestation of private libraries has been noticed for the Middle Kingdom—i.e. Ramesseum papyri—the activities of the lector priests during the Old Kingdom might have permitted the use and possession of ritual papyri (see Morales 2013, 120, n. 315; Morenz 1996, 37). As Ong points out in relation to the role of the *reciter*, it is relevant to observe the illiterateness of singers, poets and bards in early literacy cultures (Ong 1982, 59–61; similarly, Eyre 1993, 115, 120). In the case of Old Kingdom Egypt, the interaction of recitational abilities and literary knowledge in the person of the ritualist should be interpreted as a proof of the gradual development of the institution of writing in society. In fact, the etymology of the title *ẖrj-ḥbt* ("bearer of the festal papyrus scroll" or lector priest) refers to the relationship of the specialist with the (written) document: see Morales 2013, 50, n. 116, for further bibliography on the emergence of the lector specialist. In a related note, Eyre points out that "[o]ral poets, all performers, literate or not, dislike having their work copied down. They lose ownership: copyright" (Eyre 1993, 115).

47 See Assmann, 1999, 5–6, endorsed by Ong 1982, 96, who highlights the idea that in the early cultures exposed to literacy, "orality could linger in the presence of writing, even in the administrative milieu". Cf. also the views in Baines 2004, 19, who believes that the most sacred and prestigious texts

Trying to compensate for this shortcoming, scholars have discussed the absence of paratextual notations in the corpus, a question that relates to the production, use and management of the Pyramid Texts.[48] It is known for later compositions such as the Coffin Texts that paratextual notations helped to define the significance and usage of a text, although they might have not been useful for the ritualist, who would be truly informed of the text's purpose, usefulness and benefits in the ritual. In an attempt to explain the lack of paratexts, if the conventions established in the Old Kingdom for the use of Pyramid Texts were recurrent and corresponded with contemporaneous ritual practices, then no paratextual instruction would be necessary for the accomplished performer to navigate his way through the texts in the scriptural media. Another possibility for the absence of paratexts in the initial phase of the mortuary literature tradition could be that the principles of decorum and secrecy, which ruled the access to and knowledge of these texts in scriptural form, hindered any attempt to apply a system of notations for arranging the texts beyond the arcane role of the specialist dealing with the collection. Finally, the expectation that Old Kingdom manuscripts with Pyramid Texts bore paratextual notations is at this time a vexing question, since archaeological work in Saqqara has not (yet) provided us with any papyrological evidence of such antiquity. Two exceptional exemplars, however, might help us to suggest the possibility that some manuscripts displayed paratextual notations such as titles or directions:[49] pap. Ramesseum E and pap. MafS T2147. The first papyrus consists of several fragments in poor condition with a copy of a ritual text—not included in the Pyramid Text repertoire—in cursive hieroglyphs, with parallel lines along the top for the writing of a horizontal heading. The second instance was found in the temple of Pepi I and also presents a horizontal band for a heading, though it is blank (see n. 40).

For the most part, the processes of *Verschriftung* and *Verschriftlichung* have powerful implications for the structure and language of these recitations as observed in the Pyramid Texts. Investigating how the corpus emerged through the combination of recitations from different settings that were put in writing and how these varied categories of texts (*Stilmischung*) adapted to a new arrangement for Wenis can therefore elucidate issues such as the modification of the deictic relationship between text and

might not have been written down as a restriction that the forms of decorum and the primacy of oral tradition imposed on the priesthood and scribal classes; and Haring 2003, 256–259, who focuses on the resistance of the oral discourse and the capacity of the institution of writing to overcome reactionary positions and spread in society.

48 Most recently, Hays 2012, 3–4; Mathieu 2004; and Grimm 1986. For the concept of paratextuality, see Assmann 2005, 248; and Hays 2004, 178, n. 20. Cf. the rigorous approach in Aufrère 2010, 160, n. 6, who takes the hieroglyphic system as a proof that ancient Egyptians could select particular iconography to express further information about a text without the need for paratextual notation.

49 See Eyre 2013, 46, and Altenmüller 1968, 59 (sections 1 and 2). Both authors state that in religious papyri it is common to find a superior horizontal band, which in most cases is left blank.

protagonist (i.e. from first-person pronoun in oral and scriptural domains to second-/third-person in the monumental versions), the partial transformation of oral *written* discourse into literary style, the remnant presence of poetic and speech elements, or the flexibility shown by the corpus in adapting to new practices and beliefs in the realm of the private mortuary rituals in the Middle Kingdom.[50] Specifically, the configuration of the Pyramid Texts as a text-bricolage with a series of recitations from different settings and contexts[51] is what permitted the structuring of the corpus in the form of segments or sequences—also known as *Spruchfolge*[52] or "building blocks"[53]—and its capacity to adopt and adapt to the ritual practices of the First Intermediate Period and the Middle Kingdom.

3 The "Pyramid Texts" in the Early Old Kingdom

A fundamental question in the analysis of the transfer of Pyramid Texts from the oral domain to pyramid walls is the identification of the primary settings in which these recitations were originally used. In discerning cultural and religious milieux associated with Pyramid Text rituals before the reign of Wenis, one must distinguish between the oral—unrecorded—stage of the process, which unfortunately left only a few vestiges in the inscriptions of the royal assemblages, and the stage of fixation in the writing of recitations (*Verschriftung*), in which the extant evidence demonstrates the deployment of these texts in particular settings. As regards the second stage and the extant evidence, it was the German Egyptologist Kurt Sethe who first hypothesized that one could identify clear associations between the assortment of items included in offering lists of early Fifth Dynasty royal complexes and the later Pyramid Texts.[54] Later, Hartwig Altenmüller highlighted the tabular nature of these precedents that he regarded as "pictorial versions" (*Bildfassung*) of the Pyramid Texts.[55]

"Pictorial versions of Pyramid Texts" refers to non-narrative representations of a particular group of this corpus in a particular setting outside the domain of the royal tomb. One of these groups was the offering Pyramid Texts,[56] which appeared in the

---

50 See Morales 2013.

51 In Bauman/Briggs's words, the new systematization or recension of texts becomes an "emergent structure" (Bauman/Briggs 1990, 76).

52 Altenmüller 1972; Allen 1994.

53 As used in Hussein 2013, 275, n. 8 (with the term "building block" borrowed from Boltz 2005).

54 Sethe 1913, 126, ns. 1–10 (pl. 63); and later in the same fashion, Junker 1934, 15, n. 1; and Firchow 1953, 9.

55 Altenmüller 1974, 278.

56 See Hays 2012, 81–92; Hays 2010a, 127–130; and Smith 2009, 8–9, in which the authors discuss the group of offering texts as a clear instance of Pyramid Text material pre-dating the corpus of Wenis. Unfortunately, this is the only group that has been identified in the inscriptional repertoire previous

form of lists of items with a specific order in Third and Fourth Dynasty offering lists, an order that in some cases corresponded unmistakably with the series of (narrative) Pyramid Texts of the royal pyramids that alluded to the same offerings, items, and rites.[57]

In the royal context, the existence of Pyramid Texts in pictorial form was first attested at the beginning of the Fifth Dynasty, in the pyramid of Sahure at Abusir (fig. 1).[58] In spite of its fragmentary condition, the remnants of the offering list in the mortuary temple of Sahure provide us with a partial ordered list[59] of items existing in the later Pyramid Texts: PT82 PT84, PT92 PT94, PT128–129, PT139 and PT140–141 (table 1). The presence of these items in the offering list of the mortuary complex of Sahure not only evidences the existence of a non-narrative version of the Pyramid Texts for the king previous to the assemblage of Wenis,[60] but also confirms that the same repertoire of ritual texts informed the ritualists before their fixation in writing at the end of the dynasty.[61] The relationship between item and text based on a development from the canonical tabular list to the ritual narrative demonstrates that there was a common practice in which every sacred item was necessary for a particular rite.

to the reign of Wenis, and its associations have not yet been analyzed at length (cf. the study by Barta 1963, 5–90, who identified the associated materials but did not refer to the nature of the relationship between the private domain of these offering lists and the later or comtemporaneous royal Pyramid Texts).

57 The Pyramid Texts' narrative version of the tabular lists of offerings and items certainly provides further information on each rite. A phenomenon of similar characteristics has been examined in the Homeric literature, in which extensive lists and catalogues that originally had a performative and oral character were embedded into the narrative for the exhibition of the virtuoso-singer, whose memory and performative skills would enable him to display a deep knowledge of technical matters such as ship construction (Minchin 1996, 19–20).

58 Note that the earlier date of the sources from the private context with offering lists relating to the offering Pyramid Texts might indicate the private origin of the offering rites alluded to in the texts. Cf. the case of the king Pepi II providing written materials for the performance of rituals on the occasion of the funeral of Sabni's father, Mekhu: Jürgens 1995, 85, n. 95.

59 One can observe that the recitations of each fragment follow the canonical order as observed first in the offering lists and then in the Pyramid Texts. For the canonical order (type A) in the offering lists, see Barta 1963, 47–78; and cf. the development of the "genre" in Junker 1934, 71–96.

60 For the related order as observed in Wenis's pyramid, cf. W/S/N i (PT82), W/S/N i 30 (PT84), W/S/N ii 38 (PT92), W/S/N ii 44 (PT94), W/S/N iii 12–13 (PT128–129), and W/S/N iii 24–25 (PT140–141).

61 *Pace* Eyre 2002, 17–18: "The assumption that Pyramid Texts simply represent a stable and ancient oral tradition, first written down in the later Old Kingdom, belongs to the romantic intellectual climate of the late nineteenth and early twentieth centuries, and is rooted in universalist preconceptions of cultural evolution. In crude form such assumptions are neither substantiated nor sustantiable on the basis of hard evidence."

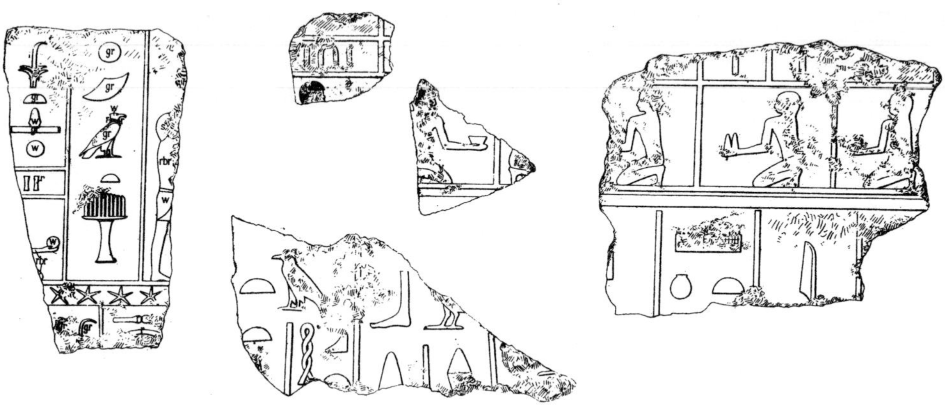

Fig. 1: Offering list fragments from the mortuary temple of Sahure (after Borchardt 1913, pl. 63).

Table 1: Offering list items of Sahure and associated Pyramid Text recitations.

Recitation	Sahure's offering list	Pyramid Texts
PT82, Pyr. 58b	*ẖ3wt*	*Ḏḥwtj jn sw ẖr=s* *pr.n=f ẖr jrt Ḥrw* *ẖ3wt dj prt-ḫrw*
PT84, Pyr. 59a	*ḥtp-nsw* 2	*Wsjr NN m-n=k jrt Ḥrw ḥtp.* *n=f ḥr=s* *ḥtp-nsw* 2
PT92, Pyr. 61c	*f3t t f3t ḥn(q)t*	*Wsjr NN m-n=k jrt Ḥrw f3 n=k* *s jr ḥr=k* *f3t t* 1 *f3t ḥn(q)t* 1
PT94, Pyr. 64a	[*š*]*bw*[62]	*d-mdw zp* 4 *dj šbw* *ḏj šbw* *šbw*
PT128, Pyr. 80d	*zẖn*	*Wsjr NN m-n=k jrt Ḥrw zẖnt=k* *ḏd-mdw zp* 4 *n NN pn f3t ḥn(q)* *t zp* 4 *zẖn* 1
PT129, Pyr. 81b	*zwt*	*Wsjr NN m-n=k zwtt jrt Ḥrw* *ḏd-mdw zp* 4 *n NN pn f3t ḥn(q)* *t zp* 4 *zwt* 1
PT139, Pyr. 86b	*s*[*r*]	*Wsjr NN m-n=k jw j.sr=f sn* *ḏd-mdw zp* 4 *n NN pn f3t ḥn(q)* *t zp* 4 *sr* 1

62 The *šbw* might allude to PT94–95 as this section is related to the two spells in Wenis.

Recitation	Sahure's offering list	Pyramid Texts
PT140, Pyr. 86d	*mnwt*	*Wsjr NN m-n=k jrt Ḥrw ẖw mn=f s* *ḏd-mdw zp* 4 *n NN pn fꜣt ḥn(q) t zp* 4 *mnwt* 1
PT141, Pyr. 86f	*t zjf*	*Wsjr NN m-n=k jrt Ḥrw jtḥt.n=f* *ḏd-mdw zp* 4 *n NN pn fꜣt ḥn(q) t zp* 4 *t zjf* 1

The same evidence can be attested in the pictorial version of offering Pyramid Texts in the mortuary temples of Sahure's successor, Neferirkare Kakai, and the later king Niuserre. The fragments of the offering list located at the mortuary complex of Neferirkare Kakai in Abusir (fig. 2) presents the list of items PT23, PT82, PT84–89, PT128–133 (table 2), which represents a section with a similar order of part of the offering group of Pyramid Texts in the pyramid of Wenis.[63]

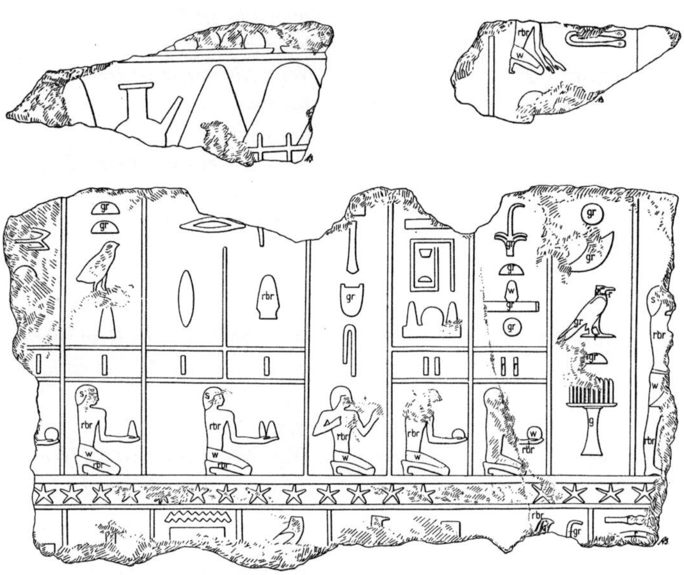

Fig. 2: Offering list fragments from the mortuary temple of Neferirkare Kakai (after Borchardt 1909, fig. 32).

63 See W/S/N i 1–4 (PT23), W/S/N ii 28 (PT82), W/S/N ii 30–35 (PT84–89), W/S/N iii 12–17 (PT128–133).

Table 2: Offering list items of Neferirkare Kakai and associated Pyramid Text recitations.

Recitation	Neferirkare's offering list	Pyramid Texts
PT23, Pyr. 16d	*zꜣṯ*	*Wsjr jṯ n=k msḏḏw NN nbw* *mdw m rn=f ḏw* *Ḏḥwtj jṯ sw n Wsjr jn mdw m* *rn n NN ḏw* *dj n=k sw m ḏrt=k ḏd-mdw zp* 4 *m sfḫḫ=k jm=f* *zꜣ jm=k sfḫḫw jm=f* *zꜣṯ*
PT82, Pyr. 58b	*ḫꜣwt*	*Ḏḥwtj jn sw ḫr=s* *pr.n=f ḫr jrt Ḥrw* *ḫꜣwt dj prt-ḫrw*
PT84, Pyr. 59a	*ḥtp-nsw* 2	*Wsjr NN m-n=k jrt Ḥrw ḥtp.* *n=f ḫr=s* *ḥtp-nsw* 2
PT85, Pyr. 59c	*ḥtp wsḫt* 2	*Wsjr NN m-n=k jrt Ḥrw ḥtp* *ḥr=s* *ḥtp wsḫt* 2
PT86, Pyr. 59d	*ḥms*	*sḫm n=k s ḫr=k* *ḏd-mdw ḥms j.gr prt-ḫrw-nsw*[64]
PT87, Pyr. 60a	*jꜥw-r t ds*[65]	*Wsjr NN m-n=k jrt Ḥrw jꜥb n=k* *s jr r=k* *jꜥw-r t* 1 *ds* 1
PT88, Pyr. 60b	*t-tw*	*Wsjr NN m-n=k jrt Ḥrw ḫw n=k* *jtj=f s* *t-tw* 1
PT89, Pyr. 60c	*t-jtḥ*	*Wsjr NN m-n=k jrt Ḥrw jtḥt.n=f* *t-jtḥ* 1
PT128, Pyr. 80d	*z[ḫn]*	*Wsjr NN m-n=k jrt Ḥrw zḫnt=k* *ḏd-mdw zp* 4 *n NN pn fꜣt ḥn(q)* *t zp* 4 *zḫn* 1
PT129, Pyr. 81b	*zw[t]*	*Wsjr NN m-n=k zwtt jrt Ḥrw* *ḏd-mdw zp* 4 *n NN pn fꜣt ḥn(q)* *t zp* 4 *zwt* 1

64 The direction "Sit down, be silent: the king's invocation" is addressed to the audience participating in the ritual performance and denotes the integration of external practices into the corpus. See analogous directions in PT460 (Pyr. §868c); PT618 (Pyr. §1746a); and sPT734 (Pyr. §§2263a-2264a), CT312 (*ECT* IV, 70e–71a); Old Kingdom mastabas (Junker 1943, 20–21); and later in the foundation ritual in Luxor and Medinet Habu (Barguet 1952, 5, pls. 1–2); stela Turin 1599 (Mekis 2011, 49–50, l. 3 "Ô tous les dieux, silence, silence"); and O. DeM 1696, verso II, 5 (Meeks 2000, 244–245).

65 For the rite of "mouth-washing", see Meulenaere 1981, 87–89, where the author suggests the reading *ꜥb* (*rꜣ*) on the restitution of the term in the Saite Period.

Recitation	Neferirkare's offering list	Pyramid Texts
PT130, Pyr. 81d	*[s]p[r]*	*Wsjr NN m-n=k sbjw jr.k* *ḏd-mdw zp* 4 *n NN pn fꜣt ḥn(q)* *t zp* 4 *spr* 4
PT131, Pyr. 82b	*ꜣ[šrt]*	*Wsjr NN m-n=k j.sšꜣw=k* *ḏd-mdw zp* 4 *n NN pn fꜣt ḥn(q)* *t zp* 4 *ꜣšrt* 1
PT132, Pyr. 82d	*m[zt]*	*Wsjr NN m-n=k jrt Ḥrw j.zꜣ=k* *jr=s* *ḏd-mdw zp* 4 *n NN pn fꜣt ḥn(q)* *t zp* 4 *mzt* 1
PT133, Pyr. 83b	*nnšm*	*Wsjr NN m-n=k jrt Ḥrw* *ḏd-mdw zp* 4 *n NN pn fꜣt ḥn(q)* *t zp* 4 *nnšm* 1

Even the scant evidence attained from the single preserved fragment of the offering list of Niuserre (fig. 3) supports the pre-existence of offering rites and performances in the royal context. Although the fragment only allows for the identification of three ritualists under a register for offering quantities, the objects represented in their hands[66] provide further information and relate the three sections with the items involved in PT129–131 (table 3): the ritual rations *zwt* (shank of meat), *spr* (four ribs), and *ꜣšrt* (roasted meat).

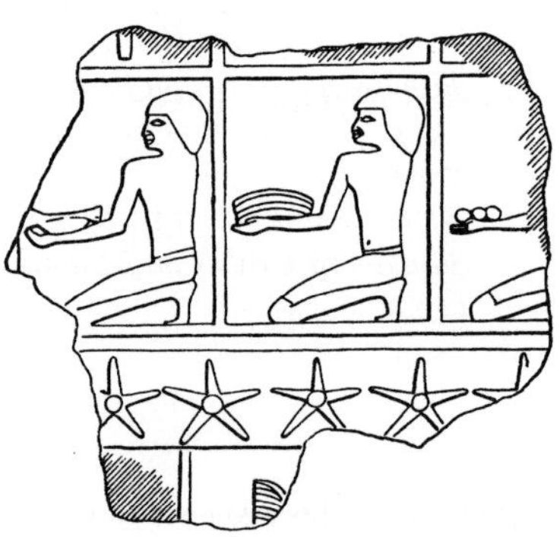

Fig. 3: Offering list fragment from the mortuary temple of Niuserre (after Borchardt 1907, fig. 59).

66 Cf. the objects of the three ritualists with the determinatives for *zwt*, *spr* and *ꜣšrt* in Sethe 1960, 46 (PT129, Pyr. §81b: ; PT130, Pyr. §81d: and PT131, Pyr. §82b:). Cf. the types of meat in Köhler/Jones 2009, 39 and 104.

Table 3: Offering list items of Niuserre and associated Pyramid Text recitations.

Recitation	Niuserre's offering list	Pyramid Texts
PT129 [Pyr. 81b]	[*zwt*]	*Wsjr NN m-n=k zwtt jrt Ḥrw* *ḏd-mdw zp* 4 *n NN pn fꜣt ḥn(q)* *t zp* 4 *zwt* 1
PT130 [Pyr. 81d]	[*spr*]	*Wsjr NN m-n=k sbjw jr=k* *ḏd-mdw zp* 4 *n NN pn fꜣt ḥn(q)* *t zp* 4 *spr* 4
PT131 [Pyr. 82b]	[*ꜣšrt*]	*Wsjr NN m-n=k j.sšꜣw=k* *ḏd-mdw zp* 4 *n NN pn fꜣt ḥn(q)* *t zp* 4 *ꜣšrt* 1

There are even extensive remains of the offering list of Pepi II from the north and south walls of his mortuary temple, which evidences the use of pictorial versions of the texts while the inner chamber of his pyramid was also inscribed with narrative versions of the same rituals.[67] In this regard, the monument of Pepi II Neferkare offers evidence that royal mortuary complexes could incorporate both traditions—offering lists and Pyramid Texts—simultaneously, in order to ensure the continual supply of prescribed provisions and items in perpetuity.[68]

In the private context, however, the attestation of the use of tabular versions of the Pyramid Texts goes beyond Sahure and can be dated to the earlier Old Kingdom. No doubt, as the rituals constituting the Pyramid Texts corpus of Wenis resulted from a long process of experiencing the rites, transferring performances and recitations into script, and monumentalizing these texts on the walls of his pyramid, the rites of the offering lists also underwent a process of "canonization" before the traditional lists of the Fourth and Fifth Dynasty appeared.[69]

Thus, before the constitution of the type A list, which incorporates ninety items that match the same number of offering rites in the Pyramid Texts, there were more limited groups of items in slab-stelae and tomb inscriptions[70] that provide the earliest

67 See offering list of Pepi II on the south wall of the main chapel in his mortuary temple, in Jéquier 1938, 56–64, pls. 61, 67–70.

68 As Assmann 2005, 346 notes: "[t]he mortuary offering and recitation are to transform this place of destitution into a place of abundance".

69 See *Listentyp A* with the example of Debeheni in Barta 1963, 47–50, and fig. 4. For differences between the initial royal examples mentioned above and their counterparts from the private context, see Smith 2009, 9 (with bibliography).

70 For the construction of the earliest examples of offering lists, see Martin 2011 (First Dynasty);

examples of some of the items necessary for the offering rites. For instance, the late Second or early Third Dynasty stela of Sisi (Helwan)[71] lists the items necessary for the rites of censing (*snṯr*: PT25, Pyr. §18d) and preparation of the offering table with two kinds of bread (*tꜣ rtḥ*: PT89, Pyr. §60c; and *tꜣ wr*: PT177, Pyr. §103a).[72] Another list of items required for the mortuary service, later incorporated into the corpus of the Pyramid Texts, was found on the Second Dynasty stela of Meri (also known as the Bankfield stela), in which one finds epitomized the rites for censing (*snṯr*: PT25, Pyr. §18d), libating (*qbḥw*: PT32, Pyr. §23b), and the preparation of the offering table with bread (*ḥṯꜣ*: PT113, Pyr. §73f).[73] This evidence constitutes the earliest proof of the existence of particular rites connected with mortuary services outside the tomb and extraneous to the royal domain, predecessors of the succeeding recitations monumentalized in the pyramid of Wenis.

The progressive development of mortuary services might have been the major reason for expanding the stock of items in the offering lists. Evidence from the Third and Fourth Dynasty indicates the performance of further rites and the incorporation of new products. Nonetheless, the sequencing of the items so far does not follow the canonical order of the type A list. I perceive this variability as an indication of the variety of local traditions in Meidum, Giza, and Saqqara, on the one hand, and the absence of formal religious sanctions at this time on the multiple services addressed to the deceased, on the other. The stela of Khabausokar[74] (fig. 4), for instance, shows that by the Third Dynasty some new items had been integrated into the mortuary service for the deceased, including the sacerdotal services for bread offering (*ḥṯꜣ*: PT113, Pyr. §73f), libation (*qbḥw*: PT32, Pyr. §23b), censing (*snṯr*: PT25, Pyr. §18d), provision of oils (*ḥꜣṯṯ ꜥš*: PT77, Pyr. §53b; and *sfṯ*: PT74, Pyr. §51a), wine jars (*jrp ꜥbš Ḫ*: PT154, Pyr. §92d), three kinds of bread (*tꜣ wr*: PT177, Pyr. §103a; *tꜣ rtḥ*: PT89, Pyr. §60c; and *šꜥt*: PT142, Pyr. §87b), and wine again (*jrp ꜥbš*: PT154, Pyr. §92d), and *jšd*-berries (PT160, Pyr. §95d).[75] In the early Fourth Dynasty, the offering lists of Rahotep and Metjen also extended the number of items related to the mortuary services. The offer-

---

Köhler/Jones 2009 (Second-Third Dynasties); Kaplony 1963, 235–241; with further comments in Kaplony 1966. I thank Jochem Kahl for bringing Kaplony's references to my attention and discussing the nature of the offering items that account for the representations of the *Speisetischszene*.

71 Grabplatten Sp. 23 in Kaplony, *Die Inschriften I*, 231, 619. For a comparative analysis of the dating criteria used with the slab stelae from Helwan and the relative date of this stela in the late Second or Third Dynasty, see Kahl 1997, 137–145, fig. 1; and Kahl/Kloth/Zimmermann 1995, 178–179, n. 11.

72 Saad 1957, 46–48, pl. 27.

73 Hays 2010a, 129–130, fig. 4; Barta 1963, 24–25; and Gardiner 1917, 259–260.

74 Murray 1905, 32–35, pl. 1.

75 In addition, the stela of Khabausokar includes four types of meat (*swt*, *jwꜥ*, *wšn* and *spr*) that might correspond to other known items used in the Pyramid Texts. See a similar case in the fragmentary offering list of Niuserre in n. 66 above. For the types of meat offerings found in the early slab stelae, see Köhler/Jones 2009, 36–56.

Fig. 4: Central panel from the tomb of Khabausokar (after Murray 1905, pl. 1).

ing list of Rahotep shows PT25 (Pyr. §18d) PT79 (Pyr. §54d) PT154 (Pyr. §92d) PT152 (Pyr. §91d) PT23 (Pyr. §16d) PT20 (Pyr. §11c) and PT177 (Pyr. §103a), an order similar to the one present in the door tablet of Metjen's false door.[76]

Most remarkable about this analogy between the construction of a list of canonical offerings required for the mortuary services and the later royal Pyramid Texts is the development of the traditional offering list by the mid-Fourth Dynasty (2550 BCE), two hundred years before the reign of Wenis. Until now, the only evidence demonstrating the existence of offering lists that agreed with the order of the Pyramid Text spells was the offering lists of Debeheni and Khafkhufu I.[77] However, by the reign of Khufu and the succeeding kings of the Fourth Dynasty, other instances of offering lists with abridged versions and similar order can be identified. The relationship between the order of the items in the offering lists of the Fourth Dynasty and the later order of the Pyramid Texts reveals that a canonization of mortuary services and rites had begun by the reign of Khufu and had taken its definitive shape by the Fifth Dynasty.

The tomb of Nefer (G 2110), completed in the reign of Khufu,[78] offers us two examples of offering lists from this time. Artists represented an offering scene in the south entrance thickness of his tomb (now Louvre B 151), in which the order of the ritual items was PT23 PT25 PT78–80 PT154 PT169 PT112 PT177 (fig. 5). Far from following the exact order of the traditional offering list type A, this abridged version again offered an example of the existence of rites (with a shortened ritual structure) in the private context connected with those alluded to in the royal Pyramid Texts. In addition, Nefer also included a more comprehensive offering list on the false door tablet in the west wall. In this case, the number of items and their order did resemble the canonical one (fig. 6). In this offering list the following sequence can be identified: PT32 PT25 PT78–80 *unknown item* PT84 PT117 PT87–90 PT95 PT117 PT164 PT117 PT94–96 PT32

76 *LD* II, 3 and 5.

77 Hays 2010a, 129–130; and Smith 2009, 9.

78 See Manuelian 2009, 154–155.

Fig. 5: Tomb of Nefer, south entrance thickness (= Louvre B 151) (after Manuelian 2009, 181).

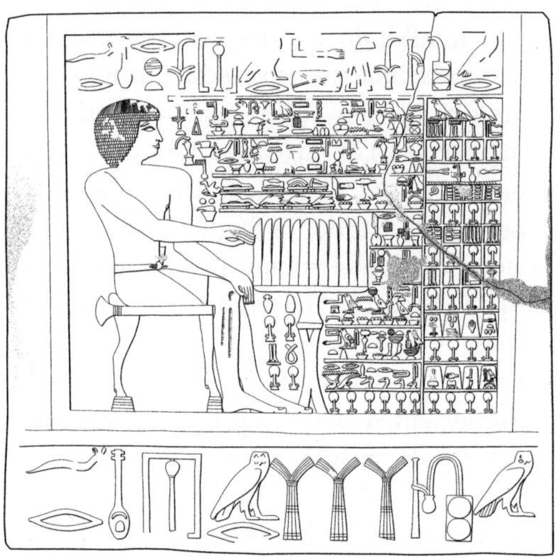

Fig. 6: Tomb of Nefer, chapel west wall, door tablet (after Manuelian 2009, 200).

PT25 PT112 PT160 PT152 PT166–168 PT154 PT146 PT149–150 PT164 PT159 PT158 PT157 PT142 PT161–162 PT169–170 *rn-jwꜣ rn-mꜣ-ḥḏ* PT138 PT136–137 PT139.

A similar distribution occurs in the tomb of Khafkhufu I (G 7140),[79] in which two major instances of lists of ritual items can be identified. In the south wall of the chapel relief, Khafkhufu I appears seated before an offering list with the following order: PT141–157 PT159 PT158 PT160–162 PT164 PT163 PT165–171.[80] In the false door tablet, however, the offering list shows a more comprehensive group with an order comparable to the canonical list A type: PT84–85 [...] PT88–89 PT92 [...] PT96 PT146 PT148 PT147 PT144 PT109 PT111–113 PT115 PT114 PT113var. PT116–126 PT129 PT127 PT130 PT128 PT131 PT134 PT136–140 PT145 PT148 PT153 PT160 PT162 PT164 PT166–167 PT169–170.[81] Other examples of the mid-late-Fourth Dynasty, such as the tomb of Seshatsekhentiu (G 2120)[82] and the anonymous owner of G 2135,[83] also denote the progressive develop-

79 For this tomb, see Simpson 1978.
80 See Simpson 1978, 14–15, fig. 31, pl. 19.
81 Simpson 1978, 15–16, fig. 32, pls. 20–21.
82 Manuelian 2009, 209–216, figs. 7.66–7.68: see offering list (now MFA 06.1894) with items relating to PT25 PT79–80 PT78 PT161–162 PT177 PT112 PT114–115 PT142 PT114 PT116 PT149 PT145 *unknown* PT146 PT154 PT160 PT152 PT166–168 PT158 PT164 PT169.
83 Manuelian 2009, 281–283, and fig. 10.15: see offering list (now Vienna ÄS 7799) with items relating to PT80 PT78 PT160 PT79 PT154 PT166.

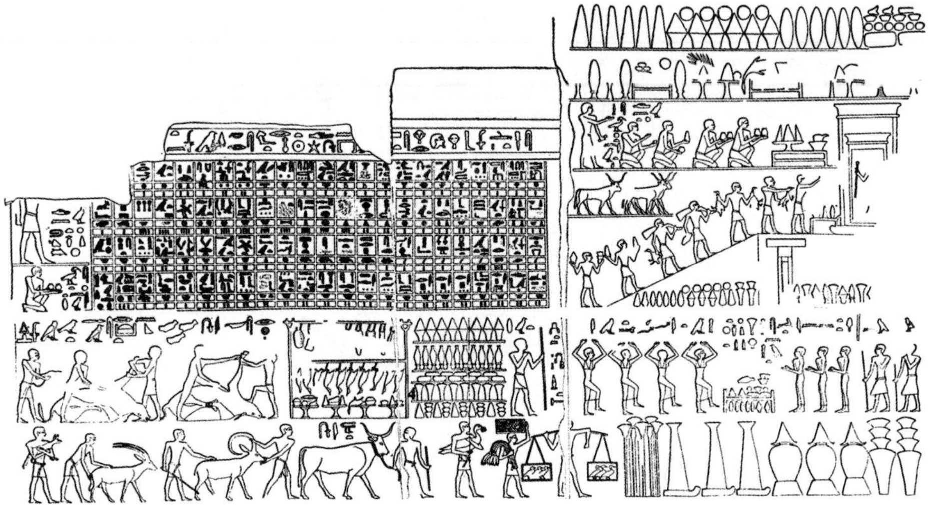

Fig. 7: Offering list of Debeheni (Giza, LG90) (after *LD* II, pl. 35).

ment of the rituals and the incorporation of further sacerdotal items (i.e. ritual activities) into the mortuary services.

The preservation of these lists testifies to the existence of particular conventions in the performance of mortuary services that were promoted to a canonical repertoire composed for the ritualists. The affiliation of a larger number of items and activities indicates the increasing interest in fixing the structure of the mortuary ritual, an idea that fostered the later composition of the Pyramid Texts corpus in the arena of royal ceremonies and beliefs. In the late Fourth and early Fifth Dynasty we find other testimonies of the canonization of the offering list, with some instances, such as the offering lists of Debeheni and Kaninesu I. In the case of Debeheni,[84] as Hays highlighted, "the third through ninetieth entries [...] correspond to the items and actions specified at the end of the eighty-eight Pyramid Texts in the same sequential order, beginning with *sṯ-ḥꜣb*".[85] The offering list of Debeheni (fig. 7) goes on with the same particular items related to the Pyramid Texts and concludes with two further items typical of the canonized offering type A list: *gsw* and *ḥꜣt-wdḥw* (Barta's items 91 and 95).[86]

The tomb of Kaninesu I (G 2155; d. late Fourth or early Fifth Dynasty)[87] provides two instances of disparate offering lists: a shorter version on the north entrance thick-

84 For this tomb, see Hassan 1948, 159–184, n. 122; Junker 1938, 50; and Junker 1934, 85–96.

85 Hays 2012, 86.

86 The sequencing of items in the offering list of Debeheni includes PT23 PT25 PT72–81 PT25 PT32 PT82–92 PT94 PT96 PT108–171 and the items *gsw* and *ḥꜣt-wdḥw*.

87 See Manuelian 2009, 367–383 (especially for the date of this tomb, 368).

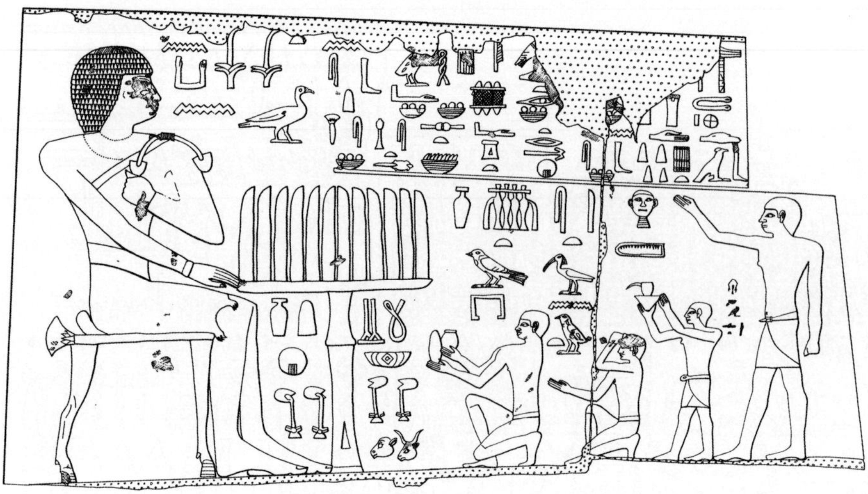

Fig. 8: Tomb of Kaninesu I, north entrance thickness (= Vienna ÄS 8006) (after Manuelian 2009, fig. 13.37).

ness, and a more extended offering list on the south wall of the chapel. The abridged version (fig. 8) on the north entrance thickness lists fifteen ritual items related to the Pyramid Texts,[88] while the offering list on the south wall[89] (fig. 9) sets forth a more canonical list of items: PT23 PT81 PT32 PT25 PT72–73 PT78–80 PT25 PT84–85 PT87–90 PT92 PT94 PT96 PT32 PT113 PT151 PT154 PT115 PT114 PT116–120 PT123 PT121 PT124–128 PT130 PT132 PT133 PT135 PT134 PT136–137 *smn* PT138–141 PT109 *jdꜣt* (Barta's item 41) PT142–145 PT149 PT151–152 PT154 PT153 PT150 PT160–161 PT163–165.

In addition to the cases of Debeheni and Kaninesu I, other offering lists of the early Fifth Dynasty demonstrate the incorporation of new items and the process of canonization of the sequence in which these items would be employed in the mortuary services. Two other examples are the offering lists found in the tombs of Nensedjerkai (G 2101) and Kanefer (G 2150).[90] The first instance is an offering list found in the south false door tablet with the sequence of items PT25 PT72–79 PT81.[91] The case of Kanefer is also remarkable as the concise sequence of twenty-four items corresponds—in contents and order—with a section of the series found in the Pyramid Texts of Wenis: PT23 PT25 PT72–81 PT32 PT82 PT84–92 PT94 PT96.[92]

88 Now Vienna ÄS 8006: see Manuelian 2009, figs. 13.36 and 13.37, with PT25 [PT32] PT79–80 PT160 PT153 PT134 PT167 PT152 PT115 PT166 PT112 PT163 PT161–162.

89 A section of the same mastaba in Vienna ÄS 8006: see Manuelian 2009, figs. 13.47 and 13.48.

90 Both tombs and their offering lists in Manuelian 2009, 117–124 (Nensedjerkai) and 307–318 (Kanefer).

91 See Manuelian 2009, 142, fig. 5.47.

92 Manuelian 2009, 351, fig. 12.80.

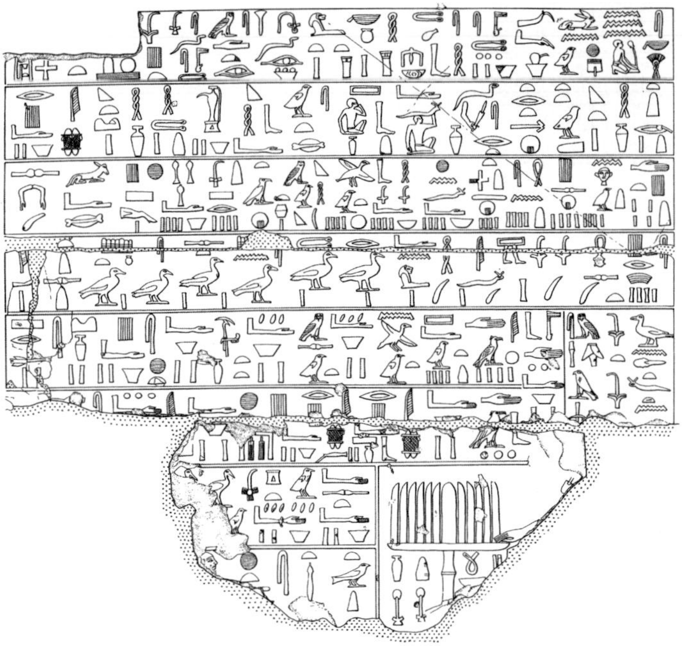

Fig. 9: Tomb of Kaninesu I, chapel south wall (= Vienna ÄS 8006) (after Manuelian 2009, fig. 13.48).

The development of the offering lists into the standard type-A list and its effect on the composition of the group of offering texts for the pyramid of Wenis becomes even more visible in the instances of the mid-Fifth Dynasty and those contemporaneous with the monumentalization of the Pyramid Texts in the inner chambers of Wenis. Two noteworthy examples for the understanding of the offering lists as reflections of ritual observances and objects, also attested to in the Pyramid Texts, can be seen in the tombs of Iymery (G 6020) and Neferbauptah (G 6010). The tomb of Iymery, which dates to the reign of Niuserre, provides us with an offering list conceptualized and composed before the reign of Wenis, with the same items and order as in the narrative version of the offering Pyramid Texts (fig. 10). Iymery's offering list contains the items for PT23 PT25 PT72–79 PT81 PT25 PT32 PT82 PT84–92 PT94 PT96 PT108–171, and completes the list with the items *gsw*, *ḥꜣt-wdḥw* and *stpt* (Barta's items 91, 95 and 94).[93] The second case is the offering list of Neferbauptah, dating to the time of Djedkare Izezi, which brings together the same sequencing of items for the ritualist (fig. 11):

93 Weeks 1994, 31–57 (especially the details about the offering list in 54–55), fig. 44.

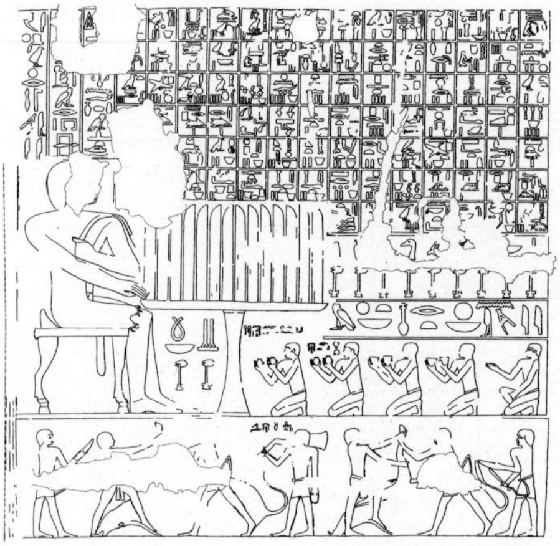

Fig. 10: Tomb of Iymery, third chamber, west wall, southern section (after Weeks 1994, fig. 44).

PT23 PT25 PT72–79 PT81 PT25 PT32 PT82 PT84–92 PT94 PT96 PT108–171, and ends with *gsw* (Barta's item 91).[94]

Other remarkable examples from the late Fifth Dynasty can be retrieved from the tomb of Khafkhufu II in Giza (G 7150),[95] and the tombs of Usernetjer,[96] Sekhemka,[97] and Iteti[98] from Saqqara. Interestingly, the instance of Iteti's offering list draws to a close our analysis of the relation between the series of items on the offering lists found in the private context and the Pyramid Texts used in the royal one. Iteti's offering list was composed during the reign of Wenis, which means it was contemporaneous with the theological endeavour to transfer into the king's monument the same rites that had previously existed in tabular form.

On a smaller scale, another setting from which Pyramid Text recitations might have originated was the domain of magical and apotropaic practices (personal and collective). The group of protective incantations has been considered some of the

94 Weeks 1994, 21–29, fig. 22.

95 Simpson 1978, 21–27, fig. 50: [...] PT94 PT96 PT108–109 [...] PT116–120 [...] PT126–130 [...] PT136–137 [...] PT147–149 [...] PT154–155 [...] PT159–161 [...] PT169–170 PT96 PT146 PT148 PT147 PT144 PT109 PT111–113 PT115 PT114 PT113 var. PT116–126 PT129 PT127 PT130 PT128 PT131 PT134 PT136–140 PT145 PT148 PT153 PT160 PT162 PT164 PT166 PT167 PT169–170.

96 Murray 1905, 19–24, pl. 23. The offering list of Usernetjer includes PT23 PT25 PT72–81 [PT25 PT32] PT82 PT84–92 PT94 PT96 PT108 PT108–171 *gsw ḥꜣt-wdḥw stpt* (Barta's items 91, 95 and 94).

97 Murray 1905, 7–10, pl. 7. The list of items in the offering list of Sekhemka includes PT23 PT25 PT72–75 [...] PT85–92 PT94 PT96 PT108–171 *gsw pẖr stpt ḥꜣt-wdḥw stpt* (Barta's items 91, 92, 94 and 95) PT126.

98 Murray 1905, 18–19, pl. 18: PT23 PT25 PT72–81 PT25 PT32 PT82 PT84–92 PT94 PT96 PT108–171 *ḥꜣt-wdḥw* (Barta's item 95) PT32 PT25.

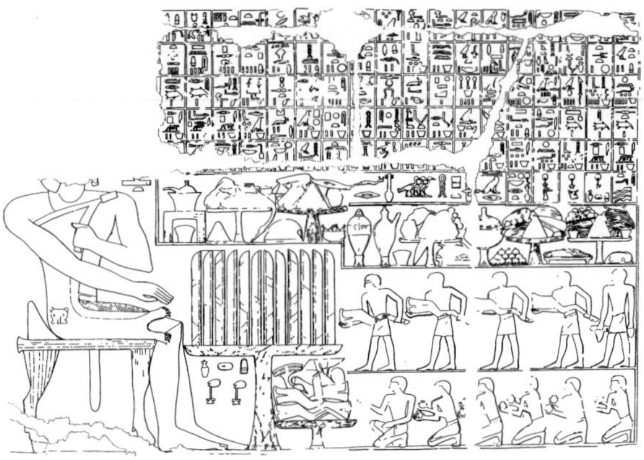

Fig. 11: Tomb of Neferbauptah, third chamber, west wall, central section (after Weeks 1994, fig. 22).

earliest recitational material to be integrated into the royal Pyramid Texts.[99] The two groups of apotropaic texts (i.e. PT232–243 and PT277–293)[100] have generally been taken as independent sets of materials embedded into the Pyramid Texts of Wenis for the protection of the king's corpse and his rebirth.[101] Originally, these incantations functioned as an agency through which to control snakes and ward off any other noxious creature that might harm the living or the dead.[102] It is clear that these recitations continued their development in other royal assemblages of the Old Kingdom,[103] and remained in use from this time to the Greco-Roman period.[104] Regret-

99 See Osing 1986, 132–136 (*Gruppe* A1: PT226–243; *Gruppe* A2: PT276–299); and Altenmüller 1984, 20.

100 Following Hays 2012, 107–108, and group K chart in 685. As indicated by Bernard Mathieu, the second group in Wenis, PT277–293, belongs to a larger set of apotropaic texts attested in other pyramids: PT276–299.

101 Ritner 2011, ix. Although Ritner supports the original independence of particular series of the group (i.e. the anti-snake spells or *Schlangensprüche* PT232–238 and PT281–287) as proof of the embedment of earlier non-Egyptian material, it is possible that the distinction observed in these spells stemmed from the variegated Egyptian settings previous to the late Fifth Dynasty in which these materials were originally located. For the 'Byblite' origin of these spells, which made it into Egypt "perhaps accompanying known timber shipments", see Rittner 2011, xi; and Steiner 2011, 8–14. Cf. the critical positions against this hypothesis in Bojowald 2012, 236–242; and Breyer 2012, 141–146. See further comments on the critical Egyptological positioning against this hypothesis, in particular by Thomas Schneider, in Morales 2013, 85, n. 224.

102 For the significance of these spells, see Sperveslage 2011, 30–37; Kousoulis 2011, 14–26; Borghouts 2007, 21–25, ns. 118–119; Leitz 2002–2003, 701–702 ("Rerek"); Meurer 2002, 269–315; Leitz 1996, 381–427; and Borghouts 1984, 707. For the metatextual heading of PT226 that appears in the Middle Kingdom and identifies the primary use of this recitation and the group to which it belongs for protective purposes, see Allen 2006a, 153; and Sledzianowski 1976, I, 24 (1): L1NY L, col. 1: *rꜣ n(j) sẖsf r(k)rk m ẖrt-nṯr*.

103 With some modifications; see Hussein 2013, 277–278, n. 11; and Hays 2012, 685.

104 See, for instance, the recent studies on the subject by Hussein 2013, passim; Hussein 2011, 220–

tably, the performative and oral nature of these incantations in their original state did not contribute to their fixation in writing beyond a few examples from the late Sixth Dynasty and First Intermediate Period.[105] Among the recitations identified by Jürgen Osing in his study of the magical texts of Niankh-Pepy, one can observe the presence of excerpts from the Middle Kingdom recitations CT473 (*Spruch* III),[106] CT885 (*Sprüche* I and IV),[107] and CT930 (*Spruch* II).[108] Interestingly, CT885 is associated with some of the Old Kingdom series of apotropaic (anti-snake) texts inscribed on the west wall of Wenis's sarcophagus chamber.[109] A remarkable aspect of the spells found in Niankh-Pepy, however, is the personal category of the ritualist, mentioned in the first person.[110] In contrast, in Wenis's sarcophagus chamber, texts of the personal category are not common and, in fact, the second or third-person structure is typical here, as it is believed that a ritualist performed the services for the king.[111] If less perfect in reliability, this disagreement between the apotropaic texts of Niankh-Pepy, with the beneficiary in the first person, and the texts of Wenis, in the second or third person, is a vestige of certain editorial work applied to the materials during the monumentalization and recontextualization of earlier personal texts into a sacerdotal setting, such as the king's sarcophagus chamber.[112] In other words, in the process of transmission from the personal setting to the royal context, the magical texts initially indicating the performer in the first person were transferred into the pyramid and changed to the second and third persons, so as to ensure that the sacerdotal class served the king in his voyage into the afterlife. In spite of the lack of evidence from the private domain in the Fourth and Fifth Dynasties, the magical texts of Niankh-Pepy seem to indicate that the apotropaic Pyramid Texts of Wenis might have had precursors of the same type as the late Old Kingdom ones. In addition, textual features of these recitations

---

222; and Hussein 2009, 89–93. The transmission of this group of Pyramid Texts evidences a sustained system of beliefs on the magical potentiality of these recitations against noxious creatures: cf. the attestation of Pyramid Texts (snake-recitations) in stela BM 190 (probably from Memphis or Saqqara), with PT226–243 (Pyr. §§225a–248a-b). Thanks are due to the research project SFB 980 for providing me with funds for a research visit to the British Museum (d. 30.09.13–10.10.13) where I could examine this stela and other objects of the Late Period that bear late versions of Pyramid Texts and Coffin Texts.

105 See the four magical spells inscribed on the wooden coffin bedframe deposited inside the coffin of Niankh-Pepy in Osing 1987, 205–210; and Hassan 1975, 21–22, pls. 15 (B) and 19–20.

106 *ECT* VI, 3ff.

107 *ECT* VII, 97p–s.

108 *ECT* VII, 131b–e, and n.

109 Hussein 2013, 278, n. 14; and Topmann 2010, 341–371.

110 For the deictic distinction of the sacerdotal/collective and personal categories in Pyramid Texts and associated materials, see mainly Hays 2012, 28–33, 52–60; Reintges 2011, 28–31; and Hays 2006, 33–40. See further bibliography in Morales 2013, 46, n. 103.

111 Hays 2009a, 48, fig. 1.

112 Hays 2006, 47–54.

remind us of particular compositions known as "water-songs" (*ḥsy-mw* or *ḥst-mw*)[113] which appeared in private tombs before the reign of Wenis for the protection of cattle against crocodiles, hippopotami and other dangers.[114]

In interim summary, the evidence on the existence of rites and objects common to both the private offering lists and the royal Pyramid Texts of the Old Kingdom reveal a clear association between the two realms. Likewise, the incorporation of magical spells that belonged to the realm of personal practices also manifest the existence of primary settings in which these recitations were first declaimed and performed, before they achieved the status of monumental inscriptions. As for the offering recitations, the earliest examples of ritual activities, such as censing, libating, provision of foodstuff and the listing of related items from the Second Dynasty onwards, show a convergence toward a collective system of rituals and services—private and royal—from which the Pyramid Texts emerged.[115] Certainly, the construction of a canonical list of rites and items used for the composition of offering lists in tabular form demanded the fixation in writing of these elements to be used as guidelines, that is, the execution of the process of *Verschriftung* or entextualization, so that they became more accessible and portable as papyri (i.e. operative copies). No doubt, the process of entextualization of the rites and the fixation of their details (e.g. items, quantities, number of services) not only gave shape to the offering lists for the tomb of particulars, but also provided the fundamental materials for the composition of the offering category of Pyramid Texts in the royal domain. In addition, it might be the case that the development of the offering lists during the Fourth Dynasty in the cemeteries of Giza reached a canonical form before being extended to other necropolises such as Saqqara, and that the traditional model resulting in the early Fifth Dynasty had an effect on the syntax of offering rites of the late Fifth Dynasty and the rest of the Old Kingdom mortuary culture.[116] Regarding the apotropaic texts, the absence of personal

---

113 Strudwick 2005, 402 (text no. 302.A), 405 (text no. 306.F); Leitz 1999, 39–40, pls. 17–18 (pap. BM EA 10042, recto 6, 10); Mathieu 1996b, 106 (n. 349: with bibliography) and 231; Morenz 1996, 60–61, 128–129; Koenig 1994, 74–75 (n. 42); Borghouts 1978, 87 (text no. 126); and Simpson 1976b, 13.

114 Ogdon 1989, esp. 59–62, n. 15.

115 For the statement that other materials, such as the references to sakhu in private tombs, constitute another bond between the mortuary traditions of both contexts (private and royal), see Enmarch 2013, 87–88, n. 25, although the author still considers this link "tantalising and indirect". Cf. the strong relationship of both contexts through reference to the same types of rites, discussed in Hays 2010a; and Smith 2009.

116 For a consideration of the social and religious regulations that might have affected the establishment of mortuary services with particular offering lists and activities in the region of Aswan (Qubbet el-Hawa), see Seyfried 2003. In addition, one must also consider the later development of the related offering formulae in the First Intermediate Period, when political and economic decentralization produced changes in the expressions of authority among the elite: see Barta 1968, 30, for the emergence of new *Bitte* in the offering formulae such as "to cross the sky like the king" (*Bitte* 30) or "to be accepted by the great god" (*Bitte* 34).

magical recitations before the end of the Old Kingdom does not preclude a consideration of the original setting from which the apotropaic Pyramid Texts derived, a group of texts that demanded editorial modification for their inception as sacerdotal services into the royal corpus.

4 Traces of the Original Forms and Style of the "Pyramid Texts" before Wenis

Theological forethought and scribal management of textual materials were essential in the composition of the royal Pyramid Texts. As an intellectual undertaking, this plan (*sẖr*) was generated by the Heliopolitan sacerdotal class for the benefit of the king's existence—initially Wenis's. It demanded selection, editing, copying, and inscription of a great many religious texts from discrete traditions, and the reconsideration of the doctrine of the afterlife affecting the king's persona.[117] As a result, each assemblage became a reflection of prevalent religious ideas and practices, a monumentalization that therefore allowed for later transformation as the theological ideas and ritual praxis of each cultural phase (*Zeitgeist*) affected the corpus and the priestly vision thereof.[118]

The process of transforming oral recitations into collections of texts probably emerged even before this plan for the monumentalization of religious texts existed.[119] As seen in the previous section, the initial process of fixation into writing (i.e. *Verschriftung* or entextualization) came about when the discourse embedded in oral and performative practices—mortuary services, magical recitations, and stage directions—had to be retained for operative and storage purposes. That stage of development of the mortuary recitations, from voice to papyrus, was not primarily associated with the more substantial undertaking for the royal pyramids, a process that came about only when the priests of Memphis decided to outmatch the previous mortuary complexes by introducing priestly and personal recitations into the inner chambers of a king.

For the construction of Wenis's assemblage, therefore, the Heliopolitan priests anchored their plan to the extant repertoire of written recitations in operative papyri, probably created, copied, and stored in temple repositories. In addition, they might even have composed or searched for further exemplars.[120] In other words, theological

117 Mathieu 2010, 78.

118 Morales 2013, 12.

119 As Hays defined it, "[t]he Pyramid Texts were not composed to decorate the walls of the tombs in which they are first attested. They were adapted to that use from texts prepared to be recited in religious performances" (Hays 2012, 251).

120 The increasing number of Pyramid Texts attested in the assemblages after Wenis seems to corroborate this possibility. For instance, the Pyramid of Wenis contains around two hundred and thirty

and scribal work with the existing collections supplied the primary stock of texts for the composition of Wenis's assemblage; textual composition might have enlarged it with new spells. This process, however, necessitated further assignments, mainly the enhancement of the recitational discourse by transforming the language and structure of the recitations from oral-compositional style to literary form (*Verschriftlichung* or textualization), and the preparation of master copies of the materials—also known as "blueprints"—to be carved onto the walls of Wenis's crypt. This major transformation of the recitations, from papyrus to wall, constitutes the second development of the corpus. Fortunately, vestiges from the earlier stage of the history of the corpus, in which its recitations were written in the oral style and fastened in operative papyri, become visible in the assemblage made for Wenis. Certainly, the gradual transference of oral materials, first to manuscripts and later to monument walls, reflects the ability of the religious discourse to adapt to new media and materiality. Therefore, I consider it essential to visualize both processes (entextualization and textualization) as a result of scribal manipulation and transmission of portentous texts in the high-cultural domain of temples.[121] Copies with the respective sections of the assemblage would be edited and personalized in the library, and only later handed over to the personnel assigned to King Wenis's pyramid, who would supervise the carving, emendation, and final approval of the corpus.[122] The modification of the texts *in situ* would have been impractical, as the nature and number of modifications would have required a prolonged stay in an adequate locale with scribal and sacerdotal experts (i.e. library, repository), to carry out editing exercises including loud reading or recitation,[123] identification of target-forms, modification, and composition.[124]

One also encounters references to the inscription of texts on tomb walls, for instance, in the private tomb of Senedjemib Inti (G 2370). Here there is evidence of necropolis personnel (scribes, draughtmen) who employed similar materials to the master copies previously mentioned as drafts (i.e. *zẖ-qdt*) for the carving of particular texts in private tombs:

> *jw rdj.n(=j) d(j).t(j) m zẖ m zẖ-qd(t) ḥr jz=f pn sḥr=sn jn qsty ḏd m ḥrj=(j) tpt-rd jm mr psš m stp-zꜣ*

texts, whereas the last assemblage for a king of the Old Kingdom, in the pyramid of Pepi II, has about six hundred and seventy-five.

121 Following Baines 2004, 26–30.

122 For the two-step process of textual transfer from papyri to the pyramid walls with emendations along the carving and a concluding revision, see Mathieu 1996a, esp. 290–293 (s.v. *première* and *seconde vérification*).

123 The practice of silent reading would also be feasible at this stage, although the use of this procedure is not clear (see note below).

124 For the multiple forms of reading (including silent and loud reading), see Contardi 2010; and Morenz 1996, 43–52, fig. 5.

> "His Majesty has had the decrees concerning it endorsed with the documentary seal. Funerary priests were appointed for him. I have had them (i.e. the decrees) put in writing in a preliminary sketch on this his tomb, and they were carved by the sculptor. The stipulations in them were recited in my face[125] according to the apportioning in the court council".[126]

The *zẖ-qdt* sketches could then designate scribal copies written in papyri and employed in the carving of texts—in this case, a royal decree by Djedkare Izezi—on the walls of a private tomb. The form of the sketched texts used for the inscription of the collectanea of the Pyramid Texts of Wenis was no doubt unconventional and more extensive and intricate than the copies prepared for Senedjemib's tomb.[127] Yet evidence for the use of sketches in the construction and decoration of private tombs before the reign of Wenis highlights the primary role of master copies in the transfer of textual (and iconographic) materials from the locale of the libraries to the site under construction, as well as the editing procedures applied along the way.

One may widen the discussion on the transference of texts in the form of customized master copies from library settings to a particular pyramid by alluding to the traces of the oral style present in the Pyramid Texts. These traits, found in the language and syntax of the Pyramid Texts, corroborate the antiquity of the recitations and their genesis outside the domain of the pyramid. What sorts of vestiges are these, then? Some of the most common features are transformation of the original deictic-structure, coalescence of dialectic variants, presence of archaisms, use of repetitive language and structures, formulaic language, wordplay, and even Semitic loans. Below I present an analysis of instances of these traces, vestiges of the oral-compositional style that resulted from the application of the processes of *Verschriftung* and *Verschriftlichung*.

An element of outstanding significance in the transformation of the oral discourse into a literary style was the modification of the deictic-structure of texts from personal structure (first person: "I") to sacerdotal structure (second and third persons: "you,

---

125 Can this statement be taken as a proof of concluding revision or recitation by priests and relatives? For the latter, see Baines 1999, 25, who believes that priests and relatives could commemorate the memory of the deceased by reading and reciting the biographical texts of the tomb in a formal ceremony. In this respect, Baines follows Helck's opinion that private inscriptions would be recited (Helck 1972, 11). The excavators of the Late Period tomb of Padihor (tomb R1) have also attested the proofreading of the hieroglyphic inscriptions upon their completion (see Coppens/Smoláriková 2009, 69–71).

126 See inscription C, lines 23–26 (biographical data on Senedjemib Inti as provided by his son Senedjemib Mehit) in Brovarski 2001, 43 (n. 94) and 102.

127 For the decoration and inscription of pyramids, see Pfirsch 1994, 293–298; and Pfirsch 1992, 35–36.

he") to adhere to the ritual conditions in which the ritualist performed for the king.[128] For this matter, one can identify modifications[129] of the texts in the form of:

1. physical recarving of personal pronouns, originally chiselled in the first person and later modified to observe the sacerdotal structure;[130]
2. vacillation, with different pronouns used in the same section;[131]
3. doubling of pronouns, the original and the adjunct following one another;[132]
4. residual *-j* and *-jj* with third weak verbs, as a vestige of the presence of first-person pronoun before editing;[133]
5. exemplar disagreement, when two or more attestations of the same text show the use of different pronouns;[134] and

128 For the modification of the performative structure in the Pyramid Texts, see n. 110 above.

129 The evidence regarding the particular modifications of the beneficiary in the Pyramid Texts of Wenis and the assemblages of the Sixth Dynasty kings has been collected from Hays 2006, 40–56, and Appendix A.

130 PT283 (Pyr. §424a) [W]; PT296 (Pyr. §439a) [W]; PT301 (Pyr. §448b) [W]; PT311 (Pyr. §495c) [W]; PT322 (Pyr. §518c) [P]; PT333 (Pyr. §542c) [P]; PT407 (Pyr. §710a) [P]; PT408 (Pyr. §714a) [P]; PT494 (Pyr. §1063c) [P]; PT495 (Pyr. §1064c) [P]; PT504 (Pyr. §1079a) [P]; PT504 (Pyr. §1083a) [P]; PT505 (Pyr. §§1090e–f) [P]; PT506 (Pyr. §1094a) [P]; PT507 (Pyr. §1104a) [P]; PT508 (Pyr. §1107a) [P]; PT509 (Pyr. §1120c) [P]; PT510 (Pyr. §§1133a-b) [P]; PT511 (Pyr. §1149b) [P]; PT513 (Pyr. §1174b) [P]; PT515 (Pyr. §1176b) [M]. The *siglum* in square brackets [ ] indicates the Old Kingdom royal assemblage in which the modification is observed.

131 In this case, the ancient editor did not identify the presence of original first-person pronouns and therefore the section retained a sacerdotal structure (i.e. second and third-person pronouns) with intact first-person pronouns. See PT263 (Pyr. §329c) [T]; PT299 (Pyr. §444c) [W]; PT304 (Pyr. §471d) [W]; PT311 (Pyr. §499a) [W]; PT327 (Pyr. §536b) [T]; PT335 (Pyr. §546a) [T]; PT346 (Pyr. §561d) [N]; PT406 (Pyr. §708a) [T]; PT469 (Pyr. §909c) [P]; PT470 (Pyr. §911b) [P]; PT477 (Pyr. §966d) [N]; PT484 (Pyr. §1023b) [P]; PT502H (Pyr. §1076) [P]; PT503 (Pyr. §1079b) [P]; PT504 (Pyr. §1086a) [P]; PT508 (Pyr. §1113c) [P]; PT510 (Pyr. §1140c) [P]; PT511 (Pyr. §1152b) [P]; PT525 (Pyr. §1246b) [M]; PT528 (Pyr. §1251a) [P]; PT555 (Pyr. §1376a) [N]; PT562 (Pyr. §1406a–b) [P]; PT565 (Pyr. §1423a) [P]; PT567 (Pyr. §1430e) [N]; PT569 (Pyr. §1440c) [P]; sPT570A (Pyr. §1443b) [P]; PT573 (Pyr. §1482) [P]; PT573 (Pyr. §1484) [M]; PT609 (Pyr. §1708a–b) [M]; PT626 (Pyr. §1770c) [P].

132 See PT269 (Pyr. §378a) [P]; PT270 (Pyr. §386a) [M]; PT336 (Pyr. §548a) [M]; PT439 (Pyr. §812c) [P]; PT467 (Pyr. §890b) [N]; PT469 (Pyr. §909a) [P]; PT505 (Pyr. §1093d) [P]; PT508 (Pyr. §1116d) [P]; PT510 (Pyr. §1135b) [P]; PT511 (Pyr. §1150c) [P]; PT513 (Pyr. §1168a) [P]; PT515 (Pyr. §1181a) [P]; sPT570A (Pyr. §1451b) [P]; PT611 (Pyr. §1726a) [N].

133 See PT266 (Pyr. §358h) [P]; PT271 (Pyr. §390a) [N]; PT302 (Pyr. §461a) [W]; PT362 (Pyr. §§606a-b) [T]; PT456 (Pyr. §856b) [N]; PT467 (Pyr. §889c) [N]; PT469 (Pyr. §906d) [P]; PT471 (Pyr. §922b) [N]; PT473 (Pyr. §927d) [N]; PT477 (Pyr. §967d) [M]; PT481 (Pyr. §1000b) [N]; PT485 (Pyr. §1036b) [P]; PT504 (Pyr. §1087a) [M]; PT509 (Pyr. §1123a) [P]; PT510 (Pyr. §1143b) [M]; PT511 (Pyr. §1159c) [N]; PT519 (Pyr. §1204a) [M]; PT527 (Pyr. §1249c) [M]; PT555 (Pyr. §1374a) [N]; PT563 (Pyr. §1416b) [N]; PT569 (Pyr. §1442c) [M]; PT571 (Pyr. §1467a) [P]; PT576 (Pyr. §1517b) [P]; PT681 (Pyr. §2037a) [N]; PT684 (Pyr. §2054) [N]. See further comments on this phenomenon in Allen 2013, 114. For the same process in the Coffin Texts, see Schenkel 2000.

134 See PT262 (Pyr. §329c) [P]; PT304 (Pyr. §471d) [T]; PT306 (Pyr. §478b) [M]; PT327 (Pyr. §536b) [N]; PT335 (Pyr. §546a) [N]; PT406 (Pyr. §708a) [N]; PT419 (Pyr. §748c) [M]; PT466 (Pyr. §883c) [M]; PT468

6. advanced noun, a noun in an advanced position only suitable for a pronoun.[135]

In addition, other features of the language of the Pyramid Texts reveal that textual editing was carried out with particular segments of the corpus. For instance, a procedure that might indicate the adaptation of the primary compositional style (speech) to the literate character of the monumentalized corpus (text) is the transformation of the genitive from direct to indirect (paradigm A (*nj*) B: *mwt Wnjs—mwt n(j)t Wnjs*).[136]

Occasionally, however, the oral-compositional structure of the Pyramid Texts surfaces through instances in which the spoken and performative nature of the text has not been altered. For instance, one observes the concatenation of interjections, imperatives, vocatives, questioning, and even directions for the ritualists, which reflect the dramatic and performative nature of the recitations:

PT214 (Pyr. §136a)[W]
136a *hꜣ Wnjs zꜣ kw šj ḏd-mdw zp* 4
"Ho, Wenis! Beware of the lake! Recitation 4 times".[137]

PT221 (Pyr. §§196a–b)[W]
196a *hj Nt hj ꞽIn hj Wrt*
196b *hj Wrt-ḥkꜣw hj Nzrt*
"Ho, Red Crown! Ho, Curl! Ho, Great One!
Ho, Great of Magic! Ho, Fiery One!"

PT296 (Pyr. §§439a–c)[W]
439a *ṯtw ṯnj šm=k ꜥḥꜥ n Wnjs*
439b *Wnjs pj Gbb hmṯ sn nj hmṯt*
439c *mt jt=k ḏꜥꜥmjw*

(Pyr. §900e) [P]; PT470 (Pyr. §911b) [N]; PT474 (Pyr. §941b) [N]; PT475 (Pyr. §947b) [M]; PT477 (Pyr. §966d) [P]; PT517 (Pyr. §§1189e-f) [P]; PT521 (Pyr. §§1225c–d) [M]; PT525 (Pyr. §1245a) [M]; PT528 (Pyr. §1251a) [M]; PT555 (Pyr. §1376a) [M]; PT565 (Pyr. §1423a) [N]; PT567 (Pyr. §1430e) [P]; PT569 (Pyr. §1440c) [M]; sPT570A (Pyr. §§1443b–1444a) [M]; PT573 (Pyr. §1482a) [M]; PT594 (Pyr. §1638a) [N]; sPT625A (Pyr. §1762b) [N]; PT626 (Pyr. §1770c) [N].

135 See PT265 (Pyr. §§355b-c) [P]; PT266 (Pyr. §§360b-d) [P]; PT321 (Pyr. §517a) [W]; PT332 (Pyr. §541c) [T]; PT344 (Pyr. §599c) [N]; PT345 (Pyr. §560c) [N]; PT349 (Pyr. §566c) [N]; PT361 (Pyr. §604c) [N]; PT406 (Pyr. §707a) [N]; PT407 (Pyr. §710b) [T]; PT471 (Pyr. §921c) [P]; PT473 (Pyr. §927a) [P]; PT477 (Pyr. §968c) [N]; PT478 (Pyr. §975a) [N]; PT480 (Pyr. §993a) [N]; PT504 (Pyr. §1087a) [N]; PT511 (Pyr. §1151a) [N]; PT515 (Pyr. §1181a) [N]; PT518 (Pyr. §1193b) [M]; PT519 (Pyr. §§1208a–b) [M]; PT520 (Pyr. §1222a) [M]; PT531 (Pyr. §1254c) [M]; PT572 (Pyr. §1473b) [P]; PT573 (Pyr. §1480a) [P]; PT587 (Pyr. §1597d) [P]; PT602 (Pyr. §1673b) [M]; PT681 (Pyr. §2036c) [N].

136 In Wenis's assemblage, see PT204 (Pyr. §118a); PT252 (Pyr. §273b); PT269 (Pyr. §380a); PT271 (Pyr. §389a and Pyr. §390b); and PT307 (Pyr. §484b). Observe the less extensive transformation of indirect genitives into direct forms: PT50 (Pyr. §37c) and PT204 (Pyr. §118c). Data retrieved from Allen 2013, 72, n. 35.

137 The ending section of Pyr. §136a might be defined as a metatextual note (guideline) addressed to the sacerdotal agent for the declaiming of this section of the recitation four times.

"*Ṯtw*-snake, where are you going? You will not go! Wait for Wenis
because it is Wenis, Geb. *ḥmṯ*-snake, brother of *ḥmṯt*-snake,
you whose father has died, Djaamiu-snake!"

PT299 (Pyr. §§444a–c)[W]
444a *ḏt r pt zpꜣ-Ḥrw r tꜣ*
444b *ṯbt Ḥrw šꜣs=f nb ḥwt kꜣ tpḥt*
444c *šnṯ nj šnṯ=j*
"Cobra to the sky! Horus's centipede to the earth!
Horus's sandal is treading on the enclosure's lord, the cavern's bull!
Shunned snake, I cannot be shunned!"[138]

In a stimulating study observing the oral-compositional style of the Pyramid Texts, Chris Reintges explains that the multiple forms of language found in the Pyramid Texts should be perceived as reflections of dialect variations, ongoing language change, and different regional backgrounds.[139] These aspects of the corpus evidence the compilation of materials of assorted antiquity from different areas and contexts.[140] Therefore, if contexts beyond the mortuary service, such as the temple cult, domestic magical practices, and popular settings with soldiers or guildmen singing,[141] supplied recitational material for the royal corpus, then it should be possible to identify some of its attributes and anchor particular sections of the Pyramid Texts to early Old Kingdom materials. Accordingly, the sundry materials revealing dialectic and contemporaneous variation (diglossia), technical language (e.g. legal, mythical, nautical), and archaisms[142] would betray this bridge between oral discourse, varied

138 Cf. Teti's version of PT299 (Pyr. §444c) with the deictic reference to the first person modified to the noun of the king to fit the sacerdotal structure: *šnṯ nj šnṯ Ttj*—"Shunned snake, Teti cannot be shunned!"
139 Reintges 2011, 36.
140 The idea that the corpus of Pyramid Texts included recitations from different periods of the Old Kingdom has been recently studied by Gundacker 2009 (*non vidi*).
141 See Reintges 2011, 15–16 (n. 12); Vachala 2010, 777; and Helck 1972, 13. These authors comment on the performative and oral nature of the *Siegeslied* embedded in the biographical text of Wenis at Abydos, whose origin is to be found in the context of "einem Chor (mit dem Refrain) und einem Vorsänger" (Vachala), some "secular traditional art, which originated at the royal court" (Reintges), or a workers' song (Helck).
142 One of the questions that has intensified the debate about the archaistic forms in the Pyramid Texts is the so-called "split-stative hypothesis", mainly adopted by Kammerzell 1990 and followed by Schenkel 1994, according to which variant spellings of statives indicate two discrete conjugation patterns, a perfect for the first person (with independent syntactic usage) and the pseudoparticiple (syntactically dependent) for the other persons. Against this hint, both Reintges 2006 and Borghouts 2001 do not take the diverse endings *k(j)*, *kj*, *t(j)* and *tj* as morphologically different, but as pronunciations of the same verb with prosodic and emphatic distinction. In addition, the attestation of an obsolete form *sḏmm=f*, a prospective passive counterpart of *sḏm.w=f* only attested in Pyramid Texts and Coffin Texts (Allen 2013, 112), and the use of the negation w (Reintges 2011, 33–34) seem to indicate a form of archaic language not used in private texts during the Fifth Dynasty.

settings, and the formal plan carried out by Memphite priests for collecting and composing a corpus for Wenis.[143]

Additionally, a recent study on the group of Pyramid Text recitations against snakes (PT232–238, PT281–287) has provided a new interpretation for some of the "untranslatable"[144] texts in the group. According to Richard Steiner, some of these recitations might have been transferred into the collection prepared for Wenis from early Northwest Semitic (Canaanite) spells,[145] a hypothesis that Robert Ritner already proposed in 1995.[146] The evidence suggests that the recitations of Semitic origin were incomprehensible to most Egyptians at the time of their inscription into the pyramid of Wenis.[147] These recitations, probably disseminated into Egypt from Byblos, might have been copied by some ritualist who knew Canaanite and added them to the esoteric materials projected by the Heliopolitan sacerdotal class. In this case, both the oral style of the recitations and their foreign origin expose a *Sitz im Leben* absolutely separate from the mortuary context of the royal pyramid. However, this hypothesis has found detractors, mainly on the basis that some of the lexemes identified for Semitic would have not existed around 2500 BCE.[148]

PT281 (Pyr. §§422a–d)[W]
422a *j z z h k w k b b h ꜣꜣꜣ b j*
422b *rw n p h tj j rw n p ṯ tj j p h tj j p ṯ tj j*
422c *m mj n j(w)nw ꜣꜣꜣ ṯ w b s j(w)f w j(w)nw hnw*
422d *nʿy nʿy nʿy nʿy*
"His whispering, the uttering of his spell: Aaa is in me.
See my mouths, see my pudenda, my mouths, my pudenda,
Who am I? Aaa, fragant perfume of the nose, I am they.

---

143 One may consider the possibility that the priests of the Memphite region initially collected these materials for previous kings in the Fourth or Fifth Dynasty, for whom they could have been used in oral form. See, for instance, the date assigned to some of the Pyramid Texts by Allen, 2001, 97, who thinks that part of the corpus might have been written by the reign of Menkaure; Baines 2004, 28, who dates the composition of the Pyramid Texts to the reign of Sahure (early Fifth Dynasty), although he takes the potential existence of the corpus to the late Second or early Third Dynasty; and Kahl 1999, 97–99, in which the author dates the origin of the group PT220–222 in the late Second Dynasty (reign of Khasekhemuy).

144 Sethe 1935, 212, comments that some of these recitations, such as PT236 (Pyr. §240), included "zunächst unverständliche Zauberworte, die in ihrem *hj.tj bj.tj* schon äusserlich an unser Hokuspokus erinnern".

145 Steiner 2011, 23, 77, and passim.

146 Ritner 1995, 3351–3352, n. 85.

147 Steiner 2011, 24–25: the Semitic segment would be embedded in the Egyptian recitation in the form of three bilingual units, PT232–238, PT281–282, and PT286–287.

148 See the possibility offered by Mathieu 2002, 191, fig. 4, who considers that we might have some palindromes here: e.g. he suggests that "Kebebehititibitiches" might be read as "*sš-Tj-bjtj-jtj-bjk*" (lit. "that is, the scribe, Thoth, the king, the sovereign, the Ibis, and the falcon"). For the positions against Steiner's reading of North West Semitic in these spells, see n. 101 above.

Go! Go! Nay-snake! Nay-snake!"[149]

PT232 (Pyr. §§236a–c)[W]
236a *m mtj m mj mtj mj mtj*
236b *ꜣꜣꜣ mwt=f zp-2 mj mtj m mtj*
236c *jꜥ.tj ḫꜣst n(=j) m ḫm w(j)*
"Come, poison! Come, poison! Look, poison! Look, poison!
You whose mother is Aaa! You whose mother is Aaa!
Look, poison! Look, poison!
Be washed away from me, foreign land! Don't ignore me!"[150]

Another set of attributes attested in the Pyramid Texts that reveal the oral nature of the corpus is related to the presence of repetitive patterns, parallel phrasing and cadence, including verbatim repetition, alliteration, dislocation, chiasmus[151] and the use of formulae such as epithets. The use of alliteration in the Pyramid Texts[152] represents a noticeable marker of oral discoursive form, providing the ritualist or reciter with routine phonetic inflection. In the repetition of particular words or *stoicheia*, one must see a mechanism to elicit alliterations that could guide and embellish the discourse:[153]

PT274 (Pyr. §§407a–b)[T]
407a *Ttj pw sḫm wr sḫm m sḫmw*
407b *Ttj pw ꜥšm ꜥšm ꜥšmw*
"Teti is the most controlling power, who controls the controlling powers;
Teti is the sacred image[154] who is most sacred of sacred images."

PT385 (Pyr. §§674b–675b)[T]
674b *ḥfnw ḥfnnt*
675a *sḏm n=f sḏm n tꜣ sḏm n jt=k Gbb*
675b *j.tm=k sḏm n=f sḏm=k ꜣbt=f jmt tpj=k*
"Male snake, female snake,
listen to him, listen to the ground, listen to your father Geb!
Should you not listen to him, you will hear his brand on your head."

149 Following Steiner's translation and interpretation of the spell written in Semitic language: Steiner 2011, 39.
150 Steiner 2011, 26, ns. 4–6.
151 For other instances of chiasmus, see PT215 (Pyr. §143b); PT230 (Pyr. §§230c, 233a); PT289 (Pyr. §430a); sPT727 (Pyr. §2254a); PT355 (Pyr. §572a); PT485 (Pyr. §1037a); PT570 (Pyr. §§1462a–b); PT613 (Pyr. §§1738c–1739b); PT667B (Pyr. §§1944d–1945a); PT675 (Pyr. §2004b).
152 See Reintges 2011, 16–18 (ex. 9: analysis of PT263); Kammerzell 2000, 193; and Firchow 1953, 217–220.
153 See the analysis of the same phenomenon in Mesopotamian literature in Teffeteller 2007, 67–70.
154 One can also point out the use of polyptota here: see also i.a. Pyr. §§181a, 235a, 417a, 481c, 803a–b, 1797c, 1913a. My thanks to Bernard Mathieu for pointing out this particular aspect in Pyr. §407b.

PT479 (Pyr. §§981a–989b)[P]

981a *wnjj ꜥꜣwj pt jznn ꜥꜣwj qbḥw n Ḥrw nṯrw*
981b *pr=f m tp hrw wꜥb=f m sḫt jꜣrw*
982a *wnjj ꜥꜣwj pt jznjj ꜥꜣwj qbḥw n Ḥrw jꜣbtj*
982b *pr=f m tp hrw wꜥb.n=f m sḫt jꜣrw*
983a *wnjj ꜥꜣwj pt jznjj ꜥꜣwj qbḥw n Ḥrw šzmt*
983b *pr=f m tp Ḥrw wꜥb=f m sḫt jꜣrw*
984a *wnjj ꜥꜣwj pt jznjj ꜥꜣwj qbḥw n Wsjr*
984b *pr=f m tp Ḥrw wꜥb=f m sḫt jꜣrw*
985a *wnjj ꜥꜣwj pt jznjj ꜥꜣwj qbḥw n Ppj pn*
985b *pr=f m tp Ḥrw wꜥb=f m sḫt jꜣrw*
986a *jj*
986b *pr jr=f pr m tpj hrw wꜥb.n=f m sḫt jꜣrw*
986c *pr Ḥrw nṯrw m tpj hrw wꜥb.n=f m sḫt jꜣrw*
987a *pr jr=f pr m tpj hrw wꜥb.n=f m sḫt jꜣrw*
987b *pr Ḥrw šzmt m tpj hrw wꜥb.n=f m sḫt jꜣrw*
988a *pr jr=f pr m tpj hrw wꜥb.n=f m sḫt jꜣrw*
988b *pr Wsjr m tpj hrw wꜥb.n=f m sḫt jꜣrw*
989a *pr jr=f pr m tpj hrw wꜥb.n=f m sḫt jꜣrw*
989b *pr Ppj pn m tpj hrw wꜥb.n=f m sḫt jꜣrw*

"The sky's door has been opened, the Cool Waters' door has been pulled open
for Horus of the gods,
that he might go forth at daybreak, having become clean in the Marsh of Reeds.
The sky's door has been opened, the Cool Waters' door has been pulled open
for eastern Horus,
that he might go forth at daybreak, having become clean in the Marsh of Reeds.
The sky's door has been opened, the Cool Waters' door has been pulled open
for Horus Shezmet,
that he might go forth at daybreak, having become clean in the Marsh of Reeds.
The sky's door has been opened, the Cool Waters' door has been pulled open
for Osiris,
that he might go forth at daybreak, having become clean in the Marsh of Reeds.
The sky's door has been opened, the Cool Waters' door has been pulled open
for this Pepi,
that he might go forth at daybreak, having become clean in the Marsh of Reeds.
So, someone has come forth at daybreak, having become clean in the Marsh of Reeds:
Horus of the gods
has come forth at daybreak, having become clean in the Marsh of Reeds.
So, someone has come forth at daybreak, having become clean in the Marsh of Reeds:
Horus Shezmet
has come forth at daybreak, having become clean in the Marsh of Reeds.
So, someone has come forth at daybreak, having become clean in the Marsh of Reeds:
Osiris
has come forth at daybreak, having become clean in the Marsh of Reeds.
So, someone has come forth at daybreak, having become clean in the Marsh of Reeds:
Pepi
has come forth at daybreak, having become clean in the Marsh of Reeds."

PT670 (Pyr. §§1977a–c)[N]
1977a *ḥw.n=f n=k ḥw ṯw m jḥ*
1977b *smꜣ.n=f n=k smꜣ ṯw m smꜣ*
1977b *qꜣs.n=f n=k qꜣs ṯw*
"He has hit for you the one who hit you as a bull;
he has killed for you the one who killed you as a wild bull,
and has tied up for you the one who tied you up."

The associated use of parallel phrasing or *Doppelung*[155] consists in the repetition of names or clauses to draw the attention of the listener to a particular segment of the discourse, and is usually connected to the phenomenon of alliteration observed above:

PT280 (Pyr. §§421a–b)[W]
421a *j.jr.tj j.jr.tj sꜣ.tj sꜣ.tj*
421b *ḥr=k ḥꜣ=k zꜣ ṯw rjj wr*
"You of the (evil) deed, you of the (evil) deed! You of the wall, you of the wall!
Your face behind you! Beware, O great mouth!"

PT287 (Pyr. §§428a–b)[T]
428a *nn nj mwt=f nn nj mwt=f*
428b *j(w).k rr m nn j(w).k rr m nn mꜣ ṯjf ṯjf mꜣ ṯjf*
"You whose mother turned him away, you whose mother turned him away,
aren't you such, aren't you such? Lion, spit out! Spit out! Lion, spit out!"

PT400 (Pyr. §§695a–696g)[T]
695a *ndfdf jrt Ḥrw ḥr bꜣt nt ḏnw*
695b *bjkwj ḫntwj prw nb ḏfꜣw wr m ꜣIwnw*
695c *ḏj=k t n Ttj ḏj=k ḥnqt n Ttj sꜣḏ Ttj*
696a *sḏꜣ=k wdḥw n Ttj*
696b *sḏꜣ=k nmt nt Ttj*
696c *ḥqr Ttj ḥqr rwtj*
696d *jbb Ttj jbb Nḫbt*
696e *hdnwt hdnwt*
696f *m jn sṯ hdn=ṯ r Ttj*
696g *tm ḥr=ṯ hdn=ṯ r Ttj*
"Horus's eye has dripped on Horus's Dnw-bush.
Falcon (Horus), foremost of the houses, lord of sustenance, great one in Heliopolis,
may you give bread to Teti, may you give beer to Teti; may you fresh Teti,
may you fresh the offering table of Teti, may you freshen the slaughterhouse of Teti.
Should Teti hunger, the dual lion will hunger,
should Teti thirst, Nekhbet will thirst.
Broom-plant goddess, broom-plant goddess,
don't fetch the scent of your broom-plant against Teti,
for you don't have to fetch the scent of your broom-plant against Teti."

155 See the phenomenon discussed in Firchow 1953, 12–20.

PT564 (Pyr. §§1421a–1422c)[P]
1421a *wꜥb wꜥb m šj jꜣrw*
1421b *wꜥb Rꜥ m šj jꜣrw*
1421c *wꜥb Ppj pn ḏs=f m šj jꜣrw*
1421d *wꜥb Šw m šj jꜣrw*
1421e *wꜥb Ppj ḏs=f m šj jꜣrw*
1422a *Šw Šw fꜣ Ppj pn jr pt*
1422b *Nwt ḏj ꜥwj=ṯ jr=f*
1422c *hpꜣ=f hpꜣ=f hnn hnn hnn hpꜣ=f hpꜣ=f*[156]
"Someone has become clean in the Lake of Reeds,
the Sun has become clean in the Lake of Reeds,
this Pepi himself has become clean in the Lake of Reeds.
Shu has become clean in the Lake of Reeds,
Pepi himself has become clean in the Lake of Reeds.
Shu, Shu, lift this Pepi to the sky!
Nut, give your arms toward him!
He will fly up, he will fly up.
Howl, howl, howl! He will fly up, he will fly up!"

The reorganization of the grammatical structure with the purpose of emphasizing a particular element of the discourse by topicalization or postposition is another phenomenon attested in the Pyramid Texts. The conscious modification of the discourse produces the dislocation of the structure (with anteposition or afterthoughts),[157] and allows the reciter to lay emphasis on the most important element of the recitation:

PT473 (Pyr. §§930d–e)[M]
930d *jm t kw jn=sn jr Mr-n-Rꜥ*
930e *jn ꜣḫjw m r=sn ꜥpr*
"Who are you? they say about Merenre,
the Akhs with their mouth equipped."

A similar mechanism for elevating the intensity of a particular element in the spell is the chiasmus, also known as "inverted parallelism", a figure of speech in which two or more clauses are related to each other through a reversal of structures in order to emphasize an idea.

PT374 (Pyr. §658c)[T]
658c *ꜥꜣ rd=k rd=k wr.j šꜣs=f mnmwt wrt*
"Your foot is big, large is your foot, and it shall traverse the Great Bed."

156 The principle of alliteration, the repetition of the same or similar sound for the construction of a phonetic routine for the ritualist, can be observed here as well.
157 For further information on both mechanisms (left-dislocation and right-dislocation) see Reintges 2011, 37–38.

PT662 (Pyr. §§1874a–c)[N]
1874a *wbnnj wbnnj ḫprr ḫprr*
1874b *j=k r Ppj jw Ppj jr=k*
1874c *jw ʿnḫ=k jr Ppj jw ʿnḫ Ppj jr=k*
"Riser, riser! Beetle, beetle!
You are related to Pepi, Pepi is related to you;
your life is related to Pepi, Pepi's life is related to yours."

PT662 (Pyr. §§1880a–b)[N]
1880a *bꜣ.n(=j) n=k bdt skꜣ.n(=j) n=k jt*
1880b *jt n wꜣg=k bdt n rnpwt=k*
"I have hoed emmer for you and have plowed barley for you:
(that is) barley for your supply, emmer for your yearly supply."

The use of formulaic expressions constitutes another marker of orality in the Pyramid Texts.[158] No doubt, the repetition of particular words such as epithets can benefit the reciter by providing him with time to remember his next lines, or pause.[159] This aspect of the spoken language offers a possible explanation for the use of a rich repertoire of epithets for the gods and the king in the Pyramid Texts.

PT487 (Pyr. §1047b)[P]
1047b[...] *Ḥrw pw mrjj Ppj pn Ḥrw zꜣ nḏ jt=f*
"[...] it is Horus the one whom Pepi loves, Horus the son who tends his father."

PT569 (Pyr. §§1442a–b)[P]
1442a *jw.n Ppj ḫr=k Rʿ*
1442b *nj-ḏr=f*
"This Pepi has come to you, Re,
limitless sun [...]"

PT576 (Pyr. §§1505a–b)[P]
1505a *Ppj pw mtwt=k Wsjr spdt*
1505b *m rn=ṯ pw n Ḥrw jmj wꜣḏ-wr Ḥrw ḫnt ꜣḫjw*
"Pepi is your seed, Osiris, which is sharp
in this identity of yours of Horus in the Great Green, Horus at the fore of the Akhs."

Another hint in the identification of oral-compositional forms in the Pyramid Texts is the attestation of wordplays, many of which are achieved by devising meaningful puns with alliteration or parallel phrasing. In the first example below, parallel

158 For a similar phenomenon in Homeric poetic literature, see further comments in Ong 1982, 58–60.
159 In this note, see Rendsburg 2000, 16, n. 15, in which the author discusses the parallel issue of using red ink in manuscripts in order to mark a section for a pause, in the same tone as scribes did with the system of *setuma* and *petuḥa* in Biblical manuscripts. In this case, the use of epithets in oral-compositional styled texts might correspond to the same demand by the reciter, who might need a short juncture before continuing with the recitation

phrasing can be observed with *m kw*, while paronomasia is achieved by playing with couplets of words constructed with *km* ("to be black"), *w3ḏ* ("to be green") and *šn* ("round"), on the one hand, and *wr*, on the other, so that the recitation focuses on the resulting identities of the deceased, here presented as *Km-wr* ("great black"), *W3ḏ-wr* ("great green"), and *Šn-wr* ("great round").

PT366 (Pyr. §§628a–629c)[T]
628a *j n=k sntj=k 'Ist Nbt-ḥwt sḏ3=sn kw*
628b *km=t wrt m rn=k n Km-wr*
628c *w3ḏ=t wrt m rn=k pw n W3ḏ-wr*
629a *m kw wr.t šnj.t m Šn-wr*
629b *m kw dbn.tj šn.t m dbn pšr ḥ3 nbwt*
629c *m kw šn.tj ꜥ3.tj m šn ꜥ3 sk*
"Your sisters Isis and Nephthys have come to you, making you sound;
you are very black in your identity of the Great Black Wall,
you are very green in your identity of the Great Green.
Look, you have become great and round, as the Great Round.
Look, you have become encircled and round, as the circuit surrounding the sk-islands.
Look, you have become round and big, as he who surrounds the Big Waters."

PT366 (Pyr. §632d)[T]
632d *Ḥr spd pr jm=k m Ḥrw jmj Spdt*
"[...] sharp Horus has emerged from you as Horus who is in Sothis."

PT600 (Pyr. §1652c)[N]
1652c *jšš.n=k m Šw tf.n=k m Tnft*
"(What) you have sneezed is Shu and (what) you have spat is Tefnut."

Upon observing these instances that reveal characteristics of the oral discourse in the Pyramid Texts, one may wonder to what rites do these verbal accompaniments written on the walls of each pyramid refer. In other words, what rituals do they represent? As observed in the previous section, the original setting of two particular groups attested in the Pyramid Texts was certainly not the pyramid. On the one hand, the association between the offering Pyramid Texts and the offering lists, mainly from the Fourth and Fifth Dynasty, contributes to the idea that such recitations were already part of the private mortuary domain before the Memphite priests of Wenis prepared the corpus for the king's pyramid. Furthermore, the link observed between apotropaic recitations in the Pyramid Texts and magical spells employed in private contexts in the Sixth Dynasty also alludes to an original setting outside the king's monument, in which these practices were commonly used, perhaps even as far away as Byblos, from which some of these recitations could have come. Now, the question is whether there were any other settings, contexts, or circles in which similar materials could have been used before their transfer into the royal assemblages.

Offering lists and snake-repelling recitations occurred in private contexts before they were attested in royal monuments, which indicates the prominent role of the former on the configuration of the royal mortuary tradition. The examination of texts, scenes and material culture in the private tombs in the Old Kingdom reveals a similar metaphysical function.[160] In fact, most of scholars understand that the words, actions and space represented in the private tombs were intended to elevate the mortal to the superhuman status of a god, a "true of voice" and, ultimately, an Akh,[161] in the same manner as the Pyramid Texts intended for the king. Therefore, it is plausible to consider that certain interactions between the repertoire of images and "voices" in the private tombs, the later outcoming Pyramid Texts for the royal individual, and the ideological principles that ruled and distinguish both views of the afterlife must have existed,[162] all part of the same religious culture of the period.

Beyond the statement, "I know every ritual by which one becomes an Akh in the necropolis", commonly attested in the biographical texts of the Fifth and Sixth Dynasty,[163] it is clear that the non-royal deceased was the beneficiary of a series of performances, ceremonies, and recitations whose effects were perpetuated in his tomb and sarcophagus. Consequently, the deceased's transition to the afterlife could be achieved by knowledge and practice[164] without the use of extensive written recitations, a medium restricted to the king.[165] Certainly, the tombs of private individuals lacked a profuse corpus of religious texts like the Pyramid Texts, but short rubrics and accompanying scenes provided symbolic and referential information associated with the concepts of death and afterlife.[166] Regarding the performance of sakhu-rites,[167]

---

160 Hays 2010a, 118–120; Wilson 1944, 210.

161 Hays 2010a, 123; Hays 2010b, 1; Altenmüller 1972, 52 (mainly for the later transmission and usage of the Pyramid Texts in the Middle Kingdom private domain); and Wilson 1944, 209–210.

162 Allen 2006b, 9.

163 See Hays 2010a, 124; Kloth 2002, 118–119; and Edel 1944, 22–30.

164 Hays 2010a, 123–126; and Assmann 2005, 352–355, 393–404.

165 Baines 1990, 11–12, n. 63. The implications of decorum are also discussed in Bauman/Briggs 1990, 77. Both anthropologists believe that the institutional structures possess mechanisms that confer legitimate authority to control texts, although the idea that "the appropriation of particular forms of discourse may be the basis of institutional power" must also be taken into consideration as it might reflect the situation with the elite of the First Intermediate Period and its access to the corpus of Pyramid Texts.

166 For the symbolic/metaphoric aspect of the scenes and texts in the private mastabas of the Old Kingdom, see Roth 2006, esp. 244–245; Bochi 2003, 161–167; and Frandsen 1997, esp. 82–93. Cf. contra: Walsem 2006, 298–299; and Walsem, 2005, 71–83, in which the author discusses the dichotomy *Sehbild – Sinnbild*, and suggests that the allegorical or metaphorical meaning of a representation might depend on the context and the intention of the observer.

167 These rites were probably performed upon the monumental superstructure of the tomb: *ꜥḥꜥ m-dp jz*. See Alexanian 2003, 29, n. 4; and Bolshakov 1997, 101, n. 42.

concise texts such as an inscription in the tomb of Tjetu[168] or a caption in the tomb of Qar[169] confirm that ritualists carried out the necessary rituals for the deceased.

Complementary to John Baines's interpretation on the origin of the oral and performative discourse of the Pyramid Texts found in the first-person speeches of gods from the late Second and early Third Dynasties,[170] here I bring forward the implication of the private mortuary context in the construction of a royal narrative for the pyramids. Although Reintges provides some import to the gods' speeches,[171] he also alludes to the interlocutive performance structure of dialogues in the private tombs of the Fourth and Fifth Dynasty as a major stimulus. He understands the role of the commoner's talk, the question-answer situations between artisans, and the reference to songs recited by retainers as models for the configuration of the oral-compositional style in the Pyramid Texts.[172] Because of the importance of the original setting for the oral discourse in the development of written forms and the Pyramid Texts, I believe that early gods' speeches (probably connected to the temple cultic environs), biographical and legal stipulations,[173] and religious allegoric references in scenes from daily life in the private tombs all contributed to the configuration of the royal corpus of religious texts. As Firchow already articulated in 1953,

> [d]ie Inschriften in den gennanten Pyramiden sind Sammlungen von Sprüchen verschiedenen Alters und Inhalts. Viele sind eigens zur Verwendung als Totentexte und erst in der Zeit ihrer Niederschrift in den Königsgräbern verfaßt oder aus älteren Teilen zusammengestellt worden.

---

168 See Simpson 1980, 12, pls. 28–29, and fig. 24: *sꜣḫt jn ẖrj-ḥꜣbt* "making glorifications by the lector priest"; *dj qbḥw* "making a libation"; and *jnt rd jn ẖrj-ḥꜣbt* "bringing the broom by the lector priest".
169 Simpson 1976a, 5, pls. 7a, 8, and figs. 22–24: *wdn j.ḫt ẖrj-ḥꜣbt* "dedicating offerings by the lector priest"; *sꜣḫt jn ẖrj-ḥbt* "making glorifications by the lector priest"; *rdjt mw* "dispensing water"; and *rdjt snṯr* "bringing the broom". Interestingly, the mastaba of Qar (G 7101) also provides us with a prominent example of the coalescence of meanings, with priests performing sakhu-rites for the deceased including, i.a., *snṯr* (PT25, Pyr. §18d), *qbḥw ṯꜣ* (PT32, Pyr. §23b), and *wnḫw* (PT81, Pyr. §57e), together with the general statement that *sꜣḫt rꜥ nb jn wt.w rꜥ nb* "daily glorification (is provided) by the embalmers every day" and *pr(t)-ḫrw m dbḥt-ḥtp* "invocation offering consisting of food requirements". For G 7101, see Simpson 1976a, 7, pl. 9, fig. 25.
170 Baines 2004, 28.
171 Reintges 2011, 10.
172 Reintges 2011, 11–16. See also the combination of different models of poetical and legal-administrative discourses that determined the composition of the Pyramid Texts in Reintges 2011, 19, fig. 1. For the development of the legal-administrative discourse as a major reason for the fixation in writing of legal aspects for the tomb owner, see Schenkel 1983, 60.
173 See Reintges 2011, 29–30, for an analysis of the intervention of the two types of biographical texts (*Idealbiographie* and *Ereignisbiographie*) in the configuration of particular aspects of the oral compositional form in the Pyramid Texts; cf. Kloth 2002, 230–235, who focuses on the association of the first-person biographical texts with the oath statements to highlight the role of legal context in the fixation of oral discourse into writing. In my opinion, a significant impact on the development of the oral discourse lies in the use of the first-person pronoun recitations of justification and good behavior of the deceased (see, for instance, the mastaba of Idu: Simpson 1976a, 20, pl. 17, fig. 33).

> Andere wiederum lassen sich als weit früher erschaffene Ritualsprüche des Königs- und Götterkultes wie als Hymnen an den Sonnengott erkennen, die mit größeren oder kleineren Veränderungen oder auch unverändert in die Sammlungen von königlichen Totensprüchen übernommen wurden. Dazu gesellen sich Zaubersprüche zur Abwehr giftiger Schlangen und anderer schädlicher Tiere, Schlachtopfersprüche, Begleitworte für das Darbringen von Opfergaben und manches sonst.[174]

Therefore, some of the rites incorporated into the inscriptional and iconographic repertoire of the private tombs before Wenis might offer us some clues to understand the *Sitze im Leben* in which the Pyramid Texts might have originated. This affiliation between the religious corpus created for the kings of the Fifth and Sixth Dynasties and the private mortuary traditions of the early and mid-Old Kingdom is evidently not forthright and categorical.[175] Some of the tabular materials seen in the form of elite offering lists and the oral and most popular recitations such as magical spells and guild songs demonstrate the bond between private mortuary cult and personal practices, on the one hand, and royal engagement on the other. Similarly, by allegoric principles, other themes from the private tombs might disguise religious and mortuary conceptions[176] that experienced a high-cultural transmission from elite monuments to the royal realm,[177] standing for aspects of resurrection and afterlife later embedded in narrative form within the Pyramid Texts.[178]

Following this hypothesis, there is a series of iconographic and textual themes in the private mastabas of the Old Kingdom that call for further examination. Among

---

174 Firchow 1953, 9.

175 These features might lead scholars to some forms of hyper interpretation: Bochi argues that as a *topos*, the references to clothing and linen in Old Kingdom tomb scenes, statues, and tomb equipment could imply generosity, moral obligation, prominent status in this world, regeneration, and a blessed afterlife (see Bochi 1996, 245–246, n. 63); consequently, she discusses the conflation of the motif in diverse contexts. I believe that our understanding of the Egyptian cultural attitude toward clothing in general should restrain us from singling out particular significances of this motif as evidence for the interaction of the private and royal realms.

176 Observe that, as Eyre points out, some of the texts and themes of the private tomb repertoire might also have a different *Sitze im Leben*: Eyre 1993, 116.

177 Two interesting examples of this approach—with some misconceptions—are Burn 2011 and Vischak 2003, both following Allen 1994. Burn highlights that tomb decoration and Pyramid Texts "appear to have been inspired by the same ideological force" (Burn 2011, 245), although his research points to modifications in the distribution of tomb scenes before the emergence of the Pyramid Texts of Wenis. In her contribution, Vischak aims at identifying similar patterns in the placement of particular themes from mastaba decoration and Pyramid Texts position in the pyramids. Although the hypothesis presented here, that there is a correlation between themes in both realms, seems to me reasonable and well grounded, Vischak carries this association further, to defend a symbolic interpretation of the cosmos present in the architectural space of both monuments, an idea that has recently been proved to be misleading. *Cf.* the critical position against this idea in Hays 2009b, 201, n. 37.

178 See Bochi 2003, 161, for the inverted phenomenon of royal influence (e.g. scenes from the "world chamber" at the sun temple of Niuserre) over private tomb decoration (e.g. Mereruka and Khentika's scenes of painting the year's seasons at an easel).

daily life, ceremonial, and cultic scenes, some particular texts and representations stand out for their exceedengly referential portrayal of religious, ritual and netherworldly matters: offering rituals, carrying chair episodes, pulling papyrus scenes, painting the year seasons, agricultural activities, boat scenes, net fishing, harper songs, and other themes constitute instances of orientational metaphors, whose cultural and religious properties embodied common ideas in both the royal and private realms.[179]

The theme of the offerings provided to the deceased has been extensively treated in the previous section. The emergence of a corresponding structure in the order and nature of the provisions of the offering lists and the offering Pyramid Texts unmasks such a cultural bond. However, other elements of the private mortuary context offer us further input in the interaction of both realms. In regard to the architecture of the mastabas, Nicole Alexanian comments on the coincidence in the use of the Pyramid Texts by Wenis and the construction of private tombs with stairs.[180] Following the author, this structure would be the locale of a particular rite, "the Ba in the sky, the corpse in the netherworld", also mentioned in the Pyramid Texts in similar terms (PT395),[181] and later in the Coffin Texts through the entire ritual sequence CT94–96 CT488–500.[182] In this ritual, the Ba was supposed to ascend to the sky, carried out by the sun's rays. In addition, this author also refers to the ascension by joining the sun-god (*ḫnm-Ꞽtn*) in another rite, which reminds us of the ritual performances achieved on top of the mastaba of Debeheni (see fig. 7) and the ascensional discourse in PT222:[183]

PT222 (Pyr. §§207c–e + 213b)[W]
207c *ḫpr=k ḥnꜤ jt=k Tm qꜣ=k ḥnꜤ jt=k Tm*
207d *wbn=k ḥnꜤ jt=k Tm j.fḫ n=k mꜣrw*
207e *tpj=k n rpwt jwnt* [...]
213a *Tm sjꜤ n=k Wnjs pn šn n=k ẖnw Ꜥwj=k*
213b *zꜣ=k pw n ḏt=k n ḏt*
"You shall evolve with your father Atum, you shall go high with your father Atum,
you shall rise with your father Atum, and release needs for you.
Head to (Nut), the Heliopolitan in the sedan chair [...][184]

179 Vachala 2010, 777; Bochi 2003, 164–165; and Frandsen 1997, 82.
180 See Alexanian 2003, 35. For details on the representation of *ḥmw-kꜣ* priests performing offering rites, see Seyfried 2003, 42–44, figs. 1–2.
181 PT305, Pyr. §474a: *ꜣḫ jr pt šꜤt jr tꜣ*. See also further remarks on the performance of rites at the roof of the tomb in Theis 2011, 160–165, as well as additional comments on the interaction of both private and royal realms.
182 See Assmann 2005, 90–94. The title of CT94, initial text in this liturgy, reads *sḥr bꜣ r ẖꜣt* "Causing the Ba to depart from the corpse" [coffin B1C: *ECT* II, 67a].
183 Alexanian 2003, 37. For the particular interpretation of this rite as a segment of a larger Middle Kingdom liturgy of offerings to the deceased in the tomb, see Assmann 2002, esp. 485–489.
184 Following Allen 2005, 40 (W155). Cf. Assmann 2002, 487, n. 70.

Atum, elevate him to you, encircle him inside your arms:
he is your son of your body, forever."

Another instance of interaction between both realms can be observed in the song of the sedan carrying chair porters (*mḥnk*).[185] The carrying chair motif is an interesting one even if the term "sedan chair" (*ẖwdt*)[186] is not present in the repertoire of the Pyramid Texts. Here, as a marker of social and religious status, the divine sedan chairs are called *wr-ḥ*ꜥ and *rp(w)t*.[187] However, as Roth has pointed out,[188] there is a link between the use of the sedan carrying chair as a marker of social status with its metaphoric reference to the journey of the tomb owner from one world to another,[189] on the one hand, and the funerary overtone of the references to the king being transported in a carrying chair during the royal jubilee and related statements, on the other.

Similarly, the motif of the deceased at an easel painting the seasons of the year in his tomb has also been understood as a reflection of royal or even divine privileges.[190] The gesture of painting "seasons" within cartouches—a motif attested in the tombs of the Sixth Dynasty officials Mereruka and Khentika[191]—alludes to the demiurgic act of creating time and cosmic *Maat*, performed by the king and the gods. In a sense, the seemingly modest appearance of the creative authority of the deceased in this type of scene also illustrates the restriction of decorum with the artistic representations of the period. This instance of dialogue between royal ideology and elite imitation could also manifest itself in the Pyramid Texts, bridging the gap between both realms and revealing common aspects of their creativity and innovation. Patricia Bochi has suggested that by creating seasons, the deceased was securing his own provision of time as the king himself does in the Pyramid Texts by cultivating and storing barley and emmer in the netherworld.[192]

PT422 (Pyr. §§760a–761)[P]
760a *ꜥḥꜥ zꜣ=k ḥr nst=k ꜥpr m jrw=k*
760b *jr=f wnt=k jr=k m bꜣḥ ḫntj ꜥnḫw*

185 See Roth 2006, 244–245 (n. 3), 248 (n. 16); and Altenmüller 1984–1985, esp. 28–29.
186 See Altenmüller 1984–1985, 21 (col. E: *Ḏꜥw—Šmꜣj* from Deir el-Gebrawi)
187 Observe the use of *wr-ḥ*ꜥ in PT81 (Pyr. §56c); PT438 (Pyr. §811a); and PT467 (Pyr. §892c), while *rp(w)t* is attested in PT222 (Pyr. §207e); PT356 (Pyr. §580a); PT423 (Pyr. §767b); PT443 (Pyr. §823d); PT549 (Pyr. §1349b); and PT691C (Pyr. §2130a).
188 Roth 2006, 253.
189 Or, as Strudwick 2005, 418 puts it, "the scene with the carrying chair is without doubt an allusion to the desire to be brought back to earth after death."
190 Bochi 2003, esp. 163–164.
191 Although there are only two attestations of this motif, it is important to consider the possibility of individual selection for the tomb decoration from particular artistic patterns in the royal artistic repertoire or the search for innovative motifs, which could largely reflect private aspirations. See Bochi 2003, 168, n. 35.
192 Bochi 2003, 164 cites as examples PT422 and PT461.

760c *m wḏt Rꜥ nṯr ꜥꜣ*
761 *skꜣ=f jt skꜣ=f bdt ḥnk=f ṯw jm*
"Your son shall take up his position on your throne, equipped with your form,
and do what you used to do before at the fore of the living,
by command of Re, the great god.
He shall farm barley, farm emmer, and endow you with them."

This idea of securing regeneration, resurrection and time by storing barley and emmer is attested in agricultural scenes and texts from the mastabas of the Old Kingdom, a parallel that promotes a cogent bond—as a religious metaphor—between the royal and private realms:[193]

Tomb of Sekhemankhptah, Saqqara
1 *mꜣꜣ kꜣt sḫt skꜣ ꜣsḫ ḥwj mḥꜥw šdt jꜥꜣw ḥwj jꜥꜣw spwt ḫꜣḫꜣ* [...]
2 *skꜣ m hb js hꜣ js*
3 *j.ḏd(=j) n=ṯn rḥw jw jt hrww jn ꜣsḫ r nfr jr=f sw*
4 *jšst pw r=f ṯꜥy srf jb*
5 *ḥꜣt jt jn ḏꜣḏꜣt*
6 *ḫꜣḫꜣ jn djwt ꜣḫ jt jn djwt*
7 *jr m(j) hꜣ=k jm=sn*
8 *mꜣꜣ=k jrt=k*
9 *wbs bdt*
10 *wꜣḥ r nfr ḥn=k m ꜥnḫ*
"Viewing the work of the fields—cultivation, reaping, pulling flax, loading donkeys, donkeys treading the threshing floors, and winnowing [...]
Cultivation with the plow. Go forward! O go forward!
I say to you: men, barley is there—he who reaps the best will get it.
What is this then, a careful man?
Measuring barley by the assessors.
Winnowing by the team of five; gleaning barley with a brush by the team of five.
You must drive them around.
Can you see what you have done?
Stacking emmer wheat.
Pile it up well, and you shall prosper in life!"[194]

Furthermore, references to seasons of the year are observed in the Pyramid Texts in the context of resurrection recitations for the king. In the first instance, it is Osiris—a counterpart of the tomb owner painting the seasons—who masters time in both realms, sky and earth. In the second, the king is fervidly solicited to continue living season after season, precisely the same *raison d'être* of the representation of seasons in the tomb of Mereruka and Khentika:

193 This category of text has also been associated with the Middle Kingdom version CT368.
194 Now MFA 04.1760: see Simpson 1976b, 10–16, pl. D.

PT219 (Pyr. §§186a–187a)[W]
186a *m rn=k jmj sꜣḥ tr=k r pt tr=k r tꜣ*
186b *Wsjr pšr ḥr=k mꜣ=k n Wnjs pn*
186c *mtwt=k prt jm=k spdt*
187a *ꜥnḫ=f ꜥnḫ Wnjs pn nj mt=f nj mt Wnjs pn*
"In your identity of one of Orion, your season is at the sky, your season is at the earth.
Osiris, turn your face and look at this Wenis,
your seed that came forth from you, efective.
He will live and this Wenis will live; he will not die and this Wenis will not die."

PT535 (Pyr. §§1289c–1290b)[P]
1289c *ꜥnḫ=f m ꜥnḫw ꜥnḫ Zkr m ꜥnḫw*
1289d *ꜥnḫ=f m ꜥnḫw ꜥnḫ Ppj m ꜥnḫw*
1290a *hꜣ Ppj pw mj ꜥnḫ ꜥnḫ=k nn m tr=k m tr=k*
1290b *m rnpwt jptn ḥtp.tj srf mrwt=k*
"He shall live with the living as Sokar lives with the living;
as he lives with the living, this Pepi shall live with the living.
O Pepi, Come! Keep alive and live here from one of your seasons to the other
in these years, being content, with the love of your warmness."

Moreover, the repertoire of texts and iconography in the mastabas provides further examples of the interactions between the royal and private circles, such as the representation of "tearing papyrus" (*zšš wꜣḏ*).[195] Altenmüller explains that the papyrus obtained through the activities of the deceased was thought to be used in the construction of a ladder to reach the (celestial) netherworld and join the goddess Hathor. This ascension occurs precisely in the region of the northern Deltic marshes (*sḫt-jꜣrw*),[196] therefore, "das *zšš wꜣḏ* den Himmelsaufstieg des Grabherrn am Ende des Tages und am Beginn der Nacht symbolisiert".[197] The deceased appears wearing a tiara that associates him with the sakhu-rites performed in the liminal region of papyrus thickets and swamps,[198] in connection with the ideas of rebirth, regeneration and transfiguration.[199] Concerning the idea of the deceased as a participant in the

195 See mainly Lapp 2013, 51–64; Woods 2009, 314–319; Altenmüller 2002a; Wettengel 1992, esp. 323–326, with commentary on previous positions regarding this type of scene; Troy 1986, 58; Harpur 1980, 53–60; and Montet 1957, 102–108. I thank Bernard Mathieu for pointing me to the existence of a block with the representation of this ritual in the mortuary complex of Ankhesenpepi II: see Callender 2011, 261, fig. 100; and Leclant/Minault-Gout 2000, fig. 8, pl. 17.
196 See Altenmüller 2002a, 26–28, and n. 74. See also the comments on the meaning of the presence of papyrus in Old Kingdom tomb equipments in Wilde 2013, 179–180. For the attestation of the "field of rushes" or "marsh of rushes" in the Pyramid Texts, see Hays 2004, 176, ns. 5–6.
197 Altenmüller 2002a, 29. The idea of connecting the "tearing papyrus" with the goddess Hathor and certain aspects of the afterlife treated in the Pyramid Texts appeared first in Junker 1940, 77–81.
198 See Altenmüller 2002a, 28, n. 74 (with reference to Altenmüller 1989, 9–21), in which the author emphasizes the role of this territory as a liminal space between both worlds and the location where the sun-god Re ascended to the sky.
199 Alexanian 2003, 35, n. 61.

ritual of “tearing papyrus”, the corpus of Pyramid Texts also emphasizes the ability of the king to traverse the marshes (*pḥww*)[200] and benefit from the purification and regeneration achieved in this territory:

Tomb of Meresankh III, Giza
1 *zšš wꜣḏ*
2 *n Ḥwt-Ḥr m pḥw*
3 *ḥnꜥ mwt=s*
4 *mꜣꜣ=sn ḫt nbt nfrt*
5 *ntt m mḥt*
“Pulling papyrus
for Hathor in the marshland
together with her mother (i.e. Hetepheres);
they view every perfect thing,
which is in the marsh.”[201]

PT267 (Pyr. §§367a–b)[W]
367a *ꜥḥꜥ j.dr tw j.ḫm-jwt*
367b *ḥms Wnjs pn m st=k ḫnjj=f m pt m wjꜥ=k Rꜥ*
“Stand up and remove yourself, you who do not know the reeds,
so that Wenis may sit in your seat. He will row to the sky in your boat, Re.”

PT512 (Pyr. §§1164a–1165b)[P]
1164a *jhj ṯz ṯw Ppj*
1164b *šzp n=k fdt=k jptw nmswt ꜥꜣbwt*
1164c *wꜥb=k m šj zꜣbj snṯr=k m šj dꜣtj*
1164d *sꜥbw=k ḥr tpj šꜣbt=k m sḫt jꜣrw*
1165a *ḫnz=k pt*
1165b *jr=k mnw=k m sḫt ḥtp mm nṯrw zjw n kꜣw=sn*
“Ho, raise yourself, Pepi!
Receive these four washing jars of yours,
become clean in the jackal lake, and wash in natron in the dual lake.
You will be cleansed on top of your water-lily in the marsh of Reeds,
you will course the sky,
and make your abode in the marsh of Offerings among the gods who have gone to their kas.”[202]

Finally, I will briefly refer to another textual category that indicates a close relationship between private and royal practices and beliefs: the ferryman texts.[203] Some

200 In this case, an ability that the ferryman of the sun-god’s boat itself does not possess; for this interpretation, see Allen 2005, 48, n. 61.
201 See Dunham/Simpson 1974, 10, fig. 4. For the text, see Strudwick 2005, 420, with further examples from the tombs of Iazen at Giza and Fetekta at South Abusir.
202 See also the same general theme in PT479 (Pyr. §§981a–989b) and PT564 (Pyr. §§1421a–1422c) discussed above.
203 See Hays 2006–2007, 46, n. 22; Krauss 1997, 67–85; Willems 1996, 192, 196 (n. 1071), 415; Depuydt

inscriptions in the private tombs deal with the control of the boat in a journey by recognizing its parts as well as the benefits of traversing the hereafter on a boat. Scholars have pointed out that these succint inscriptions reflect magical and religious aspects that found certain parallels in the royal corpus of Pyramid Texts.[204]

PT256 (Pyr. §§303c–d)[W]
303c *ẖnj sw mwt=f jtḥ sw dmj=f*
303d *hjj nwḥ=k*
"Row him, his mother! Pull him in his boat, his harbor: haul your rope!"[205]

PT613 (Pyr. §§1738a–f)[P]
1738a *jꜣ Hḏhḏ mẖnt nj mr-n-ḫꜣ*
1738b *jn mẖnt tw n Ppj pn pds ṯbwt nṯr ḏꜣ zmṯ jr sḫt ḥtp*
1738c *j r=f r=f Hḏhḏ mẖnt n mr-n-ḫꜣj*
1738d *jnt=k mẖnt tw n Ppj pn*
1738e *dꜣjj=f jm=s jr sḫt ḥtp*
1738f *d=f sw ḥr gs ʾImntj n sḫt ḥtp ḥꜣ nṯrwj ꜥꜣwj*
"O Hedjhedj, ferryman of the Winding Canal,
fetch this ferryboat for this Pepi, that the crossing god's sandal might trample over the ladder to the Marsh of Offerings,
So, Hedjhedj, ferryman of the Winding Canal,
fetch this ferryboat for this Pepi,
that he may cross in it to the Marsh of Offerings,
put himself on the west side of the Marsh of Offerings behind the two great gods."

Tomb of Kaiemankh, Giza
1 *rs r ḥr mr ʾImnt mj-nw ꜣ tp nfr pw*
2 *rs r ḥr sbꜣtj-mw ꜣ tp nfr pw*
3 *mꜣꜥ r=k j(w)=k m nj-ẖnw*
4 *ṯꜣw mrj pw jw=f ḥꜣ wḥmw mr ʾImnt wrt jrj ḥr jmj-wrt wꜣt nfrt*
5 *rs.t(j) r ḥr jrj=j ḥnꜥ jmj ḥmww ẖt=k sw mwt=f jtḥ sw dmj=f*
6 *ẖt=k ḥr mw ṯꜣ(j) pw*
7 *jmj-wrt wꜣt nfrt sbꜣtj-mw mꜣꜥ r=k*
8 *rs.t(j) r ḥr*
"Keep an eye on the sail-rope, for this is the canal of the West—it is truly good!
Keep an eye on the sail-rope, pilot. Hold a good course, one-of-the-stream.
The wind of the canal is behind the messenger, for this is the canal of the West. Keep your course to the port, the perfect way!

1992, 33–38; and Bidoli 1976, esp. 26–29.

204 In fact, the most remarkable examples of this type of texts date to the Middle Kingdom but have some antecedents in the Pyramid Texts; see, for instance, Ibi/S/S 587–596. See Hays 2012, 281, n. 1007; and Bickel 2004, 94–96, fig. 2.

205 On the expression *hjj nwḥ=k*, see Jones 1988, 170–171 (3); and Faulkner 1971, 202. See other attestations in PT1033, PT1073, CT274 (*ECT* IV, 15e), and CT659 (*ECT* VI, 280l). Cf. the expressions *šspw nwḥ* "the holder of the cord", and *dwnw nwḥ* "the stretcher of the cord" in Quirke 1990, 174–176; and Smither 1941, 74–76, pl. 9A (pap. Harageh 3, ll. 21–22).

Keep an eye on the sail-rope, for I shall work with the man with the steering oar.
Keep low to the water, o boy!
The port is the perfect way. O master of the water, hold a good course.
Keep an eye on the sail rope!"[206]

In addition, Bolshakov pointed out that the scenes of "boat jousting" offer a new perspective on the symbolic significance of tomb decoration. In his opinion, the meaning behind the boatmen's contests alludes to provisions and offerings to be secured for the deceased and to complement the households for the mortuary cult.[207] Furthermore, Altenmüller observed that the scenes of boats in the non-royal tombs of the Old Kingdom referred to the need to ensure access to the day and night boats for the deceased, in a similar fashion to that which particular recitations of the Pyramid Texts stipulate for the king.[208] A similar tradition associated with the use of the ferryman texts is the use of net-texts, whose themes and function not only reflect points of contact between the private and royal realms, but also seem to reveal the continuities of the mortuary literature in the Old and Middle Kingdoms.[209]

Finally, it is also noticeable that there are similarities between certain aspects of the legal royal and private compositions and the idea of the deceased's justification (*mꜣ*ᶜ*-ḫrw*) in the Pyramid Texts.[210] The conception of death as a transference to the netherworld through the presentation of the deceased before a tribunal,[211] the significance of vindication for the integration of the deceased within the realm of the gods,[212] and the role of the divine tribunal of Osiris[213] are only a few examples derived from the corpus.

206 See Strudwick 2005, 417; Kanawati 2001, 42–43, pl. 37; and Junker 1940, 57–61, pls. 3–7.
207 Bolshakov 1993, 36–39.
208 See Altenmüller 2002b, esp. 275, 281, 284, in which the author discusses the relationship of the two types of non-royal boats, Henet and Shabet, with the solar bark used by the king, Mesketet.
209 See Bidoli 1976, 11, n. 3, in which the author associated early representations of *Fangnetz*-activities in the area of marshes, depicted in Old Kingdom tombs, with the religious themes as found in the later (Middle Kingdom) recitations CT473–480. In addition, Bidoli points out the presence of a list of the parts of a ferryboat in the Sixth Dynasty tomb of *Kꜣ.j-m-*ᶜ*nḫ* (*id.*, 28), which connects it with the Middle Kingdom net and ferryman's texts. For a recent study on the continuities of Old Kingdom ferryman texts and Middle Kingdom ferryman and net texts, see Hays 2006–2007, esp. 45–47 (with further bibliography on ferryman texts in ns. 15 and 29).
210 See n. 172 above. See also Mathieu 1997a, 289–304; and Mathieu 1997b, 11–28, for the hypothesis that the eastern section of the pyramid antechamber and the serdab façade [W/A/E] constituted the place for judgement.
211 See i.a. PT263 (Pyr. §§340a–b), PT265 (Pyr. §§365a–b), PT266 (Pyr. §361a), PT374 (Pyr. §658b), PT440 (Pyr. §816d), PT517 (Pyr. §1190a), PT609 (§1708c), and PT1046.
212 See, for instance, PT71J^{N} (Pyr. §§49+9), PT260 (Pyr. §316d), PT265 (Pyr. §§354a–b, 356c–d, 357c–d), PT266 (Pyr. §§361b–c), PT473 (Pyr. §§929a, 935a–b); and PT689 (Pyr. §2089a).
213 For the mythical allusion, see Mathieu 2011, esp. 150–151; and Mathieu 1998, 71–78.

5 Conclusion

In summary, this study argues for a re-interpretation of long-standing assumptions on the origin and development of the Pyramid Texts. It is widely acknowledged that the Old Kingdom Pyramid Texts represent the birth of the ancient Egyptian mortuary literature tradition. This intellectual shift, with apparent traces of Memphite-Heliopolitan formulation, culminated at the demise of king Wenis with the monumentalization of ritual voices and deeds on the walls of his pyramid. Underneath the magnitude of Wenis's novelty, however, ancient Egyptians perceived not only the sudden reshaping of the mortuary tradition for the royalty but also a progressive transformation of the process of tradition or *Überlieferungsgeschehen*, the act of tradition itself.[214]

The genesis of this gradual process of religious transformation therefore occurred long before the reign of Wenis, when the sacerdotal class envisaged the incipient use of operative papyri as *aides-mémoire* for the performance of rituals and the control of restricted knowledge. By entextualizing oral recitations, the priests of the early Old Kingdom transferred the ritual discourse to the scriptural media. As the relationship between the original setting of a recitation and its textual counterpart was not severed, the ritual discourse was not decontextualized. Also, the materiality of the new media (papyrus scroll) opened up the possibility to distinguish, arrange and store texts by categories. By the reign of Wenis, the decision to transfer scriptural materials to the walls of the king's pyramid entailed the decontextualization and monumentalization of the recitations and their subsequent detachment from their primary settings. In this study, I have attempted to suggest ways to bridge the gap between the form of the Pyramid Texts in the royal monuments and their original settings. For example, examination of the offering lists from the earliest instances in the first dynasties evidences a clear association with the offering Pyramid Texts. Such a nexus indicates that the setting of the mortuary cult and the provision of offerings was involved in supplying recitational material for the later royal corpus.[215]

In suggesting that this setting of mortuary cultic service provided these materials for the Pyramid Texts, one could argue that the commitment of the basic ritual instructions in stelae and tomb walls of the early Old Kingdom corresponded to a private process of monumentalization of the oral and scriptural discourse. However, this implementation—which would predate the monumentalization of Wenis's pyramid by several centuries—would be restricted by the tenets of decorum inherent to the domain of high-cultural production. Consequently, in the following dynasties the confines of the private offering lists and other categories of non-royal inscriptions would never attain the dimension of the Memphite creation for Wenis and its ritual corpus as *scriptio continua*.

214 Morales 2013, 50–52.
215 Smith 2009, 6–7.

Above all, the analysis of the oral-compositional style of the Pyramid Texts corroborates the antiquity of the corpus and its interaction with earlier forms of inscriptional evidence. Alongside this, some scholars have opted to stress the association of mid-Old Kingdom temple inscriptions with the Pyramid Texts.[216] Based on the correlation of other textual traditions that do not belong to the context of the temple, I suggest that the royal corpus incorporated not only mortuary service and temple materials, but also other types of recitations associated with magical practices, guilds' ceremonies, local festivities, and even arcana. Additionally, there is also ample evidence in mastaba decoration and its scanty inscriptional testimonies to demonstrate the correspondence of afterlife beliefs between both domains.

In sum, within their own world view, the high officials, priests and courtiers of the late Fifth Dynasty witnessed the culmination of a theological plan for perpetuating mortuary rituals that ensured the beneficial afterlife of the king. In all probability, however, these individuals were not intrigued by the nature of the texts used in the royal corpus, as most of them were familiar with the old recitations "by which anyone becomes an Akh".

Abbreviations

ECT	de Buck, Adrian/Gardiner, Alan H. (1936–1961), *The Egyptian Coffin Texts 1–4*, Chicago.
LD	Lepsius, Carl Richard (1897–1913), *Denkmäler aus Ägypten und Äthiopien*, Leipzig.
MafS	*Mission archéologique française de Saqqarah*

Bibliography

Alexanian, Nicole (2003), "Himmelstreppen und Himmelsaufstieg. Zur Interpretation von Ritualen auf Grabdächern im Alten Reich", in: Heike Guksch, Eva Hofmann and Martin Bommas (eds.), *Grab und Totenkult im Alten Ägypten*, Munich, 27–40.

Allen, James P. (1994), "Reading a Pyramid", in: Catherine Berger, Gisèle Clerc and Nicolas Grimal (eds.), *Hommages à Jean Leclant*, vol. 1: Études pharaoniques (Bibliothèque d'Étude 106.1), Cairo, 5–28.

Allen, James P. (2001), "Pyramid Texts", in: Donald B. Redford (ed.), *The Oxford Encyclopedia of Ancient Egypt*, vol. 3, Oxford, 95–97.

Allen, James P. (2005), *The Ancient Egyptian Pyramid Texts* (Writings from the Ancient World 23), Atlanta.

Allen, James P. (2006a), *The Egyptian Coffin Texts VIII: Middle Kingdom Copies of Pyramid Texts* (Oriental Institute Publications 132), Chicago.

216 Hays 2012, 22, n. 119 (with further positions in this respect).

Allen, James P. (2006b), "Some Aspects of the Non-Royal Afterlife in the Old Kingdom", in: Miroslav Bárta (ed.), *The Old Kingdom Art and Archaeology*. Proceedings of the Conference Held in Prague, May 31–June 4, 2004, Prague, 9–17.

Allen, James P. (2013), *The Ancient Egyptian Language. An Historical Study*, Cambridge.

Altenmüller, Hartwig (1968), "Zum Beschriftungssystem bei religiösen Texten", in: Wolfgang Voigt (ed.), *XVIII Deutscher Orientalistentag vom 1. bis 5. Oktober 1972 in Lübeck. Vorträge* (Zeitschrift der Deutschen Morgenländischen Gesellschaft Supplement 2), Wiesbaden, 58–67.

Altenmüller, Hartwig (1972), *Die Texte zum Begräbnisritual in den Pyramiden des Alten Reiches* (Ägyptologische Abhandlungen 24), Wiesbaden.

Altenmüller, Hartwig (1974), "Zur Ritualstruktur der Pyramidentexte", in: Wolfgang Voigt (ed.), *XVIII Deutscher Orientalistentag vom 1. bis 5. Oktober 1972 in Lübeck. Vorträge* (Zeitschrift der Deutschen Morgenländischen Gesellschaft Supplement 2), Wiesbaden, 8–17.

Altenmüller, Hartwig (1984), "Pyramidentexte", in: *Lexikon der Ägyptologie*, vol. 5, 14–24.

Altenmüller, Hartwig (1984–1985), "Das 'Sänftenlied' des Alten Reiches", in: *Bulletin de la Société d'Égyptologie Genève* 9/10, 15–30.

Altenmüller, Hartwig (1989), "Nilpferd und Papyrusdickicht in den Gräbern des Alten Reiches", in: *Bulletin de la Société d'Égyptologie Genève* 13, 9–21.

Altenmüller, Hartwig (2002a), "Der Himmelsaufstieg des Grabherrn. Zu den Szenen des *zšš wꜣḏ* in den Gräbern des Alten Reiches", in: *Studien zur Altägyptischen Kultur* 30, 1–42.

Altenmüller, Hartwig (2002b), "Funerary Boats and Boat Pits of the Old Kingdom", in: *Archiv Orientální* 70, 269–290.

Assmann, Jan (1988), "Schrift, Tradition, Kultur", in: Wolfgang Raible (ed.), *Zwischen Festtag und Alltag* (ScriptOralia 6), Tübingen, 25–49.

Assmann, Jan (1999), "Cultural and Literary Texts", in: Gerald Moers (eds.), *Definitely: Egyptian Literature*. Proceedings of the Symposion "Ancient Egyptian Literature: History and Forms", Los Angeles, March 24–26 1995 (Lingua Aegyptia Studia Monographica 2), Göttingen, 1–15.

Assmann, Jan (2000), *Religion und kulturelles Gedächtnis*, Munich.

Assmann, Jan (2002), *Altägyptische Totenliturgien. Totenliturgien in den Sargtexten des Mittleren Reiches* (Supplemente zu den Schriften der Heidelberger Akademie der Wissenschaften, Philosophisch-historische Klasse 1), Heidelberg.

Assmann, Jan (2005), *Death and Salvation in Ancient Egypt*, Ithaca/London.

Aufrère, Sydney H. (2010), "Priestly Texts, Recensions, Rewritings and Paratexts in the Late Egyptian Period", in: Philip S. Alexander, Armin Lange and Renate J. Pillinger (eds.), *In the Second Degree: Paratextual Literature in Ancient Near Eastern and Ancient Mediterranean Cultures and Its Reflections in Medieval Literature*, Leiden/Boston, 159–180.

Baines, John (1990), "Restricted Knowledge, Hierarchy and Decorum: Modern Perceptions and Ancient Institutions", in: *Journal of the American Research Center in Egypt* 27, 1–23.

Baines, John (1999), "Forerunners of Narrative Biographies", in: Anthony Leahy and William J. Tait (eds.), *Studies on Ancient Egypt in Honour of H. S. Smith* (Occasional Publications 13), London, 23–37.

Baines, John (2004), "Modelling Sources, Processes, and Locations of Early Mortuary Texts", in: Susanne Bickel and Bernard Mathieu (eds.), *D'un monde à l'autre. Textes des Pyramides & Textes des Sarcophages* (Bibliothèque d'Étude 139), Cairo, 15–41.

Bakker, Egbert J. (1999), "How Oral is Oral Composition?", in: E. Anne Mackay (ed.), *Signs of Orality. The Oral Tradition and Its Influence in the Greek and Roman World* (Mnemosyne Supplementum 188), Boston/Cologne, 29–47.

Barguet, Paul (1952), "Le rituel archaïque de fondation des temples de Medinet-Habou et de Louxor", in: *Revue d'Égyptologie* 9, 1–22.

Barta, Winfried (1963), *Die altägyptische Opferliste von der Frühzeit bis zur griechisch-römischen Epoche* (Münchner Ägyptologische Studien 3), Berlin.

Barta, Winfried (1968), *Aufbau und Bedeutung der altägyptischen Opferformel* (Ägyptologische Forschungen 24), Glückstadt.

Barta, Winfried (1981), *Die Bedeutung der Pyramidentexte für den verstorbenen König* (Münchner Ägyptologische Studien 39), Berlin.

Bauman, Richard/Briggs, Charles L. (1990), "Poetics of Performance as Critical Perspectives of Language and Social Life", in: *Annual Review of Anthropology* 19, 58–88.

Berger-El Naggar, Catherine (2004), "Des Textes des Pyramides sur papyrus dans les archives du temple funéraire de Pépy Ier", in: Susanne Bickel and Bernard Mathieu (eds.), *D'un monde à l'autre. Textes des Pyramides & Textes des Sarcophages* (Bibliothèque d'Étude 139), Cairo, 85–90.

Bickel, Susanne (1994), *La cosmogonie égyptien avant le Nouvel Empire* (Orbis Biblicus et Orientalis 134), Fribourg/Göttingen.

Bickel, Susanne (1997), "Héliopoli.e. le tribunal des dieux", in: Catherine Berger and Bernard Mathieu (eds.), *Études sur l'Ancien Empire et la nécropole de Saqqâra dédiées à J.-Ph. Lauer*, vol. 1 (Orientalia Monspeliensia 9), 113–122.

Bickel, Susanne (2004), "D'un monde à l'autre: le thème du passeur et de sa barque dans le pensée funéraire", in: Susanne Bickel and Bernard Mathieu (eds.), *D'un monde à l'autre. Textes des Pyramides & Textes des Sarcophages* (Bibliothèque d'Étude 139), Cairo, 91–117.

Bidoli, Dino (1976), *Die Sprüche der Fangnetze in den altägyptischen Sargtexten* (Abhandlungen des Deutschen Archäologischen Instituts Kairo 9), Glückstadt.

Billing, Nils (2011), "Monumentalizing the Beyond: Reading the Pyramid Before and After the Pyramid Texts", in: *Studien zur Altägyptischen Kultur* 40, 53–66.

Blackman, Aylward M. (1938), "The Use of the Egyptian Word *ḥt* 'House' in the Sense of 'stanza'", in: *Orientalia* 7, 64–66.

Bochi, Patricia A. (1996), "Of Lines, Linen, and Language: A Study of a Patterned Textile and Its Interweaving with Egyptian Beliefs", in: *Chronique d'Égypte* 71, 221–253.

Bochi, Patricia A. (2003), "The Enigmatic Activity of Painting the Seasons at an Easel: Contemplative Leisure or Preemptive Measure?", in: *Journal of the American Research Center in Egypt* 40, 159–169.

Bojowald, Stefan (2012), "Review: Steiner, R.C., Early Northwest Semitic Serpent Spells in the Pyramid Texts (Harvard Semitic Studies 61)", in: *Bibliotheca Orientalis* 69 (3/4), 236–242.

Bolshakov, Andrey O. (1993), "The Scene of the Boatmen Jousting in Old Kingdom Tomb Representations?", in: *Bulletin de la Société d'Égyptologie Genève* 17, 29–39.

Bolshakov, Andrey O. (1997), *Man and His Double in Egyptian Ideology of the Old Kingdom* (Ägypten und Altes Testament 37), Wiesbaden.

Boltz, William G. (2005), "The Composite Nature of Early Chinese Texts", in: Martin Ern (ed.), *Text and Ritual in Early China*, Seattle, 50–78.

Borchardt, Ludwig (1907), *Das Grabdenkmal des Königs Ne-User-Reᶜ* (Ausgrabungen der Deutschen Orient-Gesellschaft in Abusir 1), Leipzig.

Borchardt, Ludwig (1909), *Das Grabdenkmal des Königs Nefer-Ir-Ke'-Reᶜ* (Ausgrabungen der Deutschen Orient-Gesellschaft in Abusir 5), Leipzig.

Borchardt, Ludwig (1913), *Das Grabdenkmal des Königs Sꜣ-ḥu-reᶜ II. Die Wandbilder* (Ausgrabungen der Deutschen Orient-Gesellschaft in Abusir 7), Leipzig.

Borghouts, Joris F. (1978), *Ancient Egyptian Magical Texts* (Nisaba Series 9), Leiden.

Borghouts, Joris F. (1984), "The Victorious Eyes: A Structural Analysis of Two Egyptian Mythologizing Texts of the Middle Kingdom", in: *Studien zu Sprache und Religion Ägyptens. Zu Ehren von Wolfhart Westendorf*, vol. 2, Göttingen, 703–716.

Borghouts, Joris F. (2001), "On certain Uses of the Stative", in: *Lingua Aegyptia* 9, 11–35.

Borghouts, Joris F. (2007), *Book of the Dead [39]: From Shouting to Structure* (Studien zum Altägyptischen Totenbuch 10), Wiesbaden.

Breyer, Francis (2012), "Zu den angeblich semitischen Schlangensprüchen der Pyramidentexte", in: *Orientalistische Literaturzeitung* 107 (3), 141–146.

Brovarski, Edward (2001), *The Senedjemib Complex. Part 1. The Mastabas of Senedjemib Inti (G2370), Khnumenti (G2374), and Senedjemib Mehi (G2378)* (Giza Mastabas Series), Boston.

Burkard, Günter (1980), "Bibliotheken im alten Ägypten", in: *Bibliothek. Forschung und Praxis* 4, 79–115.

Burn, John (2011), "An Ecological Approach to Determine the Potential Influence that the Pyramid Texts has upon the Sixth Dynasty Tomb Decorations", in: Miroslav Bárta, Filip Coppens and Jaromír Krejčí (eds.), *Abusir and Saqqara in the Year 2010*, vol. 1, Prague, 233–245.

Callender, Vivienne G. (2011), *In Hathor's Image I. The Wives and Mothers of Egyptian Kings from Dynasties I–VI*, Prague.

Cervelló-Autuori, Josep (2011), "The Sun-Religion in the Thinite Age: Evidence and Political Significance", in: Renée F. Friedman and Peter N. Fiske (eds.), *Egypt at Its Origins*. Proceedings of the Third International Conference "Origin of the State: Predynastic and Early Dynastic Egypt", London, 27 July–1 August 2008 (Orientalia Lovaniensia Analecta 205), Leuven, 1125–1149.

Contardi, Federico (2010), "Egyptian Terms used to Indicate the Act of Reading: An Investigation about the Act of Reading in the Egyptian Society", in: Frederick M. Fales and Giulia F. Grassi (eds.), *Camsemud 2007. Proceedings of the 13th Italian Meeting of Afro-asiatic Linguistics Held in Udine, May 21–24, 2007* (History of the Ancient Near East Monographs 10), Padua, 261–269.

Coppens, Filip/Smoláriková, Květa (2009), *Lesser Late Period Tombs at Abusir. The Tomb of Padihor and the Anonymous Tomb R3* (Abusir 20), Prague.

Depuydt, Leo (1992), "Der Fall des Hintersichschauers", in: *Göttinger Miszellen* 126, 33–38.

Dunham, Dows/Simpson, William K. (1974), *The Mastaba of Queen Mersyankh III. G 7530–7540*, vol. 1 (Giza Mastabas Series), Boston.

Edel, Elmar (1944), "Untersuchungen zur Phraseologie der Ägyptischen Inschriften des Alten Reiches", in: *Mitteilungen des Deutschen Archäologischen Instituts, Abteilung Kairo* 13, 1–90.

Enmarch, Roland (2013), "Mortuary and Literary Laments: A Comparison", in: Roland Enmarch and Verena Lepper (eds.), *Ancient Egyptian Literature. Theory and Practice* (Proceedings of the British Academy 188), Oxford, 83–99.

Eyre, Christopher J. (1993), "Why was Egyptian Literature?", in: *Atti.* Sesto Congresso Internazionale di Egittologia, vol. 2, Turin, 115–120.

Eyre, Christopher J. (2002), *The Cannibal Hymn: A Cultural and Literary Study*, Liverpool.

Eyre, Christopher J. (2013), *The Use of Documents in Pharaonic Egypt* (Oxford Studies in Ancient Documents), 2013.

Eyre, Christopher J./Baines, John (1989), "Interactions between Orality and Literacy in Ancient Egypt", in: Karen Schousboe and Mogens T. Larsen (eds.), *Literacy and Society*, Copenhagen, 91–119.

Faulkner, Raymond O. (1971), "*Hy (m) nwḥ* 'make a rope fast'", in: *Journal of Egyptian Archaeology* 57, 202.

Firchow, Otto (1953), *Untersuchungen zur ägyptischen Stilistik II: Grundzüge der Stilistik in den altägyptischen Pyramidentexten* (Deutsche Akademie der Wissenschaften zu Berlin 21), Berlin.

Foley, John M. (2004), "Textualization as Mediation: The Case of Traditional Oral Epic", in: Raimonda Modiano, Leroy Searle and Peter L. Shillingsburg (eds.), *Voice, Text, Hypertext: Emerging Practices in Textual Studies*, Seattle/London, 101–120.

Frandsen, Paul J. (1997), "On Categorization and Metaphorical Structuring: Some Remarks on Egyptian Art and Language", in: *Cambridge Archaeological Journal* 7 (1), 71–104.

Gardiner, Alan H. (1917), "An Archaic Funerary Stele", in: *Journal of Egyptian Archaeology* 4, 256–260.

Gardiner, Alan H. (1955), "A Unique Funerary Liturgy", in: *Journal of Egyptian Archaeology* 41, 9–17.

Gestermann, Louise (2005), *Die Überlieferung ausgewählter Texte altägyptischer Totenliteratur ("Sargtexte") in spätzeitlichen Grabanlagen* (Ägyptologische Abhandlungen 68), Wiesbaden.

Grapow, Hermann (1936), *Sprachliche und schriftliche Formung ägyptischer Texte* (Leipziger Ägyptologische Studien 7), Glückstadt.

Grimm, Alfred (1986), "Titel und Vermerke in den Pyramidentexten", in: *Studien zur Altägyptischen Kultur* 13, 99–106.

Gundacker, Roman (2009), *Studien zu Genese und innerer chronologischer Schichtung der Pyramidentexte*, PhD Thesis, University of Vienna.

Haring, Ben (2003), "From Oral Practice to Written Record in Ramesside Deir el-Medina", in: *Journal of the Economic and Social History of the Orient* 46 (3), 249–272.

Harpur, Yvonne M. (1980), "*zšš wꜣḏ* Scenes of the Old Kingdom", in: *Göttinger Miszellen* 38, 53–60.

Hassan, Selim (1948), *Excavations at Giza*, vol. 4: *1932–1933*, Cairo.

Hassan, Selim (1975), *Excavations at Giza*, vol. 2: *1937–1938*, Cairo.

Hays, Harold M. (2004), "Transformation of Context: The Field of Rushes in Old and Middle Kingdom Mortuary Literature", in: Susanne Bickel and Bernard Mathieu (eds.), *D'un monde à l'autre. Textes des Pyramides & Textes des Sarcophages* (*Bibliothèque d'Étude* 139), Cairo, 175–200.

Hays, Harold M. (2006), *The Typological Structure of the Pyramid Texts and Its Continuities with Middle Kingdom Mortuary Literature*, PhD Thesis, University of Chicago.

Hays, Harold M. (2006–2007), "The Mutability of Tradition: The Old Kingdom Heritage and Middle Kingdom Significance of Coffin Texts Spell 343", in: *Jaarbericht Ex Oriente Lux* 40, 45–59.

Hays, Harold M. (2009a), "Old Kingdom Sacerdotal Texts", in: *Jaarbericht Ex Oriente Lux* 41, 47–94.

Hays, Harold M. (2009b), "Unreading the Pyramids", in: *Bulletin de l'Institut Français d'Archéologie Orientale* 109, 195–220.

Hays, Harold M. (2010a), "The Death of the Democratization of the Afterlife", in: Nigel Strudwick and Helen Strudwick (eds.), *Old Kingdom, New Perspectives: Egyptian Art and Archaeology 2750–2150 BCE*, Oxford, 115–130.

Hays, Harold M. (2010b), "Funerary Rituals (Pharaonic Period)", in: Jacco Dieleman and Willeke Wendrich (eds.), *UCLA Encyclopedia of Egyptology*, Los Angeles, 1–14 http://escholarship.org/uc/item/1r32g9zn#page-3 (last accessed 10.07.2013).

Hays, Harold M. (2012), *The Organization of the Pyramid Texts: Typology and Disposition* (Probleme der Ägyptologie 31), vol. 1–2, Leiden/Boston.

Helck, Wolfgang (1972), "Zur Frage der Entstehung der ägyptischen Literatur", in: *Wiener Zeitschrift für die Kunde des Morgenlandes* 63/64, 6–26.

Honko, Lauri (2000), "Text as Process and Practice: The Textualization of Oral Epics", in: Lauri Honko (ed.), *Textualization of Oral Epics* (Trends in Linguistics. Studies and Monographs 128), Berlin/New York, 3–54.

Hornung, Erik (1999), *The Ancient Egyptian Books of the Afterlife*, Ithaca/London.

Hussein, Ramadan B. (2009), *The Saite Pyramid Texts Copies in the Memphite and Heliopolitan Shaft--Tombs: A Study of Their Selection and Layout*, PhD Thesis, Brown University.

Hussein, Ramadan B. (2011), "Notes on the Saite Copies of Pyramid Texts Spells in the Memphite and Heliopolitan Shaft-Tombs", in: Zahi Hawass, Khaled A. Daoud and Ramadan B. Hussein (eds.), *Scribe of Justice: Egyptological Studies in Honour of Shafik Allam* (Supplément aux Annales du Service des Antiquités de l'Egypte 42), Cairo, 217–233.

Hussein, Ramadan B. (2013), "Recontextualized – The Pyramid Texts 'Serpent Spells' in the Saite Contexts", in: *Études et Travaux* 26, 273–290.

Jéquier, Gustave (1938), *Fouilles à Saqqarah: Le monument funéraire de Pepi II*, vol. 2: *Le temple*, Cairo.

Jones, Dilwyn (1988), *A Glossary of Ancient Egyptian Nautical Titles and Terms*, London/New York.

Junker, Hermann (1934), *Giza II: Bericht über die von der Akademie der Wissenschaften in Wien auf gemeinsame Kosten mit Dr. Wilhelm Pelizaeus unternommenen Grabungen auf dem Friedhof des Alten Reichs bei den Pyramiden von Giza*, Vienna.

Junker, Hermann (1938), *Giza III: Bericht über die von der Akademie der Wissenschaften in Wien auf gemeinsame Kosten mit Dr. Wilhelm Pelizaeus unternommenen Grabungen auf dem Friedhof des Alten Reichs bei den Pyramiden von Giza*, Vienna.

Junker, Hermann (1940), *Giza IV: Bericht über die von der Akademie der Wissenschaften in Wien auf gemeinsame Kosten mit Dr. Wilhelm Pelizaeus unternommenen Grabungen auf dem Friedhof des Alten Reichs bei den Pyramiden von Giza*, Vienna.

Junker, Hermann (1943), *Zu einigen Reden und Rufen auf Grabbildern des Alten Reiches* (Akademie der Wissenschaften in Wien 221.5), Vienna/Leipzig.

Jürgens, Peter (1995), *Grundlinien einer Überlieferungsgeschichte der altägyptischen Sargtexte* (Göttinger Orientsforschungen, Reihe Ägypten IV/31), Wiesbaden.

Kahl, Jochem (1996), *Steh auf, gib Horus deine Hand: die Überlieferungsgeschichte von Altenmüllers Pyramidentexte – Spruchfolge D* (Göttinger Orientsforschungen, Reihe Ägypten IV/32), Wiesbaden.

Kahl, Jochem (1997), "Zur Datierung der frühen Grabplatten mit Opfertischszene", in: *Studien zur Altägyptischen Kultur* 24, 137–145.

Kahl, Jochem (1999), *Siut-Theben. Zur Wertschätzung von Traditionen im alten Ägypten* (Probleme der Ägyptologie 13), Leiden/Boston/Cologne.

Kahl, Jochem (2007), *"Ra is My Lord". Searching for the Rise of the Sun God and the Dawn of Egyptian History* (Menes 1), Wiesbaden.

Kahl, Jochem/Kloth, Nicole/Zimmermann, Ursula (1995), *Die Inschriften der 3. Dynastie. Eine Bestandaufnahme* (Ägyptologische Abhandlungen 56), Wiesbaden.

Kammerzell, Frank (1990), "Funktion und Form. Zur Opposition von Perfekt und Pseudopartizip im Alt- und Mittelägyptischen", in: *Göttinger Miszellen* 117–118, 181–202.

Kammerzell, Frank (2000), "Das Verspeisen der Götter: Religiöse Vorstellung oder poetische Fiktion?", in: *Lingua Aegyptia* 7, 183–218.

Kanawati, Naguib (2001), *Tombs at Giza*, vol. 1: *Kaiemankh (G4561) and Seshemnefer (G4940)* (Australian Centre for Egyptology, Reports 16), Warminster.

Kaplony, Peter (1963), *Die Inschriften der ägyptischen Frühzeit*, vol. I (Ägyptologische Abhandlungen 8), Wiesbaden.

Kaplony, Peter (1966), *Kleine Beiträge zu den Inschriften der ägyptischen Frühzeit* (Ägyptologische Abhandlungen 15), Wiesbaden.

Kessler, Dieter (1984) "Per-anch", in: *Lexikon der Ägyptologie*, vol. 4, 929.

Kloth, Nicole (2002), *Die (auto-)biographischen Inschriften des ägyptischen Alten Reiches: Untersuchungen zu Phraseologie und Entwicklung* (Studien zur Altägyptischen Kultur, Beihefte 8), Hamburg.

Koenig, Yvan (1994), *Magi.e. magiciens dans l'Égypte ancienne*, Paris.

Köhler, Christiana/Jones, Jana (2009), *Helwan II. The Early Dynastic and Old Kingdom Funerary Relief Slabs* (Studien zur Archäologie und Geschichte Altägyptens 25), Rahden.

Kousoulis, Panagiotis (2011), "Stop, O Poison, that I may find Your Name According to Your Aspect: A Preliminary Study on the Ambivalent Notion of Poison and the Demonization of the Scorpion's Sting in Ancient Egypt and Abroad", in: *Journal of Ancient Egyptian Interconnections* 3 (3), 14–26.

Krauss, Rolf (1997), *Astronomische Konzepte und Jenseitsvorstellungen in den* Pyramidentexten (Ägyptologische Abhandlungen 59), Wiesbaden.

Lapp, Günther (2013), "*zšn dwjt* und *zšš wꜣḏ*", in: *Göttinger Miszellen* 239, 51–64.

Leclant, Jean/Minaul-Gout, Anne (2000), "Fouilles et travaux en Égypte et au Soudan", in: *Orientalia* 69, 141–170.

Leclant, Jean/Berger-El Naggar, Cathérine/Pierre-Croisiau, Isabelle (2001), *Les textes de la pyramide de Pépy 1er* (Mémoires publiés par les membres de l'Institut Français d'Archéologie Orientale 118), Cairo.

Leitz, Christian (1996), "Die Schlangensprüche in den Pyramidentexten", in: *Orientalia* 65, 381–427.

Leitz, Christian (1999), *Magical and Medical Papyri of the New Kingdom. Hieratic Papyri in the British Museum VII*, London.

Leitz, Christian (2002–2003), "Rerek", in: *Lexikon der ägyptischen Götter und Götterbezeichnungen*, vol. 4, 701–702.

Lloyd, Alan B. (ed.) (2010), *A Companion to Ancient Egypt*, vol. 1, Oxford.

Manuelian, Peter Der (1996), "Presenting the Scroll: Papyrus Documents in the Tomb Scenes of the Old Kingdom", in: Peter Der Manuelian (ed.), *Studies in Honor of William Kelly Simpson*, vol. 2, Boston, 561–588.

Manuelian, Peter Der (2009), *Mastabas of Nucleus Cemetery G 2100*, vol. 1: *Major Mastabas G 2100–2220* (Giza Mastabas Series 8), Boston.

Martin, Geoffrey Th. (2011), *Umm el-Qaab VII: Private Stelae of the Early Dynastic Period from the Royal Cemetery at Abydos* (Archäologische Veröffentlichungen 123), Wiesbaden.

Mathieu, Bernard (1996a), "Modifications de texte dans la pyramide d'Ounas", in: *Bulletin de l'Institut Français d'Archéologie Orientale* 96, 289–311.

Mathieu, Bernard (1996b), *La poésie amoureuse de l'Égypte ancienne. Recherches sur un genre littéraire au Nouvel Empire* (Bibliothèque d'Étude 115), Cairo.

Mathieu, Bernard (1997a), "La signification du serdab dans la pyramide d'Ounas. L'architecture des appartements funéraires royaux à la lumière des Textes des Pyramides", in: Catherine Berger and Bernard Mathieu (eds.), *Études sur l'Ancien Empire et la nécropole de Saqqâra dédiées à J.-Ph. Lauer*, vol. 2 (Orientalia Monspeliensia 9), 289–304.

Mathieu, Bernard (1997b), "L'huissier, le juge et le greffier. Une hypothèse sur la fonction du serdab dans les pyramides à textes", in: *Méditerranées. Revue de l'Association Méditerranées* 13, 11–28.

Mathieu, Bernard (1998), "Un épisode du procès de Seth au tribunal d'Héliopolis (Pyr. §957a–959e)", in: *Göttinger Miszellen* 164, 71–78.

Mathieu, Bernard (2002), "Les formules conjuratoires dans les pyramides à textes: quelques réflexions", in: Yvan Koenig (ed.), *La magi.e. Égypte: à la recherche d'une définition*. Actes du colloque organisé par le musée du Louvre les 29 et 30 septembre 2000, Paris, 185–206.

Mathieu, Bernard (2004), "La distinction entre Textes des Pyramides et Textes des Sarcophages est-elle légitime?", in: Susanne Bickel and Bernard Mathieu (eds.), *D'un monde à l'autre. Textes des Pyramides & Textes des Sarcophages* (Bibliothèque d'Étude 139), Cairo, 247–262.

Mathieu, Bernard (2010), "Mais qui est donc Osiris? Ou la politique sous le linceul de la religion", in: *Égypte Nilotique et Méditerranéenne* 3, 77–107.

Mathieu, Bernard (2011), "Seth polymorphe: le rival, le vaincu, l'auxiliaire", in: *Égypte Nilotique et Méditerranéenne* 4, 137–158.

Mathieu, Bernard (in press), "Linguistique et archéologie: l'usage du déictique de proximité (*pn/tn/nn*) dans les Textes des Pyramides", in: *forthcoming*.

Meeks, Dimitri (2000), "Mots sans suite ou notations rituelles? (O. DeM 1696 et O. Petrie 36)", in: Robert J. Demarée and Arno Egberts (eds.), *Deir el-Medina in the Third Millennium AD. A Tribute to Jac. J. Janssen* (Egyptologische Uitgaven 14), 235–249.

Mekis, Tamás (2011), "L'équipement funéraire de la prêtresse thébaine Nestaneteretten", in: *Chronique d'Égypte* 86, 41–81.

Mercer, Samuel A. B. (1956), *Literary Criticism of the Pyramid Texts*, London.

Meurer, Georg (2002), *Die Feinde des Königs in den Pyramidentexten* (Orbis Biblicus et Orientalis 189), Freiburg.

Minchin, Elizabeth (1996), "The Performance of Lists and Catalogues in the Homeric Epics", in: Ian Worthington (ed.), *Voice into Text. Orality and Literacy in Ancient Greece* (Mnemosyne Supplementum 157), Leiden, 3–20.
Montet, Pierre (1957), "Hathor et les papyrus", in: *Kêmi* 14, 102–108.
Morales, Antonio J. (2013), *The Transmission of the Pyramid Texts into the Middle Kingdom: Philological Aspects of a Continuous Tradition in Egyptian Mortuary Literature*, PhD Thesis, University of Pennsylvania.
Morenz, Ludwig D. (1996), *Beiträge zur Schriftlichkeitskultur im Mittleren Reich und in der 2. Zwischenzeit* (Ägypten und Altes Testament 29), Wiesbaden.
Müller, Maya, (1984), "Musterbuch", in: *Lexikon der Ägyptologie*, vol. 4, 244–246.
Murray, Margaret A. (1905), *Saqqara Mastabas I* (Egyptian Research Account 10th year), London.
Nagy, Gregory (1998), "Homer as 'Text' and the Poetics of Cross-Reference", in: Christine Ehler and Ursula Schaefer (eds.), *Verschriftung und Verschriftlichung. Aspekte des Medienwechsels in verschiedenen Kulturen und Epochen* (ScriptOralia 94), Tübingen, 78–87.
Oesterreicher, Wulf (1993), "Verschriftung und Verschriftlichung im Kontext medialer und konzeptioneller Schriftlichkeit", in: Ursula Schaefer (ed.), *Schriftlichkeit im frühen Mittelalter* (ScriptOralia 53), Tübingen, 267–292.
Oesterreicher, Wulf (1998), "Textzentrierung und Rekontextualisierung. Zwei Grundprobleme der diachronischen Sprach- und Textforschung", in: Christine Ehler and Ursula Schaefer (eds.), *Verschriftung und Verschriftlichung. Aspekte des Medienwechsels in verschiedenen Kulturen und Epochen* (ScriptOralia 94), Tübingen, 10–39.
Oesterreicher, Wulf (2005), "Autonomisation du texte et recontextualisation. Deux problèmes des sciences du texte", in: Adolfo Murguía (ed.), *Sens et Références. Sinn und Referenz. Mélanges Georges Kleiber*, Tübingen, 199–222.
Ogdon, Jorge R. (1989), "Studies in Ancient Egyptian Magical Thought IV: An Analysis of the 'Technical' Language in the Anti-Snake Magical Spells of the Pyramid Texts", in: *Discussions in Egyptology* 13, 59–71.
Ong, Walter J. (1982), *Orality and Literacy. The Technologizing of the Word*, London/New York.
O'Rourke, Paul F. (2001), "Drama", in: Donald Redford (ed.), *The Oxford Encyclopedia of Ancient Egypt*, vol. 1, Oxford, 407–410.
Osing, Jürgen (1986), "Zur Disposition der Pyramidentexte des Unas", in: *Mitteilungen des Deutschen Archäologischen Instituts, Abteilung Kairo* 42, 131–144.
Osing, Jürgen (1987), "Sprüche gegen die *jbhꜣtj*-Schlange", in: *Mitteilungen des Deutschen Archäologischen Instituts, Abteilung Kairo* 43, 205–210.
Pfirsch, Luc (1992), "Les bâttisseurs des pyramides de Saqqara", in: *Saqqara. Aux origines de l'Égypte pharaonique*, Dijon, 34–37.
Pfirsch, Luc (1994), "À propos des constructeurs de Teti, Pépi Ier et Mérenrê", in: Catherine Berger, Gisèle Clerc and Nicolas Grimal (eds.), *Hommages à Jean Leclant*, vol. 1: Études pharaoniques (Bibliothèque d'Étude 106.1), Cairo, 293–298.
Quirke, Stephen (1990), *The Administration of Egypt in the Late Middle Kingdom*, New Malden.
Quirke, Stephen (2001), *The Cult of Ra. Sun-worship in Ancient Egypt*, London.
Reintges, Chris H. (2006), "The Older Egyptian Stative Revisited", in: *Lingua Aegyptia* 14, 115–134.
Reintges, Chris H. (2011), "The Oral-compositional Form of Pyramid Text Discourse", in: Fredrik Hagen, John Johnston, Wendy Monkhouse, Kathryn Piquette, John Tait and Martin Worthington (eds.), *Narratives of Egypt and the Ancient Near East. Literary and Linguistic Approaches* (Orientalia Lovaniensia Analecta 189), Leuven, 3–54.
Rendsburg, Gary A. (2000), "Literary Devices in the Story of the Shipwrecked Sailor", in: *Journal of the American Oriental Society* 120 (1), 13–23.
Ricke, Herbert (ed.) (1950), *Bemerkungen zur altägyptischen Baukunst des Alten Reichs* (Beiträge zur ägyptischen Bauforschung und Altertumskunde 5), Cairo, 1–128.

Ridley, Ronald T. (1983), "The Discovery of the Pyramid Texts", in: *Zeitschrift für ägyptische Sprache und Altertumskunde* 110, 74–80.

Ritner, Robert K. (1995), "Egyptian Magical Practice under the Roman Empire: The Demotic Spells and their Religious Context", in: Wolfgang Haase and Hildegard Temporini (eds.), *Aufstieg und Niedergang der römischen Welt*, vol. 2: *Principat*, Berlin, 3333–3379.

Ritner, Robert K. (2011), "Foreword: An Egyptological Perspective", in: Richard C. Steiner, *Early Northwest Semitic Serpent Spells in the Pyramid Texts* (Harvard Semitic Studies 61), Winona Lake, ix-xi.

Roeder, Hubert (1993), "Themen und Motive in den Pyramidentexten", in: *Lingua Aegyptia* 3, 81–119.

Roth, Ann Macy (2006), "Multiple Meanings in Carrying Chair Scenes", in: Martin Fitzenreiter and Michael Herb (eds.), *Dekorierte Grabanlagen im Alten Reich. Methodik und Interpretation* (Internet-Beiträge zur Ägyptologie und Sudanarchäologie 6), London, 243–253. http://www2.hu-berlin.de/nilus/net-publications/ibaes6/publikation/ibaes6_pub_text.pdf (last accessed: 26.04.2016).

Saad, Zaki Youssef (1957), *Ceiling Stelae in Second Dynasty Tombs from the Excavations at Helwan* (Annales du Service des Antiquités de l'Égypte, Supplément 21), Cairo.

Schenkel, Wolfgang (1983), "Wozu die Ägypter die Schrift gebrauchten", in: Aleida Assmann, Jan Assmann and Christof Hardmeier (eds.), *Schrift und Gedächtnis: Beiträge zur Archäologie der literarischen Kommunikation* (Archäologie der literarischen Kommunikation Serie 1), Munich, 45–63.

Schenkel, Wolfgang (1994), "*sḏm.t*-Perfekt und *sḏm.tj*-Stativ: Die beiden Pseudopartizipien des Ägyptischen nach dem Zeugnis der Sargtexte", in: Heike Behlmer (ed.), *Quaerentes Scientiam: Festgabe für Wolfhart Westendorf zu seinem 70. Geburtstag überreicht von seinen Schülern*, Göttingen, 157–182.

Schenkel, Wolfgang (2000), "Die Endungen des Prospektivs und des Subjunktivs (*sḏm=f*, *sḏm.w=f*, *sḏm.y=f*) nach Befunden der Sargtexte", in: *Lingua Aegyptia* 7, 27–112.

Schott, Siegfried (1950), "Bemerkungen zum altägyptischen Pyramidenkult", in: Herbert Ricke (ed.), *Bemerkungen zur altägyptischen Baukunst des Alten Reichs* (Beiträge zur ägyptischen Bauforschung und Altertumskunde 5), Cairo, 131–252.

Sethe, Kurt (1908–1910), *Die altaegyptischen Pyramidentexte*, vols. 1–3, Leipzig.

Sethe, Kurt (1913), "Die Inschriften", in: Ludwig Borchardt, *Das Grabdenkmal des Königs š3-ḥu-reʿ II. Die Wandbilder* (Ausgrabungen der Deutschen Orient-Gesellschaft in Abusir), Leipzig.

Sethe, Kurt (1933), *Urkunden des Alten Reichs* (Urkunden des ägyptischen Altertums 1), Leipzig.

Sethe, Kurt (1935), *Übersetzung und Kommentar zu den altägyptischen Pyramidentexten*, vol. 1: *Spruch 213–260 (§§ 134 a–322 a/b)*, Hamburg.

Sethe, Kurt (ed.) (1960), *Die altägyptischen Pyramidentexte. Nach den Papierabdrücken und Photographien des Berliner Museums*, vol. 1–3, Leipzig.

Seyfried, Karl-Joachim (2003), "Dienstpflicht mit Selbstversorgung. Die Diener des Verstorbenen im Alten Reich", in: Heike Guksch, Eva Hofmann and Martin Bommas (eds.), *Grab und Totenkult im Alten Ägypten*, Munich, 41–59.

Simpson, William K. (1976a), *The Mastabas of Qar and Idu. G 7101 and 7102* (Giza Mastabas Series 2), Boston.

Simpson, William K. (1976b), *The Offering Chapel of Sekhem-Ankh-Ptah in the Museum of Fine Arts*, Boston.

Simpson, William K. (1978), *The Mastabas of Kawab, Khafkhufu I and II* (Giza Mastabas Series 3), Boston.

Simpson, William K. (1980), *Mastabas of the Western Cemetery: Part I* (Giza Mastabas Series 4), Boston.

Sledzianowski, Bernd (1976), *Textkritische und überlieferungsgeschichtliche Untersuchung zu den Pyramidentexten der Unas-pyramide (Sprüche 226–243)*, PhD Thesis, University of Göttingen.

Smith, Mark (2009), "Democratization of the Afterlife", in: Jacco Dieleman and Willeke Wendrich (eds.), *UCLA Encyclopedia of Egyptology*, Los Angeles, 1–16. http://repositories.cdlib.org/nelc/uee/1147 (last accessed: 25.12.2013).
Smither, Paul C. (1941), "A Tax-Assesor's Journal of the Middle Kingdom", in: *Journal of Egyptian Archaeology* 27, 74–76.
Sperveslage, Gunnar (2011), "Zu den Schlangensprüche in den Pyramidentexten", in: *Sokar* 23, 30–37.
Spiegel, Joachim (1953), "Die religionsgeschichtliche Stellung der Pyramidentexte", in: *Orientalia* 22 (2), 129–157.
Spiegel, Joachim (1956), "Das Auferstehugsritual der Unaspyramide", in: *Annales du Service des Antiquités de l'Égypte* 53, 339–439.
Spiegel, Joachim (1971), *Das Auferstehungsritual der Unas-Pyramide. Beschreibung und erläuterte Übersetzung* (Ägyptologische Abhandlungen 23), Wiesbaden.
Steiner, Richard C. (2011), *Early Northwest Semitic Serpent Spells in the Pyramid Texts* (Harvard Semitic Studies 61), Winona Lake.
Strudwick, Nigel (2005), *Texts from the Pyramid Age* (Writings from the Ancient World 16), Leiden/Boston.
Svenbro, Jesper (1988), *Phrasikleia. Anthropologie de la lecture en Grèce ancienne* (Textes à l'Appui: Histoire Classique), Paris.
Teffeteller, Annette (2007), "Orality and the Politics of Scholarship", in: Craig Cooper (ed.), *Politics of Orality. Orality and Literacy in Ancient Greece* (Mnemosyne Supplementum 280), Leiden, 67–86.
Theis, Christoffer (2011), *Deine Seele zum Himmel, dein Leichnam zur Erde. Zur idealtypischen Rekonstruktion eines altägyptischen Bestattungsrituals* (Studien zur Altägyptischen Kultur, Beihefte 12), Hamburg.
Topmann, Doris (2010), "PT-Sequenzen in Spruch 885 der Sargtexte", in: *Studien zur Altägyptischen Kultur* 39, 341–371.
Troy, Lana (1986), *Patterns of Queenship in Ancient Egyptian Myth and History* (Acta Universitatis Upsaliensis, Boreas 14), Uppsala.
Urban, Greg (1996), "Entextualization, Replication, and Power", in: Michael Silverstein and Greg Urban (eds.), *Natural Histories of Discourse*, Chicago, 21–44.
Vachala, Břetislav (2010), "Das Vorfeld der Literatur in den Privatgräbern des AR: Ein Fall aus Abusir", in: Miroslav Bárta, Filip Coppens and Jaromír Krejčí (eds.), *Abusir and Saqqara in the Year 2010*, vol. 2, Prague, 767–777.
Verner, Miroslav (2001), *The Pyramids. The Mystery, Culture, and Science of Egypt's Great Monuments*, New York.
Vischak, Deborah (2003), "Common Ground between Pyramid Texts and Old Kingdom Tomb Design: The Case of Ankhmahor", in: *Journal of the American Research Center in Egypt* 40, 133–157.
Walle, Baudouin van de (1965), "Spectacle et religion dans l'Égypte ancienne", in: Guy Dumur (ed.), *Histoire des spectacles. Encyclopedia de la Pléiade*, vol. 19, Paris, 102–134.
Walsem, René van (2005), *Iconography of Old Kingdom Elite Tombs* (Mededelingen en Verhandelingen Ex Orient Lux 35), Leiden/Leuven.
Walsem, René van (2006), "Sense and Sensibility. On the Analysis and Interpretation of the Iconography Programmes of Four Old Kingdom Elite Tombs", in: Martin Fitzenreiter and Michael Herb (eds.), *Dekorierte Grabanlagen im Alten Reich* (Methodik und Interpretation, Internet-Beiträge zur Ägyptologie und Sudanarchäologie 6), London, 277–332.
Weeks, Kent R. (1994), *Mastabas of Cemetery G 6000: Including G 6010 (Neferbauptah), G 6020 (Iymery), G 6030 (Ity), G 6040 (Shepseskafankh)* (Giza Mastabas Series 5), Boston.
Wettengel, Wolfgang (1992), "Zu den Darstellungen des Papyrusraschelns", in: *Studien zur Altägyptischen Kultur* 19, 323–340.

Wilde, Keike (2013), "Grabbeigaben und ihre symbolische Bedeutung anhand eines Konvolutes aus Giza (Mastaba D 208)", in: *Zeitschrift für ägyptische Sprache und Altertumskunde* 140, 172–187.

Willems, Harco (1996), *The Coffin of Heqata (Cairo JdE 36418)* (Orientalia Lovaniensia Analecta 70), Leuven.

Wilson, John A. (1944), "Funeral Services of the Egyptian Old Kingdom", in: *Journal of Near Eastern Studies* 3 (4), 201–218.

Woods, Alexandra A. (2009), "*zšš w3ḏ* Scenes of the Old Kingdom Revisited", in: Nigel Strudwick and Helen Strudwick (eds.), *Old Kingdom, New Perspectives. Egyptian Art and Archaeology, 2750–2150 BC*, Oxford, 314–319.

Zinn, Katharina (2011), "Temples, Palaces and Libraries: A Search for an Alliance between Archaeological and Textual Evidence", in: Rolf Gundlach and Kate Spence (eds.), *Palace and Temple. Architecture–Decoration–Ritual* (Königtum, Staat und Gesellschaft früher Hochkulturen 4.2), Wiesbaden, 181–202.

Zumthor, Paul (1983), "L'intertexte performanciel", in: *Texte. Revue critique et de théorie littéraire* 2, 49–59.

Sara Campanelli

Family Cult Foundations in the Hellenistic Age

Family and Sacred Space in a Private Religious Context

1 Introduction

The private foundation consists of an endowment of funds (capital or real estate) whose income is intended to realize an ongoing purpose freely chosen by the founder.[1] A private foundation can be defined as a cult foundation when its purpose is exclusively or mainly cultic in nature.[2] In this paper, "cult" refers to a wide range of ritual performances, including rituals directed to the gods, to the dead, or to both.[3]

Evidently, what distinguishes the foundation from an isolated act of euergetism is its potentially permanent continuance. For this reason, the will of the founder had to be embedded into a system of measures and regulations that ensured its continuity beyond the individual's lifetime.

Within this regulatory system, an essential role was played by the designation of a social group (or, more rarely, an individual) as the body responsible for the continuing fulfilment of the purpose established by the founder and for the management of the funds allocated to it. The chosen group was often also the recipient of the various possible benefits connected with the foundation (e.g. banquets, money distribution, contests, building activities, and the use of landed properties).[4]

The body in charge of the foundation not only ensured its operation, but also represented the social context in which the founder intended it to operate. Therefore, a classification of Greek foundations based on the different social interlocutors

This article emerged from the Heidelberg Collaborative Research Center 933 "Material Text Cultures. Materiality and Presence of Writing in Non-Typographic Societies". The CRC 933 is financed by the German Research Foundation (DFG).

1 Cf. Laum 1914, 1, 1f.; Lupu 2005, 81f.

2 This clarification is needed because almost all Greek foundations include a cultic component, even when their main purpose is different in nature: cf. e.g. the foundation of Eudemos from Hellenistic Milet, which is dedicated to the education of Milesian children, but also includes sacrifices and a procession (*Milet* I 3, 145 = Laum 1914, 2, no. 129, ll. 30–36; 69–77); cf. also the agonistic foundation of C. Iulius Demosthenes from Oinoanda (Hadrianic period), which establishes a wide range of cult activities along with musical, poetic and rhetorical contests and sports performances (text and translation in Wörrle 1988, 4–17, cf. ll. 68–87; for the organization of the festival as a whole see 227–258).

3 This broad definition of "cult" reflects the classification of the foundations made by Laum 1914, 1, 60.

4 For the individuals and social groups known from the documents as managers and beneficiaries of foundations see Laum 1914, 1, 159–166.

DOI 10.1515/9783110417845-005,

chosen by the founders, is a fruitful way to understand the intention of the founders themselves and the significance they attributed to their initiative. A private individual who decided to create a foundation might intend it either for public or for private enjoyment: family cult foundations are foundations that limit participation in the established rituals, and management of the funds allocated thereto, to the family[5] of the founder.[6]

In light of this, it is easy to understand why the (family cult) foundation as an institution offers many opportunities for analysis. Although each aspect is closely connected to the others, studies on this subject can focus on specific perspectives of investigation, preferring, for example, the juridical or financial point of view;[7] alternatively, it is possible to focus on the social or ideological background of such an institution.[8] Moreover, when the foundation has a cult purpose, the adoption of a religious perspective is also valid;[9] family cult foundations, in particular, involve further fields of research, relating to the history of the Greco-Roman family as well as to family and household religion.[10]

---

5 The term "family" is meant here in an extensive sense, including kin relationships that go beyond the nuclear family, since some family cult foundations also allow members of the extended family to participate therein.

6 This definition of "foundation", when applied to the "family dossiers", seems not to be as problematic as argued by Carbon/Pirenne-Delforge 2013, 67f. The different conception and structure of the private foundations, depending on the different social interlocutors chosen by the founders, are briefly discussed by Campanelli 2012b, 71–75 through two Hellenistic examples: the foundation of Kritolaos from Amorgos (*IG* XII 7, 515: public enjoyment) and the foundation of Epikteta from Thera (family cult foundation: see below).

7 See e.g. Mannzmann 1962, which deals with theoretical issues connected with the juridical form of the foundation and interprets it in terms of *homologia* between the founder and the group appointed by him as the body responsible for the foundation; the *homologia* is meant as a bilateral agreement which takes on the juridical form of the contract; for criticism of this interpretation see Modrzejewski 1963. Among the more recent studies discussing financial and administrative aspects of foundations are Sosin 2001, Gabrielsen 2008, Migeotte 2010, Harter-Uibopuu 2011, and Migeotte 2012.

8 Such are the main concerns of Schmitt-Pantel 1982 and Veyne 1976, 241–251 respectively.

9 See Purvis 2003, who analyzes in detail three private cult foundations of the Classical period, contextualizing them in the general framework of the elective cults; see also Hupfloher 2012, who discusses the same pieces of evidence as Purvis, together with other documents from the Classical age onwards, but mainly concentrates on "Kultgründungen" by private individuals rather than on foundations which meet the above definition, where the emphasis is on the continuity of the established (cult) purposes over time.

10 The distinction between household religion and family religion has been conveniently pointed out by Faraone 2008, 211–213: the former is meant in a "locative" sense, referring to the cult practices that took place at the house and aimed at protecting it; the latter is meant in a "genetic" sense, referring to the rituals that were intended to cement kin relationships, based on descent from common ancestors, and to define the identity and position of the family within society. Gherchanoc 2012, 159–168 discusses family cult foundations in the framework of family rituals, celebrations and sociability; an interpretation of family cult foundations as a Hellenistic development of the traditional gentilicial cults is

This paper will present a comparative analysis of five inscriptions from the south-Aegean Doric area which record foundations that arose in a family context during the Hellenistic age. The analysis will concentrate on an aspect that has hardly been touched on in studies of this topic, namely, the physical spaces as an integral part of the foundation system.[11] The term "space" is used here to include not only the family sanctuaries and tombs, which in most cases were clearly built concurrently with the enactment of the cult, but also the estate that was the source of income intended for funding cult activities.

Art, monumental visibility, and ritual as a cluster of factors that interact to translate family identity into language of signs have been pointed out in some studies,[12] but the intention here is to look more specifically into how the spaces were meant to be used in material and conceptual terms and to see how they contributed to defining group identity. This kind of analysis partially overlaps with the developing field of studies of the "religious landscape", conceived as the result of a complex interaction of environmental, territorial, ritual, normative, and socio-economic factors.[13] In this case, the "religious landscape" falls into the elusive sphere of the private space outside the house walls.

Being essentially regulatory in nature, the texts under examination do not provide actual descriptions of the cult places or attached properties, but the very fact

suggested by Graf 1995, 112f.; some references to the foundation of Diomedon from Cos (see below) are included by Brulé 2005, 33, fn. 25; 45 (reprinted in Brulé 2007, 420f.) in his discussion of the domestic cult of Zeus; in general, however, family cult foundations are scarcely considered in studies of family and household religion, which mostly concentrate on literary sources, usually focusing on Attic documentation; the references given here by way of example only concern general works, including the post-Classical period: Jost 1992, 245–261; Price 1999, 89–107; Mikalson 2005, 133–159; Faraone 2008. Continuity and change are the main parameters adopted to interpret the post-Classical developments of the family, but the focus is generally on its civic role and/or the possible influence of non-Greek elements, social mobility, institutional rearrangements of its structure and legal aspects: cf. e.g. van Bremen 2003; Modrzejewski 2011, 359–415; however, family cult foundations are taken into account by Pomeroy 1997b, 108–113 (even though some of her conclusions are questionable: see below); studies of the juridical and socio-political condition of women during the Hellenistic and Imperial periods make extensive use of epigraphic evidence concerning (family cult) foundations: cf. van Bremen 1996 (212–216 on Epikteta's foundation); Stavrianopoulou 2006 (see in particular 226–236; on Epikteta's foundation notably 292–302).

11 Wittenburg 1990 dedicates some pages (139–147) to the architectural and decorative features of the sanctuary belonging to Epikteta's family in Thera, but he does not intend to provide an in-depth examination of this topic (as explicitly stated on p. 139, fn. 1) and, above all, he does not compare this sanctuary to evidence provided by the cult places from other family cult foundations.

12 Cf. Stavrianopoulou 2006, 290, 294–302 (with specific reference to the foundation of Epikteta from Thera).

13 For this perspective of investigation see e.g. Cole 2004 and Brulé 2012; see also the contributions included in Olshausen/Sauer 2009; for the problems of definition connected with the notion of "religious landscape" and for an attempt to circumscribe its scope see Horster 2010, 436–438.

that some of them pay a lot of attention to regulating the use of the spaces points to their centrality in the foundation system. From a methodological point of view, this kind of analysis will show how and to what extent inscriptions can be used as sources for reconstructing the architectural layout of places known only from epigraphic records.[14] Indeed, archaeological remains of structures connected with family cult places established by foundations are only preserved in one instance from Cos;[15] for this reason, the mention of buildings, open spaces, and land throughout the texts concerned represents the main piece of evidence for the architectural and spatial arrangement of family sanctuaries, tombs, and related properties.

Actual reconstruction hypotheses would require systematic comparisons between the architectural terminology provided by the inscriptions and the archaeological evidence for structures that show analogies with those under examination. This kind of analysis, however, goes beyond the purposes of this paper, which aims, above all, at identifying the significance of the spaces to the family groups who enjoyed them. References to archaeological evidence, however, will occasionally be included for the sake of comparison, to help clarify the function and architectural features of structures mentioned in the texts being examined.

The first section of this paper will address, albeit selectively, the main features of family cult foundations, placing the discussion in the framework of a survey of the main theories proposed over time concerning the origin and function of this institution. The second section will focus on the cult communities and pay particular attention to the membership requirements which are of an exclusively family nature. The third section will concentrate on the arrangement, functions and meanings of the physical spaces mentioned throughout the texts. The expression "sacred space", which appears in the title of both the paper as a whole and the third paragraph, might turn out to be problematic, given that the notion of "sacred land" is controversial, and its very existence has been questioned.[16] The scholarly debate mainly concerns the legal status of properties belonging to sanctuaries (or, more precisely, to the gods) placed under the administration of the polis or sub-polis groups, which seem to have treated

---

14 Methodological issues connected with this kind of analysis are concisely discussed by Hammerstaedt 2009, including an extensive bibliography on the relationships between epigraphy and architecture.

15 According to the convention adopted by Sherwin-White 1978, 5, fn. 3, in this paper the spelling "Cos" refers to the island as a whole, whereas "Kos" refers to the main city of the island.

16 For the debate on the notion of "sacred land" see the exhaustive overview of the scholarship provided by Papazarkadas 2011, 1–13.

them as if they were public properties.[17] Although such issues do not directly affect the subject of this paper, which concerns properties connected with private sanctuaries or tombs, attention will be paid to both the terminology that refers to the sacredness of the spaces and the regulations that concern their use. What will be borne in mind is the wide range of possible sacred spaces characterized by different degrees of sacredness and by different statutes—themselves liable to change over time—as is generally pointed out by those scholars who recognize the existence of sacred land as an autonomous category.[18] Using this varied framework, the notion of "sacred" can be applied to the buildings, open spaces and landed properties which will be examined in this paper. The conclusions will summarise the results of the analysis and point out some clearly recognizable features that characterize the spaces involved in family cult foundations, which might delineate a specific "religious landscape".

2 Scholarly Theory, Epigraphic Evidence, and Cultic Aspects

The existence of family cult foundations as a particular type of foundation is recognized for the first time by B. Laum in his massive work, *Stiftungen in der griechischen und römischen Antike*, published in 1914.[19] At that time the main epigraphic evidence for family cult foundations of the Hellenistic age was already known: the foundation of Diomedon from Cos (fig. 1), that of Poseidonios from Halicarnassus (fig. 2), and the "Testament of Epikteta" from Thera (fig. 3).[20]

Considering the chronology and geographic range of these inscriptions, Laum points out the early appearance of the foundation as an institution in the family context of the Doric insular area, in which Halicarnassus can also be included because of its proximity to and its close cultural ties with the Dodecanese. He connects the

---

17 Cf. e.g. the texts collected by Velissaropoulos-Karakostas 2011, 2, 31–36; a particularly controversial point is the expression τὰ τεμένη τὰ δημόσια (Arist., *Oec.* 2.2.3a) where the two categories of "sacred" and "public" seem to overlap: see Migeotte 2006.

18 Cf. Parker 1983, 160–168; Dignas 2002, 13–35; Cole 2004, 40–50, 57–65; Horster 2004, 7–54 (with discussion especially of the earliest archaeological, epigraphic, and literary evidence for sacred land as an autonomous and well defined category); Horster 2010, notably 444–455.

19 Cf. Laum 1914, 1, 10, 15, 68–71, 158f., 224–227, 243–245.

20 Foundation of Diomedon: *IG* XII 4, 1, 348 (= Laum 1914, 2, no. 45; end of the 4th century B.C.–about 280 B.C.); foundation of Poseidonios: the latest edition is J.-M. Carbon in Carbon/Pirenne-Delforge 2013, 99–114 (= Laum 1914, 2, no. 117; 3rd century B.C.; between 280 and 240 B.C., as proposed by Carbon on the basis of palaeographic analysis, Carbon/Pirenne-Delforge 2013, 99f.); foundation of Epikteta: Wittenburg 1990, 22–37 (= Laum 1914, 2, no. 43; between 210 and 195 B.C.); from now on, when quoted, these inscriptions will refer to the editions mentioned here (*IG*; Carbon; Wittenburg).

origin of such a Hellenistic innovation to a weakening of family bonds, which led to the creation of family cult communities gathered around a shared fund in order to entrust them with the continuation of the funerary cult, which was no longer ensured by the spontaneous acts of piety of the descendants, as it had been in earlier times.

Fig. 1: Foundation of Diomedon. Four-side engraved marble pillar now kept in storage within the Castle of Knights, Kos city (photos S. Campanelli, reproduced with the kind permission of the 22nd Ephorate of Prehistoric and Classical Antiquities, Greece).

In fact, Laum considers the funerary cult as the main purpose of family cult foundations, attaching little importance to the gods who were worshipped along with the ancestors and/or dead family members. In his view, the rise of this new form of family aggregation found favourable conditions among the conservative Doric aristocracy, who attributed to ancestry an important role in defining family identity.

The idea that disintegration of the family served as premise for the origin of the Greek foundation is further developed by E.F. Bruck,[21] and then brought into question by W. Kamps,[22] who reaffirms the crucial role played by the family in the origin of the foundation, but stresses its evolution rather than its disintegration as Bruck

---

21 Bruck 1926, notably 190–276; Bruck 1955; for a similar view cf. Nilsson 1955–1961, 2, 116f.; further bibliography and arguments against this theory in Stavrianopoulou 2006, 291; 301f.

22 Kamps 1937.

does. By evolution, Kamps means a change in the Hellenistic period from a broad clan structure identifying itself in the cult of shared ancestors to a more restricted family unit that based its new identity on cults founded *ex novo* by family members. In his view, this change did not entail a complete interruption of the traditional ancestral cult, but rather its reshaping within a more restricted kin solidarity, whose new course was marked by the foundation of cults dedicated to recently deceased individuals, namely the founder and members of his family, rather than (or not only) to the ancestors. The act of creating a foundation, therefore, caused a break in the traditional form of ancestry cult, which had been spontaneously handed down from generation to generation.

Fig. 2: Foundation of Poseidonios. Two-side engraved stele rejoined from more than twenty pieces and partially damaged by fire; now kept in the British Museum © The Trustees of the British Museum.

These newly introduced private cults gradually came to feature the heroization of the deceased, according to the chronological development which Kamps describes in analyzing the three foundations from Cos, Halicarnassus and Thera. In the first stage, represented by the foundation of Diomedon, there is no heroization and the traditional ancestry cult is still present; the ancestors are placed under the protection of Heracles, "héros locale à Cos",[23] who is the main recipient of the established rituals and is worshipped together with other "génies"[24] peculiar to the ancestral cult, the Moirai and *Pasios*, the latter meant as

23 Kamps 1937, 152; this and the next quotation from Kamps' article show how he pays little attention to the importance of the gods, whose significance actually goes beyond that of mere ancestral heroes and demons, see below.

24 Kamps 1937, 152.

Fig. 3: Foundation of Epikteta. Four engraved marble plaques attached to a base for three statues; now kept in the Museo Maffeiano, Verona, Italy (photo G. Stradiotto; with the kind permission of the Direzione Musei d'Arte di Verona).

a form of the domestic Zeus. An innovation, however, is represented by the fact that Heracles bears the epithet *Diomedonteios*, stemming from the name of the founder himself and pointing, therefore, to a privileged relationship between him and this god as a way to immortalize his memory without involving his actual heroization. In Kamps' view, an actual cult of Diomedon might have been introduced in later generations of his descendants, who would have regarded him as their ancestral hero.

Traditional elements are still present also in the foundation of Poseidonios, where it is explicitly stated that the cult conforms to the customs of the ancestors (l. 7: καθάπερ καὶ οἱ πρόγονοι); but besides the gods, whom Kamps relates partly to the ancestral milieu (Zeus *Patroos*, Moirai), and partly to the Anatolian context (Apollo of Telmessos, Mother of the Gods), the Τύχη Ἀγαθή of Poseidonios' parents and the Ἀγαθὸς Δαίμων of Poseidonios and his wife are also worshipped by the cult community. This is a step towards heroization of the deceased family members, although here it is not yet direct, but mediated by the individual demons.

A direct heroization is made manifest in the foundation of Epikteta, by the means of which a cult of family heroes is established. It consists of the founder's husband and two sons who all predeceased her; Epikteta is to be included among them after her death; they are recipients of annual sacrifices along with the Muses, the goddesses who are worshipped together with the heroes by the cult community.

Kamps argues that this cultic evolution of family cult foundations goes hand in hand with the development of their structure and function, which became increasingly complex from a juridical and administrative point of view. This study of Kamps has been presented in some detail because it is so far the only one specifically devoted to family cult foundations of the Hellenistic age as a whole,[25] and some of its conclusions are still valid. A good example of how his interpretation has affected later studies is provided by the case of an inscription from Cos published in the middle of the nineteenth century,[26] but neither included in Laum's collection nor discussed in Kamps' article. The inscription is on a boundary stone delimiting the estate conse-

---

25 Wittenburg 1990 is specifically dedicated to the analysis of Epikteta's foundation and does not include close comparisons with the other family cult foundations, which are scarcely mentioned: cf. 75, fn. 18; 76, fn. 19; 93, fn. 10; 99; Gherchanoc 2012, 159–168 (see above fn. 10) only summarizes the content of the inscriptions concerned in order to highlight the value of ritual sharing for family sociability and identity; there is a brief overview of family cult foundations in Parker 2010, 118–120 where the aim is to contextualize a recently discovered foundation inscription from Hellenistic Lycia that shows similarities with family cult foundations (see below); see also Lupu 2005, 86f., where family cult foundations are briefly discussed as a type of sacred law. During the preparation of this paper, two new works appeared: Paul 2013, addressing the family cult foundations from Cos in the framework of Coan cults and sanctuaries (108–117, 232–235, 247f.); Carbon/Pirenne-Delforge 2013, focusing on the priesthoods and cult offices in the foundations of Diomedon, Poseidonios and Epikteta, but including quite a broad comparative analysis among the three texts.

26 *Ed. pr.* Ross 1845, no. 309; *IG* XII 4, 1, 355; henceforth, when quoted, this inscription will refer to the *IG* edition.

crated to the Twelve Gods and Charmylos, the hero of the *Charmyleoi* (fig. 4). Applying Kamps' theory, S. Sherwin-White categorizes this document as pertaining to family cult foundations, on the grounds that Charmylos is not a mythical ancestor after whom the cult group of the *Charmyleoi* is named, but the historical founder of a family cult community like Diomedon,[27] even if in this case the actual foundation deed is not preserved.

Fig. 4: The boundary stone referring to Charmylos' foundation embedded into the façade of the medieval Church of the Stavros, Pyli, Cos (photos S. Campanelli).

Although the chronological development proposed by Kamps has undeniable merits, it is too linear to account for all the data, derived both from later epigraphic findings, as well as from the texts he analyzes. Among the later findings, another inscription from Cos[28] can be added to family cult foundations because it concerns the establishment of a private cult by a certain Pythion, together with an anonymous priestess who was probably his wife;[29] even though the inscription dates to the 2nd

27 Sherwin-White 1977, 207–217; but it must be noted that a similar interpretation had already been suggested by Herzog 1928, 31.
28 *Ed. pr.* Fraser 1953; *IG* XII 4, 1, 349; henceforth, when quoted, this inscription will refer to the *IG* edition.
29 Cf. Fraser 1953, 44; 59–61.

century B.C., there is no trace of heroization or funerary cult, since the foundation is intended for the worship of Artemis, Zeus *Hikesios* and the *Theoi Patrooi*. This calls into question the idea that the cult of the gods was involved in family cult foundations only as a function of the funerary cult that was their main purpose. As for the chronology relating to the development of the heroization, it must be pointed out that the inscription considered as the most ancient in this cluster of documents, the foundation of Diomedon, does not allow us to completely exclude a possibility that a cult of the founder was already practised at this stage, since the restoration of his name, albeit extremely dubious, has been proposed for ll. 26 and 36 in a very damaged part of the text that deals with ritual regulations.[30] In any case, heroization is explicitly stated in the aforementioned foundation of Charmylos, which dates to the end of the 4th or the beginning of the 3rd century B.C. and is therefore coeval with the foundation of Diomedon, whereas it is missing in the later foundation of Pythion.

Kamps identifies the cult directed both to dead family members and to gods as a feature of family cult foundations in the Hellenistic age, but underestimates the role played by the divine component. Since this paper does not intend to adopt a strictly religious perspective, it can only touch upon the fact that the choice of the gods within family cult foundations seems to have been very carefully considered, also involving deities of the civic *pantheon* who are reshaped within the private context of the family. In Cos, Heracles is something more than a local hero; he is the mythical forefather of the Doric population of the island and, accordingly, he is closely connected to the definition of the renewed civic body resulting from the μετοικισμός of 366 B.C. which involved a complex institutional as well as cultic rearrangement.[31] Thus, Diomedon performed a complex cultic operation: drawing on the strong genealogical components which characterize Coan Heracles, he bound him to his own lineage (also through the epithet *Diomedonteios*) and in this way cemented the double ancestral connection of his family, that stemming from their blood relation and that stemming

30 The fact that Kamps does not consider this possibility might have depended on the editions of the text he uses: indeed, the highly uncertain restoration θυόντω | δὲ [τῶι Ἡρακλεῖ καὶ Διομέδ]ọ[ν]τι μόσχον (ll. 25f.) appears for the first time in Segre 1993, ED 149 and is then accepted in *IG*, whereas in the editions quoted by Kamps this point is restored in a different way or not at all: cf. Dareste/Haussoullier/Reinach 1898, 94–103, no. 24b; Laum 1914 (see above, fn. 20); *Syll.*³ 1106; Herzog 1928, 28–32, no. 10; at l. 36, however, the restoration Διομέδ]οντι is already present in Herzog's edition and the reading has recently been improved by the editors of *IG* (Διομέ]δ̣οντι).

31 On the mythology concerning Coan Heracles and on the civic cult of Heracles on Cos see Campanelli 2011, 648f., 654f., 668–669, 673–677 (with particular reference to the sanctuary of the god located in the Harbour Quarter of the city of Kos); at greater length see Paul 2013, 95–108, 116–117, 288f. on the connection between Heracles *Kallinikos* and Coan citizenship (cf. the sale of his priesthood, *IG* XII 4, 1, 320, ll. 35–37, where individuals who have obtained citizenship are required to sacrifice to the god); on this point cf. also Feyel 2009, 248; for the Coan μετοικισμός, see Sherwin-White 1978, 40–75; Carlsson 2010, 216–218.

from their Coan citizenship as well as from their shared Doric ancestry.[32] A similar process must have occurred in the choice of the Twelve Gods by Charmylos, as these deities seem to be connected to the civic unity after the μετοικισμός.[33]

Even more complex involvements can be hypothesized behind the choice of deities bearing the epithet *Patroos*, such as Zeus *Patroos* in Poseidonios' foundation and the *Theoi Patrooi* in that of Pythion: since the adjective *patroos* indicates all that is inherited from the father,[34] it also summarizes all of the religious memberships handed down from father to son and, accordingly, can equally refer to the traditional household gods and to those worshipped by sub-polis groups, like the phratries, into which one was admitted only through legitimate descent; at the widest social level, finally, this epithet can refer to the ancestral gods of the *father*land.[35] For this reason, it is often difficult to know exactly which social sphere a god named *Patroos* belongs to,[36] and this is also true for the "paternal" gods involved in family cult foundations. In this respect, it can be observed that the *Theoi Patrooi* are known in Cos under this collective denomination from inscriptions of the Hellenistic and Imperial periods, especially the latter, but the nature of these deities is controversial, as they might be the ancestral gods of the whole Coan state or the ancestral gods of some civic subdivision.[37] However, the latter hypothesis is more likely, at least as far as the Hellenistic period is concerned, as is supported by a decree of the Coan deme of Isthmos that deals with a private donation for the benefit of the local tribal cults; in case of violations, a fine was to be paid to the *Theoi Patrooi*, who were probably also recipients of

32 On Heracles and other civic (Aphrodite and perhaps Dionysus) and household (Moirai, *Pasios*) deities who are part of the private *pantheon* founded by Diomedon, see Campanelli 2011, 655–657, 671, 675–677; Paul 2013, 53, 108, 111, 115, 295–297; civic cult of Aphrodite: 79–95, 285–288, 294f.; civic cult of Dionysus: 117–127; other deities must also have been worshipped, but their names are lost due to the bad condition of the stone on side A, ll. 25–36, where most of the ritual regulations appeared; the attempts at restoration made by Herzog and other editors are mere conjectures (cf. Herzog 1928, 31), and they have been almost entirely rejected by the editors of *IG*.

33 On the cult of the Twelve Gods on Cos, see Paul 2013, 40–42, 158f., 282f.

34 Cf. Liddell/Scott/Jones 1968, s.v. πατρῷος; Chantraine 2009, s.v. πατήρ; cf. also Ammon., *Diff.* 383: πάτρια πατρῴων καὶ πατρικῶν διαφέρει. Πατρῷα μὲν γὰρ τὰ ἐκ πατέρων εἰς υἱοὺς χωροῦντα˙ πατρικοὶ δ' ἢ φίλοι ἢ ξένοι˙ πάτρια δὲ τὰ τῆς πόλεως ἔθη.

35 On the gods named *Patrooi* and on their worship at gradually wider levels of Greek society (from the household to the whole polis) see Parker 2008, who pays particular attention to the intermediate groups (phratries and "phratry-like bodies"), but also collects the epigraphic evidence referring to the other social contexts where these cults are present; on Zeus *Patroos* and on the concept of ἱερὰ πατρῷα, with particular reference to the gentilicial group of the *Klytidai* on Chios, see Brulé 1998, especially 312–317 (republished in Brulé 2007, 392–398); see also Brulé 2003, 113–115 (republished in Brulé 2007, 445–448).

36 Cf. Parker 2008, 204.

37 Cf. Sherwin-White 1978, 331f.; Buraselis 2000, 33f., 46–51, 154–162 (catalogues of the dedications to the *Theoi Patrooi* from the 1st century B.C. onwards); Paul 2013, 290.

a sacrifice.[38] It is clear that in this case they are gods worshipped by the three tribes of the deme of Isthmos.[39] It is significant that the inscription of Pythion comes from this deme,[40] whose very ancient and illustrious religious traditions continued after the μετοικισμός.[41] Thus, it is likely that Pythion wanted to stress the tribal membership of his family by including the "Paternal Gods" of his place of origin in his private *pantheon*,[42] but it is also possible that he wanted to indicate various levels of religious membership, including the ancestral religious traditions of his own family, through the potentially all-embracing nature of the *Theoi Patrooi*.[43] Another of the three deities involved in Pythion's foundation seems to be connected to sub-polis groups: Zeus *Hikesios*, who is one of the gods attested in Cos by several inscriptions on small stones probably marking out the respective cult places of groups which bear names with the typical patronymic ending in *–dai*; the stones date from the 4th to the 2nd century B.C. and are from the area of the *Asklepieion*.[44] The fact that some of the deities concerned bear the epithet *patroos* beside their own name[45] confirms the extra-family involvements of this divine attribute on Cos, at least in the Hellenistic period.

On the other hand, Pythion might have re-marked the familial nature of his foundation by placing Artemis beside the *Theoi Patrooi* and Zeus *Hikesios*, since she is well known as the protector of childbirth, nurturer of children, and the goddess who presides over the female life cycle.[46] Moreover, Artemis seems to be the main recipi-

---

38 *IG* XII 4, 1, 100, ll. 26–28; on this inscription see Paul 2013, 231f.

39 For the civic organization of the Coan state after the μετοικισμός and for its internal subdivisions, see Sherwin-White 1978, 153–174; Grieb 2008, 147–153; Carlsson 2010, 218–221.

40 Cf. Fraser 1953, 35–37, 56.

41 Cf. Pugliese Carratelli 1957 and 1963–1964, 147–158; for criticism of the historical and political implications of his reconstruction, see Sherwin-White 1978, notably 74f., 196f.

42 The eventuality that an individual chose to privately honour gods of the "phratry-like body" to which he belonged is admitted by Parker 2008, 204 (with reference to the foundation of Pythion).

43 That the *Theoi Patrooi* of Pythion's foundation are to be meant as the family ancestral gods is argued by Paul 2013, 233.

44 Cf. Parker 2008, notably 202; 211 (nature of these "phratry-like groups"); Paul 2013, 185f., 290f.

45 Cf. inscriptions (g), (h) and (u) in the list provided by Parker 2008, 202.

46 A parallel can be suggested with an Attic inscription (*IG* II^2 4547) included by Purvis 2003 in her study of private cult foundations of the Classical age; Artemis *Lochia* appears among the *kourotrophoi* deities listed therein as the recipients of the cult; such a cluster of deities fits the bond between mother and son, which seems to be the forming principle of this foundation. Interpretation of this text, however, is controversial: see the exhaustive examination of Purvis 2003, 15–32 (notably 18f. on the *kourotrophoi* deities). The cult of Artemis *Lochia* is also attested in Cos (cf. *IG* XII 4, 1, 72, ll. 16f.): however, based on the hypothetical number of missing letters (about eight, according to his calculation), Fraser 1953, 38f. rules out the possibility that this was the lost epithet of Artemis in Pythion's foundation, where a gap follows the name of the goddess at l. 2 (Ἀρτέμιτο[ς- - -]ας); as alternatives he suggested Πατρώι]ας (but considers it too short) and Εἰλειθυί]ας: both would fit the context, even though neither of the two is attested in Cos; among the epithets of Artemis known on the island, other possibilities might be: Περγαί]ας (cf. *IG* XII 4, 1, 346, l. 8; *IG* XII 4, 2, 526), considered too short by

ent of the established cult, as she is the first deity mentioned at l. 2 of the inscription and the only one involved in the manumission of Makarinos, the slave whom Pythion freed by consecrating him to the goddess in order to entrust him with the care of the sanctuary he had founded (ll. 3–12).[47]

In any case, the possibility that Pythion intended to link his family microcosm to "paternal" memberships wider than the household cannot be ruled out. This would match the vitality of Coan gentilicial groups which claimed shared ancestry.[48] Their persistence on an island like Cos, where family cult foundations particularly flourished, encourages a radical review of Kamps' theory as far as the development of the family structure is concerned.[49]

The cult, privately dedicated to deities like Heracles and the *Theoi Patrooi*, who recall the plurality of social contexts to which each family actually belonged,[50] points to a will for self-representation which the founders seem to have put into practice by

---

Fraser, but re-evaluated by Cucuzza 1997, notably 17f., though not very convincingly; Ἀγροτέρ]ας (cf. *IG* XII 4, 1, 358, ll. 18f.). In general, on Artemis as *kourotrophos* and on the involvement of the goddess in female rites of passage and sexuality see Cole 2004, 198–230; for the cult of Artemis on Cos see Paul 2013, especially 140–145, 214–216, 263–265, 294, 298 (Artemis' connection with birth and childhood), 307.

47 The consecration of slaves to deities was a common form of manumission and often imposed cultic duties on the freed people; the same procedure occurs in the foundation of Diomedon, see below; cf. Darmezin 1999, 159–161, 195f. (foundations of Diomedon and Pythion); 219–222 (cultic duties); Zelnick-Abramovitz 2005, 86–99, 228 with fn. 98, 232f.

48 Cf. Pugliese Carratelli 1963–1964, 147–158; Sherwin-White 1978, especially 158–169, considers the gentilicial components traceable in Hellenistic Cos (e.g. groups' names ending in *–dai* which claim a shared ancestry) as remains of the kinship-based organization which characterized the phase predating the μετοικισμός; Vallarino 2009, 193–199 argues that in Hellenistic Cos gentilicial criteria are traceable in the composition of the official cult college of the *hieropoioi*; the lineage κατ'ἀνδρογένειαν seems to have played some role within the Ἀσκλαπιαδᾶν τὸ κοινὸν Κώιων καὶ Κνιδίων as late as the 4th century B.C., when membership had already been extended to physicians who did not belong to the original Coan *genos*: cf. Pugliese Carratelli 1991; *gene* claiming descent from Asclepius and Heracles survived in the Imperial period: cf. Pugliese Carratelli 1994.

49 Cf. Wittenburg 1990, 94f.

50 The idea of Greek religion as a network of multifaceted experiences, all of which are embedded in the all-embracing polis system, corresponds to the model of the so-called "polis religion" theorized by Sourvinou-Inwood 1988 (= Buxton 2000, 38–55); 1990 (= Buxton 2000, 13–37); cf. also Aleshire 1994; the network which joined public cults with family ones, passing through "cercles de sociabilité intermédiaires", is highlighted by Brulé in his study of the cults of Zeus *Ktesios* and Zeus *Herkeios*, the two most typical forms of the domestic Zeus, who were nonetheless also worshipped by both sub-polis groups and the *polis* as a whole: Brulé 2005 (= Brulé 2007, 405–428); a similar perspective can be found in the aforementioned work of Parker on the *Theoi Patrooi* (Parker 2008); recent studies have attempted to attenuate the idea of an all-embracing function of the polis in the religious sphere by valuing the peculiarities of family cults (Faraone 2008; Boedeker 2008) or the role of individual initiatives in the religious field (Purvis 2003, 1–13); for a review of the model of "polis religion" see Kindt 2012, 12–35.

placing their family identity in the framework of the fatherland's religious traditions. The conformity of the established cults to fatherland and ancestral *mores* is explicitly stated in the texts through expressions like κατὰ τὰ πάτρια in the foundation of Diomedon (ll. 42; 92f.) and καθάπερ καὶ οἱ πρόγονοι in the foundation of Poseidonios (l. 7).

The impulse to define family identity, which can be traced by examining, albeit partially, the composition of the *pantheon* established through family cult foundations, might find its historical explanation in the framework of the profound sociopolitical changes which affected the Hellenistic age, as argued relatively recently by A. Wittenburg and S. Pomeroy.[51] Mainly on the basis of Epikteta's foundation, they connect the origin and the ultimate aim of family cult foundations to the need for family self-preservation. Wittenburg argues that the ancient *élite* of Thera, to which Epikteta's family is likely to have belonged, experienced a sociopolitical marginalization throughout the 3rd century B.C., due to the heavy interference of Ptolemaic rulers in the internal issues of the island despite its formal independence. The author also connects the influx of foreign people to Ptolemaic presence; such people included parvenus who widely displayed their newly acquired influence, to the detriment of the established local aristocracy. In his view, the constitution of family associations, like the one founded by Epikteta, could be intended as a conservative reaction aimed at preserving the prestige of the families that had been divested of their legitimate political weight.[52] As a working suggestion, Wittenburg attempts to apply a similar sociopolitical interpretation to other places where family cult foundations arose, namely Cos and Halicarnassus.

In more general terms, Pomeroy sees family cult foundations as a strategy to prevent family disintegration due to wars, social mobility and mortality. She points out, in particular, that the lack of direct heirs affected the structure of Epikteta's family and related inheritance matters, as well as raised concerns about the continuity of the funerary cult. Following the death of her husband and two sons, Epikteta restored the male line by resorting to collateral branches of the family in order to create a male cult community. Despite the fact that some women were admitted therein, the male core would more efficiently ensure the continuing accomplishment of the funerary duties, since the patrilocal marriage tradition meant that women left their birth families to join those of their husbands. In Pomeroy's view, a similar lack of male descendants

51 Wittenburg 1990, 51–54, 57–70; Wittenburg 1998; Pomeroy 1997a, 206–209 and 1997b, 108–113.

52 Unlike Wittenburg, Stavrianopoulou 2006, 299f. with fn. 119, believes that the epigraphic evidence for members of Epikteta's entourage who held public office is sufficient to indicate that this kinship group was politically active; in Stavrianopoulou's view, the fact that the foundation text does not refer to the civic merits of the deceased men of Epikteta's own family is due to the completely private nature of this foundation: according to her, in the Hellenistic age the polis was the only entity entitled to give recognition for merit to its prominent citizens.

was a factor in the foundation of Diomedon, where the "commemorative activities"[53] were performed by the founder's freedman Libys and his offspring rather than by the descendants of Diomedon, who are only briefly mentioned throughout the text. Pomeroy concludes that the involvement of distant relatives or non-family members in commemorative celebrations and inheritance matters indicates a diminished value of the actual kin ties in the Hellenistic age.

Pomeroy's argument concerning the foundation of Diomedon is difficult to accept, because the text clearly distinguishes between Libys and his offspring and the descendants of Diomedon, who were in fact charged with guaranteeing freedom of these manumitted slaves (ll. 6–9). The lines referring to the duties with which the freedmen were entrusted are badly preserved (ll. 11–23), but they are likely to be concerned simply with the care of the sanctuary and the provisions for ritual performances (coverings for the dinner-couches, cleaning and crowning the images and so on). In contrast to Pomeroy's argument, the freedmen are no longer mentioned after the initial part of the text, where they first appear among the goods consecrated by Diomedon to Heracles *Diomedonteios* (ll. 1–5). The rest of the inscription lists regulations involving the descendants of Diomedon, who represented the actual cult community, as will be discussed below. In fact, they were in charge of the crucial ritual activities, such as the sacrifices (ll. 9–11: θυόντω δὲ τὰ ἱερὰ τοὶ ἐγ [Διο]μέδον⟦δον⟧ τος καὶ ἀεὶ τοὶ ἐξ αὐτῶṿ γεν[ό]μενοι). This instruction is placed just before the lines dealing with the freedmen's duties in order to highlight that the role of the latter did not include the sacrificial performance.

The expression τοὶ ἐγ Διομέδοντος καὶ τοὶ ἐξ αὐτῶν γενόμενοι shows that Diomedon actually had his own direct descendants, and the great importance attached to legitimate descent can be seen clearly in the ritual through which newborn children were introduced into the cult community (ll. 51–55).[54] The text specifies the prerogatives of the descendants in the male line (οἱ κατ' ἀνδρογένειαν: ll. 86f.; 154f.): the performance of the sacrifice for *Pasios* and the Moirai was reserved only for them (ll. 153–155) and, under particular conditions, they had the right to celebrate their wedding in the sanctuary of Heracles *Diomedonteios* (ll. 86–111). Among the legit-

---

53 Pomeroy 1997a, 209; 1997b, 113; actually, there is no explicit trace of commemorative activities in the text; these can be supposed to have taken place based on the fact that the foundation includes the cult of the ancestors (cf. ll. 130–140), but the regulations concerning it are completely lost (ll. 20f.?); note also the possibility that Diomedon was a recipient of the cult, see above. According to the preserved text, however, the rituals established by Diomedon mainly consisted of various sacrifices and banquets in honour of Heracles and the other deities worshipped along with him; for the ritual aspects of Diomedon's foundation, see Campanelli 2011, 657–661; Paul 2013, 109–115.

54 As for its meaning, this ritual can be compared to the ceremonies through which newborn children were introduced to the domestic hearth and young people to the paternal phratries or similar bodies to which their fathers belonged: cf. Campanelli 2011, 654; on these rituals and their connection with legitimate descent, see Gherchanoc 2012, 35–48, 150–158.

imate descendants, furthermore, the three officials of the community (*epimenioi*)[55] were appointed annually (ll. 144–146), whereas the illegitimate children were entitled neither to this office nor to the priesthood of Heracles (ll. 146–149),[56] which was to be handed down from eldest son to eldest son (ll. 23–25).[57] All of this speaks in favour of patrilineal descent as the basic requirement for membership in the cult community.

Given that the foundation of Diomedon provides no clues for supposing a generalized crisis of family ties in the Hellenistic age, the involvement of the extended family in Epikteta's foundation is probably to be understood as the consequence of accidental factors, like the high male mortality rate, which affected the structure of her family, as Pomeroy herself argues in the initial part of her discussion. In other words, the more or less extended size of these family cult communities might have depended on the individual family histories.[58] These considerations seem to accord with observations made in the studies that investigate the effect of demographic factors and socioeconomic constraints on synchronic and diachronic developments of the family.[59]

---

55 On the roles and meaning of this office in the foundations of Diomedon, Poseidonios and Epikteta, see Carbon/Pirenne-Delforge 2013, especially 83–95; the most innovative conclusion the authors come to concerns the fact that the temporary nature of the office, implied in the literal meaning of the term *epimenios*, might be mirrored in the three foundation texts by the limited span of time in which these officials carried out their task, that is, the period slightly before and during the annual celebrations: this matches evidence on *epimenioi* in the public sphere, showing the way in which family cult foundations drew on and reshaped the civic model; other equally important remarks are made on the twofold nature, ritual and administrative, of *epimenios*' office.

56 For the interpretation of this passage, see Velissaropoulos-Karakostas 2008, 254–256.

57 These lines are very damaged, but their general content can be easily understood by comparing them to the foundation of Poseidonios (ll. 19f.: ἱερατευέτω τῶν ἐκγόνων τῶν ἐκ Ποσειδωνίου ὁ πρεσβύτατος ὢν κατ' ἀνδρογένειαν).

58 For literary evidence on the possible family rearrangements due to various eventualities and involving different degrees of kinship, see the cases discussed by Cox 1998, 141–167 (with specific reference to 5th and 4th century Athens); family strategies are mirrored by funerary contexts: see Marchiandi 2011, 35–46, on the more or less extended family structures emerging from the prosopographic analysis of the individuals buried within the 5th and 4th century Attic *periboloi*; van Bremen 1996, especially 193–204, 237–272, shows how family strategies can be at least partially reconstructed through the evidence for family members acting together in civic euergetism (with particular reference to the role of women).

59 See the comparative overview of Huebner 2011; see also Zuiderhoek 2011, who interprets euergetism, including foundations for public enjoyment, as a strategy for public recognition adopted by civic *élites* of the Hellenistic period to cope with the rapid turnover due to the high mortality rate, which led notable families to quickly disappear from the political scene; as the author does not provide much evidence, for the moment this interpretation must remain an interesting working hypothesis.

What deserves attention within Pomeroy's discussion is the concept of family self-preservation as the motivation that led Epikteta to a formal family rearrangement in order to ensure the proper settlement of the inheritance and the continuity of funerary practices.[60] This view, though only suggested by Pomeroy, matches Wittenburg's interpretation of Epikteta's foundation as the way in which the old established *élite* of Thera reacted to its sociopolitical marginalization. This is not the place to attempt to apply Wittenburg's reconstruction to the parallel contexts of Cos and Halicarnassus, since such an analysis would require an in-depth examination of the local institutional arrangements, the identity of the internal groups holding the power and the relationship between local autonomy and interference from the Hellenistic kingdoms.[61] This paper will analyze the means by which this intention of family self-preservation or at least self-representation comes to light in the texts. Some useful elements in this regard have already emerged during the survey of scholarly theory, but the intention now is to further the analysis by focusing on two aspects which seem to be especially indicative for the above purpose: (1) the family acting as a cult community and (2) the sacred space with the twofold function of cult place and source of income.

3 The Cult Community

The act of foundation formally turned the founder's family into an exclusive cult community, which was both responsible for the continuity of the cult and entrusted with the management of the funds allocated for this purpose.[62] In most cases the cult community is identified in a way that clearly shows its family nature. In the foundation of Diomedon, those in charge of the cult activities were named τοὶ ἐγ [Διο]μέδον⟦δον⟧τος καὶ ἀεὶ τοὶ ἐξ αὐτῶṿ γεν[ό]μενοι (ll. 9–11).[63] Whereas this definition emphasized

60 It should be noted, however, that the creation of the male cult community is not a personal initiative of Epikteta, but satisfies the express will of her deceased son Andragoras (ll. 22f.): see below.

61 For Cos and its relationships with the Hellenistic kingdoms, see Sherwin-White 1978, expecially 82–145; Carlsson 2010, 208–214; on the Coan government after the *metoikismos*, see Carlsson 2010, 221–243; on the relationship between Halicarnassus and the Ptolemies, see Isager 2004; Pedersen 2004; Konuk 2004, among others.

62 Before Laum 1914 (see vol. 1, especially 167 and 224, where he classifies family groups among the "Vereine, die um der Stiftung willen ins Leben gerufen werden"), the foundations of Diomedon, Poseidonios and Epikteta were considered "Familienvereine" in the first works on Greek associations: cf. Poland 1909, especially 87f.; see also Wittenburg 1998, 451; the foundation of Diomedon is mentioned by Maillot 2013, 208 (cf. also 222, fn. 4) among the associations of Cos, but she considers it the only family association on the island.

63 Cf. the slight variation at ll. 136–139 (τοὺ[ς ἐ]γ Διομέδοντος ⟦I- - - - -Ω .⟧[γ]εγενημένους καὶ τοὺς ἐγγ[όν]ους αὐτῶν) and 145f.; the concept of legitimate descent down the male line is more specifically stressed by the expression τοὶ κατ' ἀνδρογένειαν (ll. 86f.; 154f.).

the descent as the basic requirement for the membership, the expression τοὶ τῶν ἱερῶν κοινωνεῦντες (ll. 7; 81; 87f.), which alternates with the previous one throughout the text,[64] highlights the shared base around which the community is gathered, namely the "sacred things". The wide range of meanings included in the concept of *hiera* is well known: it can refer to cult, rituals, cult furniture, sacrificial victims and offerings, gifts for the gods, gods' *simulacra* and symbols, cult places, funds devoted to cult purposes, or cultic matters.[65] In the text under examination *hiera* is likely to include, beyond the rituals, all that was dedicated to Heracles,[66] namely the precinct, buildings, garden, and slaves that are objects of the initial dedication to the god (l. 1: Δ̣[ιομέδων ἀνέθηκ]ε), and the cult furnishings listed in the inventory of ll. 120–130,[67] which are equally dedicated to Heracles (l. 120: ἀνέθηκα) and are said to be sacred to him (ll. 127–129: ὥστε πάντα ταῦτα ἱερὰ εἶναι τοῦ̣ Ἡρακλεῦς). Although not explicitly stated, the other buildings and lands mentioned throughout the text must also be considered *hiera* because they were sources of income used for cult purposes.[68] This finds confirmation in the appearance of the designation "those who share the sacred things" mostly in connection with management matters, such as the protection of the freedmen (ll. 6–9) and the use of the spaces belonging to the sanctuary (ll. 80–90).[69]

64 Cf. the variation οἷ[ς] μέτεστι τῶν ⟦των⟧ ἱερῶν (ll. 52f.); Paul 2013, 116 conveniently points out that κοινωνεῖν and μετεῖναι differ slightly in meaning: the former seems to entail some deliberative power over the *hiera*, whereas the latter seems to indicate merely the birth right to participate in cult; these nuances of meaning fit the contexts where κοινωνεῖν and μετεῖναι are used respectively throughout the text; Paul also notices that μετεῖναι, in turn, differs from μετέχειν, used in ll. 146–149 to indicate participation in the cult granted, within certain limits, to the illegitimate children who did not strictly belong to the eligible group; for the terminology concerning the rights of participation in cult and in cult management see Paul 2013, 196–203 (with particular reference to the tribal cult of Apollo and Heracles in the deme of Halasarna).
65 For the wide range of meanings covered in the adjective ἱερός and its neuter plural ἱερά, see Casabona 1966, 6–17; Rudhardt 1992, 22–30; Horster 2004, 50–52; on the sacred implying divine ownership and on the various categories of "sacred things", see Parker 1983, 151–175; Connor 1988 provides an in-depth examination of the expression ἱερὰ καὶ ὅσια (broadly speaking: "sacred and profane things"), pointing out its first appearance in the financial sphere.
66 Rudhardt 1992, 30 includes in his definition of ἱερός all that was transferred to the gods through the act of dedication.
67 The list of objects dedicated to Heracles continues at ll. 155–157 (ἀνέθηκα δὲ καί); for the ritual function of some of these objects, see Campanelli 2011, 658.
68 See below.
69 In contrast, Paul 2013, 340, fn. 73, classifies Diomedon's foundation among the Coan texts where *hiera* has the meaning of "cult": accordingly, she translates the expression τοὶ τῶν ἱερῶν κοινωνεῦντες into "ceux qui ont le culte en commune"; cf. also 108–111 and 114 for her translation of ll. 1–43 and 86–115; for the differentiated use of the two definitions ("the descendants of Diomedon and those who descend from them" and "those who share the sacred things") throughout the text, see Campanelli 2011, 652f. The inscription also provides evidence for the use of the term *hiera* to indicate both the sacrificial performance (l. 9: θυόντω δὲ τὰ ἱερά; ll. 65f.: ἐχθυσεῦ̣νται τὰ ἱερά) and the "sacred offerings" which were part of the provisions that members of the cult community had to give as an "entrance tax" at the

This point has to be borne in mind because of its consequences for the sanctuary's role as a family asset.

A definition of the cult community based on the concept of legitimate descent also occurs in the foundation of Poseidonios, but in this case the inclusion of the descendants down the female line is specified (l. 2–4: τοῖς ἐξ αὐτοῦ γινομένοις καὶ οὖσιν, ἔκ τε τῶν ἀρσένων καὶ τῶν θηλειῶν); the second mention of the cult group, however, indicates a more extended membership (ll. 12–14: τοῖς ἐξ ἑαυτοῦ καὶ τοῖς ἐκ τούτων γινομένοις, ἔκ τε τῶν ἀρσένων καὶ τῶν θηλειῶν, καὶ τοῖς λαμβάνουσιν ἐξ αὐτῶν), including οἱ λαμβάνοντες ἐξ αὐτῶν, probably to be understood as relatives by marriage, such as the husbands of Poseidonios' daughters.[70] The verb λαμβάνω to indicate "receive in marriage", "take as one's wife" is epigraphically attested, for example in a series of honorary decrees from Tenos (2nd century B.C.) which record a marriage agreement: Μήδειος μὲν ἐγδιδόμενος [τ]ὴν θυγατέραν Φιλίππην Σουνιάδης δὲ λαμβάνων (*IG* XII 5, 864, ll. 5f.; "Medeios giving his daughter Philippe away and Souniades receiving her ...").[71] Unlike Poseidonios' foundation, however, in this case the object of λαμβάνω is explicitly stated. A recently discovered foundation inscription from Hellenistic Lycia includes the sons-in-law of the founder in the limited group of ten male relatives who are entitled to participate in the annual banquets connected with two sacrifices in memory of the founder Symmasis and his wife Mamma.[72] The fact that the sons-in-law and their descendants are included among the relatives (cf. A, ll. 34; 46f.; B, l. 38: ἀνχιστεῖς) shows that the notion of ἀγχιστεία could vary

---

birth of a child (ll. 51–55: εἰσαγώγιον δὲ διδότω ὧι κα γένηται παιδίον, οἷ[ς] μέτεστι τῶν ⟦των⟧ ἱερῶν, χο[ῖ]ρον, ἱερά, λιβανωτόν, σπονδάν, στέφανον; cf. above, fn. 54); the priest, too, was probably required to provide *hiera* (cf. ll. 36–38: ἱερ̣[ὰ δὲ παρεχέτω ἄρτον ποτὶ τ]ὰν ἀρ̣τοφαγ[ίαν καὶ οἶνον καὶ μέλι ποτὶ τὰ]ν̣ σ̣πονδᾴ[ν]); for the offerings or sacrificial provisions called *hiera*, see Paul 2013, 341–344.

70 Cf. Kamps 1937, 157f.; Carbon/Pirenne-Delforge 2013, 81f.

71 Cf. *IG* XII 5, 863, ll. 4f.; 865, ll. 7–9; on these decrees see Stavrianopoulou 2006, 40–43 (with translation of 863, 1–12); for the terminology of the marriage agreements (including λαμβάνω) in the literary sources see Vérilhac/Vial 1998, 232–265.

72 Text and translation in Parker 2010, 104–108 and Arnaoutoglou 2012, 217–220; even though these celebrations were commemorative in nature (cf. B, ll. 19–21: εὐωχηθήσονται ἐν ταύτῃ τῇ ἡμέρᾳ ἄγοντες ἐπώνυμον ἡμέραν Συμμασιος καὶ Μαμμας), they began before the death of Symmasis and Mamma as is clear from the fact that portions of the sacrificial victims were reserved for them (cf. A, ll. 5–18; B, 38–47): this may be another similarity to Poseidonios' foundation, which is classified by Dareste/Haussoullier/Reinach 1898, 128–133 and 145, among the *donationes inter vivos*; the sacrifice for the Ἀγαθὸς Δαίμων Ποσειδωνίου καὶ Γοργίδος (ll. 35f.), especially if it was already performed during Poseidonios' and Gorgis' lifetime, can be compared to the sacrifice for the ἥρως Συμμασιος καὶ Μαμμας (B, 31f.): the "hero of Symmasis and Mamma" is close to the notion of personal demon, according to the convincing interpretation of Parker 2010, 116; in contrast, Arnaoutoglou 2012 simply translates "the hero Symmasis and Mamma": in this case, however, it would be difficult to understand why the term "hero" is in the singular; on the Ἀγαθὸς Δαίμων in Poseidonios' foundation, see Carbon/Pirenne-Delforge 2013, 97; Carbon in Carbon/Pirenne-Delforge 2013, 107f.

depending on context.[73] Although this foundation shares the creation of a family cult group and the link between funerary cult and divine cult (in this case, of Helios) with other family cult foundations, it is not included in the dossier of inscriptions being examined here because it involves a society of metalworkers along with the founder's family.[74]

Unlike Symmasis' foundation and Epikteta's foundation, which will be addressed later, that of Poseidonios does not include a list of community members designated by name, but their number must have been defined at a certain point, as we can infer from the use of the perfect participle εἰληφότες instead of the present participle λαμβάνοντες at l. 23. Here the cult community is mentioned again at the beginning of the decree by which the foundation is ratified and enacted (ll. 22f.: ἔδοξεν Ποσειδωνίωι καὶ τοῖς ἐκγόνοις τοῖς ἐκ Ποσειδωνίου καὶ τοῖς εἰληφόσιν ἐξ αὐτῶν).[75] Like Diomedon's foundation, the officials charged with ritual as well as administrative duties (*epimenioi*) were appointed annually from among the community members (ll. 23–33; 40–47). Despite this extended membership, the priesthood and the enjoyment of the properties allocated by the founder were reserved for the eldest son of Poseidonios and then handed down through the male line (ll. 18–20; l. 20: κατ' ἀνδρογένειαν).

In Epikteta's foundation the complete denomination of the cult community is κοινὸν τοῦ ἀνδρείου τῶν συγγενῶν, but the abbreviated form κοινόν appears more frequently throughout the text.[76] This definition summarizes the male basis of the community and family ties existing among its members.[77] The exclusivity of the κοινόν is highlighted by the fact that the inscription includes a list of community members, who can be distributed among the genealogical trees of three related families (ll.

73 The relatives by marriage are not included in the basic definition of ἀγχιστεία provided by Harrison 1968, 143–149; cf. Parker 2010, 115.

74 On all these aspects of Symmasis' foundation, see Parker 2010 and Arnaoutoglou 2012.

75 Laum 1914 (cf. above, fn. 20) divides the text of Poseidonios' foundation into three parts: "Veranlassung" (ll. 1–11); "Stiftung" (ll. 12–22); "Beschluß des Familienverbandes" (ll. 22–52); the text itself highlights its tripartite structure: cf. ll. 49–50 (ἀναγράψαι δὲ καὶ τὸν χρησμὸν καὶ τὴν ὑποθήκην κ[αὶ] τὸ δόγμα ἐν στήλῃι λιθίνηι); the presence of *paragraphoi* (and a punctuation mark at l. 22) graphically dividing the three sections is noticed by Carbon in Carbon/Pirenne-Delforge 2013, 103f.; Carbon/Pirenne-Delforge 2013, notably 68–73, 83–85, 88, and 95 conveniently pay attention to the composite structure of Diomedon's, Poseidonios' and Epikteta's inscriptions; on the foundation inscriptions consisting of more than one document (e.g. excerpt from the founder's testament or his/her ἐπαγγελία followed by the endorsement of the body responsible for managing the endowment) cf. Laum 1914, 1, 3; a slightly more complex situation can be envisaged for Diomedon's inscription, which does not present a well-defined structure, but is the result of unsystematic additions over time: cf. Campanelli 2011, 652, 657.

76 The denomination of the cult community with all its variations appears in the following lines: 22f., 26f., 30f., 40f., 52f., 56, 61, 71, 74, 76f., 132, 143, 145f., 165, 168, 177, 202, 205, 213, 215, 216f., 220, 222, 228, 233f., 236, 243, 248, 251, 254f., 258, 263, 278, 285f.

77 Cf. Wittenburg 1990, 97f.

81–93).[78] After the men, a number of women are listed (ll. 94–106), including, among others, the wives of those men previously mentioned and their children, the heiresses[79] with their husbands and children, and Epikteta's daughter Epiteleia; the role and position of these women within the community are unclear, but they must have been entitled to belong to the cult group in a different way than the men: whereas the latter are listed first and under the definition of συγγενεῖς, which recalls the official name of the community, the list of women is introduced by the verb πορεύομαι (literally "to go", "to walk", "to proceed to") which might be simply referring to their presence at the annual celebration for the Muses and heroes without actually participating in the management of the foundation;[80] the idea of "being admitted into (the cult community in order to attend the rituals)" might have stemmed from the literal use of πορεύομαι and especially of its compound εἰσπορεύομαι to indicate the right/prohibition to enter a sanctuary or the required conditions to gain access to a sacred place.[81] A development of this basic meaning can be seen in the participles πορευόμενοι and especially συπορευόμενοι and εἰσπορευόμενοι when referring to groups of "Festteilnehmer" or members of cult associations defined as "those who go (together)", or "participate (together) in" given feasts or cults.[82] This meaning fits the right to attend the annual celebration granted to the women listed in ll. 94–106 of Epikteta's inscription. These women, moreover, are likely to have been admitted at a second phase, due to an autonomous decision by Epikteta, independently of the will of her second deceased son who, before dying, had asked his mother to create

---

78 Cf. Wittenburg 1990, 63–66; Stavrianopoulou 2006, 295f.

79 On the heiresses in Epikteta's foundation see Velissaropoulos-Karakostas 2011, 1, 250–252.

80 Cf. Wittenburg 1990, 98f.; 101; the author includes a comparison with the foundations of Diomedon and Poseidonios where the women were probably allowed to participate in the banquets; nevertheless, at least as far as Diomedon's foundation is concerned, the question of female participation seems to be more complex than Wittenburg assumes: cf. Campanelli 2011, 660f. and below in this paper.

81 Cf. e.g. the foundation of Pythion: ἁγνὸν εἰσπορεύεσθαι (l. 15: purity requirements for gaining access to the sanctuary; see below); Andania Mysteries regulations (1st century B.C.): ἃ μὴ δεῖ ἔχοντας εἰσπορεύεσθαι (Gawlinski 2012, 72f. l. 37: items people were not permitted to carry into a special area within the sanctuary); a good parallel would have been an inscription from Philadelphia (Lydia, 2nd/1st century B.C.) concerning purity and "moral" requirements for participating in a cult that was newly established or renewed by a certain Dionysios (*TAM* V 3, 1539), but the verb πορεύομαι appears there as the result of a restoration, albeit likely: πορευ]όμενοι εἰς τὸν οἶκον τοῦτον (ll. 14f.: people allowed to enter the *oikos*/to participate in cult); εἰς τὸν οἶκον τοῦτον μ[ὴ εἰσπορευέσθω] (l. 32: entry into the *oikos*/participation in the cult denied to men and women who do not respect the instructions of Dionysios); the nature of this cult community (private cult association or household?) and the meaning of the term *oikos* (building, name of the cult group or both?) are controversial: cf. Stowers 1998; cf. Petzl's commentary on l. 5 in *TAM* edition. According to a suggestion of Stavrianopoulou 2006, 300f., fn. 122, on the other hand, the verb πορεύομαι ("mitlaufen") recalls the idea of a procession which reproduces the pecking order within the cult community founded by Epikteta.

82 Cf. Poland 1909, 73; for example, the κοινὸν τῶν συμπορευομένων παρὰ Δία Ὑέτιον (*IG* XII 4, 1, 121; Cos, ca. 200 B.C.).

the κοινὸν τοῦ ἀνδρείου τῶν συγγενῶν (ll. 22f.).[83] Membership in the community was based on a hereditary principle, which was, however, limited to the male line, since daughters were admitted only as long as they were under their father's guardianship (cf. ll. 95–97; 135–139). Just as in the foundations of Diomedon and Poseidonios, annual *epimenioi* were appointed from among the συγγενεῖς (ll. 65f.), and the same must have been true for other community officials mentioned throughout the text.[84]

The prominent position of Epikteta's own family within the cult community emerges from the fact that the priesthood of the Muses and heroes was reserved for the son of Epikteta's daughter Epiteleia and then handed down from eldest son to eldest son among Epiteleia's descendants (ll. 57–61), similar to the foundations of Diomedon and Poseidonios again; moreover, Epikteta bequeathed to Epiteleia herself the sanctuary of the Muses (*Mouseion*) and the funerary precinct with the annual obligation to pay the agreed part of the income deriving from her inheritance in order to allow the cult community to celebrate the annual feast in honour of the Muses and heroes (ll. 29–41): a similar procedure, which takes into account the rights of inheritance,[85] occurs in Poseidonios' foundation, where the firstborn of the descendants was entitled to enjoy the properties allocated by the founder, provided he paid for the family community's annual religious feast.

The term συγγενεῖς occurs in two other inscriptions, from Thera and Halicarnassus, which probably refer to foundations as far as it is possible to infer from the small preserved parts of the texts. The first[86] was thought to be related to Epikteta's foundation because of the onomastic concurrences between the two texts, but this hypothesis has not been generally accepted.[87] Beyond the term συγγενεῖς (l. 12), which is likely to refer to a cult community based on kinship, other references to descent might be the participles [γ]εγονότες (l. 17) and γινόμενοι (l. 20).

The inscription from Halicarnassus is first recognized as a foundation text by L. Robert.[88] Its family nature can be concluded from the mention of the συγγενεῖς (ll. 3; 4?; 5) and οἱ ἐκ τοῦ γένους (l. 10); the terms γέρα and [σ]πλάγχνων (ll. 8f.) very likely refer to the privileges and/or obligations of a priest, similar to the regulations in the foundations of Diomedon, Poseidonios and Epikteta;[89] similarly to these again, one

83 Cf. Wittenburg 1990, 83–84.

84 For the offices and authorities occurring in the text see below.

85 Cf. Wittenburg 1990, 75; Stavrianopoulou 2006, 141, 151–154, 126–130.

86 *IG* XII Suppl. 154 (2nd/1st century B.C.).

87 Hiller von Gaertringen 1914, 131–133; Kamps 1937, 166–168; *contra* Wittenburg 1990, 65, fn. 9.

88 Robert 1937, 466–468; the first editor, Bérard 1891, 550f., no. 22, interprets the inscription as a rent contract and considers the term συγγενεῖς as referring to a tribe of Halicarnassus: this is why the text is not included in Laum's collection.

89 Foundation of Diomedon: ll. 39–41 (γ[έ]ρη δὲ λαμβανέτω τοῦ ἱερέο̣[υ] ἑκάστου σκέλος καὶ τὸ δέρμα); ll. 101–103 (ἐ̣[φιέτω δὲ] καὶ ὁ ἱερεὺς εἰς τοὺς γάμους τὰ γέρ[η τῶι τὸ]ν γάμον ποιοῦντι); cf. also ll. 36–39 (sacrificial supply to be provided by the priest); foundation of Poseidonios: ll. 38–40 (ὁ δὲ

may suppose that the expression οἱ ἐκ τοῦ γένους, occurring after the mention of γέρα and [σ]πλάγχνων, refers to the principle of primogeniture in handing down the priesthood, if the incomplete sentence ὅταν δέ τι[–] σι τῶν ἐκ τοῦ γένους (ll. 9f.) can be compared to ll. 59–61 of Epikteta's inscription, where the succession in the priesthood is regulated in the case that something befalls the designated priest, Andragoras, son of Epiteleia (εἰ δὲ τί κα πάθῃ οὗτος, ἀεὶ ὁ πρεσβύτατος ἐκ τοῦ γένους τοῦ Ἐπιτελείας).

In Charmylos' foundation the family group is named *Charmyle(i)oi* after the founder, who therefore has an eponymic function.[90] The text provides no further information about the cult community because its function is limited to marking out the property consecrated to the Twelve Gods and to Charmylos as a hero.

An actual definition of the cult community is not present in Pythion's foundation, but the family statute of the sanctuary he founded can be inferred from ll. 15f., where it is stated that the sanctuary had to be owned in common by all his children (τὸ δὲ ἱερὸν ἔστω τῶν υἱῶν πάντων κοινόν). This fits the aforementioned fact that Pythion and the priestess with whom he acted are likely to have been husband and wife, as does the fact that the inscription makes reference to their joint testament (ll. 11f.).[91] There is no apparent reason to suppose, as Fraser does, that the expression υἱοὶ πάντες includes the offspring of the freedman Makarinos as well:[92] like in the case of Libys and his descendants in the foundation of Diomedon, Makarinos was essentially charged with the care of the sanctuary (cf. l. 6); "the other sacred and profane things" with which Makarinos was entrusted probably refer to activities relating to this practical sphere; even though the text does not specify what these consisted of, it refers to a "sacred tablet" where the requirements must have been explained in detail.[93] The task of ministering to "those who perform the joint sacrifice" (ll. 6–8: ὅπως ἐπιμέληται

ἱερ̣[ε]ὺς ‹λ›αμβανέτω ἑκάστου ̣ἱερείου κωλῆν καὶ τεταρτημ̣[ο]ρ̣ίδα σπλάγχνων, καὶ τῶν ἄλλων ἰσόμοιρος ἔσ̣τ̣ω̣); foundation of Epikteta: ll. 194–197 (οἱ δὲ ἐπιμήν[ιοι] οἱ θύοντες τὰς θυσίας ταύτας ἀποδωσο[ῦ]ντι τῶι κοινῶι τός τε ἐλλύτας πάντας κ[αὶ] τῶν σπλάγχνων τὰ ἡμίση, τὰ δὲ λοιπὰ [ἐ]ξοῦντι αὐτοί).

90 For the form *Charmyle(i)oi*, see Sherwin-White 1977, 214f.; for other cases of family groups named after the founder of the family tomb, cf. the μνῆμα Βουσελιδῶν known from D. XLIII 79 and intended for the burial of all of the descendants of the Athenian Bouselos, including members of his extended family (συγγενεῖς): cf. Marchiandi 2011, 24f.; 40f.; cf. also the family tomb built by Makrinos Philadelphos for the burial of the *Philadelphoi* (Ritti 2004, 274f., no. 9) and the funerary foundation of Makedonikos, which involved the members of the founder's family named *Makedonikoi*, along with a cult association (*Altertümer von Hierapolis* 153 = Laum 1914, 2, no. 181); both cases are from Hierapolis in Phrygia and date to the Imperial period.

91 Cf. above, fn. 29.

92 Fraser 1953, 44f.; the inscription, moreover, makes no reference to the existence of any children of Makarinos, unlike the inscription of Diomedon where the ἔγγονα of Libys are mentioned (ll. 4f.; 11f.).

93 Ll. 9–11: ἐπιμελέσθω καὶ Μακαρῖνος καὶ τῶν ἄλλων ἱερῶν καὶ βεβάλων καθάπερ καὶ ἐν τᾶι ἱερᾶι δέλτωι γέγραπται; for an interpretation of the phrase *hiera kai bebala* see Campanelli 2012a, who argues that a set of requirements concerning protection and use of Pythion's sanctuary was summarized therein.

τοῦ ἱερο[ῦ] καὶ τῶν συνθυόντων πάντων διακονῶν καὶ ὑπηρετῶν ὅσσωγ κα δῆι ἐν τῶι ἱερῶι)[94] also points to the exclusion of Makarinos from the cult community, because the task includes assisting sacrifice-performers without actually being a member of the group of the συνθύοντες. The verb συνθύω fits the shared sacrificial action as the way to cement socio-political and religious ties as well as kinship bonds, whether real or fictitious. The correlated substantive συνθύτης applies to private religious associations of the Hellenistic age gathered around the θυσία.[95] The terms συνθύτης, συνθυσία and συνθύω also appear in inscriptions concerning interstate relationships, where the joint sacrifice had the function of confirming existing alliances or συγγένεια bonds among cities.[96] In the context of household religion, which is our main concern here, the verb συνθύω occurs in a passage of Isaeus' judicial speech "On the Estate of Kiron".[97] The action of "sacrificing together" (συνθύειν) represents an argument for the legitimacy of the kinship between an Athenian named Kiron, who died without leaving sons, and the speakers, who claimed to be his natural grandsons and, accordingly, to have a right to his inheritance. The close collaboration between grandfather and grandsons in sacrificing to the domestic Zeus *Ktesios* is expressed in the text through three verbs with the prefix συν–, corresponding to as many shared ritual actions. Kiron attributed such a sense of family intimacy to this joint sacrifice that neither slaves nor freemen from outside the family were allowed to participate in it.[98]

Returning to Pythion's foundation, we can conclude that in this case the identity of the cult community passes through the sacrificial solidarity (συνθύειν) which matches the regulation concerning the sharing of the sanctuary by all the children.

---

94 The correct interpretation of ll. 7f. is due to Klaffenbach 1955, 123 who understands the two terms ΔΙΑΚΟΝΩΝ and ΥΠΗΡΕΤΩΝ as present participles in singular nominative referring to Makarinos (διακονῶν and ὑπηρετῶν: "ministering to" and "serving"), instead of plural genitives of the substantives διάκονος and ὑπηρέτης as Fraser thinks (διακόνων and ὑπηρετῶν); in fact, it would be very unlikely that Pythion founded a sanctuary in order to allow these two different categories of subordinate cult ministers to sacrifice therein; moreover, the functions of *diakonos* and *hyperetes* tend to overlap: both are classified by Poland 1909, 391f. among the sacrificial ministers; the two offices never appear together, for example, in the lists of cult personnel from north-western Greece: cf. Baldassarra 2010; the two terms are considered as synonyms in the literary sources: cf. Balsassarra/Ruggeri (2010) (with particular reference to the lists of cult personnel from north-western Greece); in some contexts, however, a slight difference in function between *diakonos* and *hyperetes* seems to be detectable: cf. Collins 1990, notably 167.

95 Cf. Poland 1909, 34; Baldassarra 2010, 355, T10; 364; Kloppenborg/Ascough 2011, 283f., no. 56 (with reference to other attestations).

96 Cf. Jones 1998; Weiß 1998.

97 VIII 15f.; this passage is often quoted as evidence for domestic sacrificial practice: cf. e.g. Nilsson 1955–1961, 1, 403f.; Jost 1992, 246; Mikalson 2005, 134f.; Parker 2005, 15f.; Faraone 2008, 216; Boedeker 2008, 230f.; Gherchanoc 2012, 74f.

98 For an exhaustive commentary on the whole passage see Ferrucci 2005, 167–174.

In this respect, the role of Makarinos can be compared to that of the domestic slaves: Kiron excluded them from participation in the sacrifice to Zeus *Ktesios* probably as a personal choice, but we cannot rule out the possibility that in other cases they were allowed to participate in domestic rituals, possibly acting as subordinate cult ministers.[99] This fits the meaning of the verbs διακονέω and ὑπερητέω, both referring to Makarinos' tasks within the joint sacrifice.[100]

Finally, there is no reason to suppose that a third group of individuals, namely, the ἐπιμελόμενοι, was involved in the foundation, as Fraser does on the basis of ll. 12–14 where "people who take care of the sanctuary and increase it" are mentioned.[101] Actually, these lines contain a blessing generically addressed to all those who will look after the sanctuary over time: such expressions are connected to the hope for the continued fulfilment of the founders' will, which is the common denominator of all foundations.[102]

4 The Sacred Space

As for Diomedon's foundation, the interpretation proposed above for the "sacred things" shared among the community members (τοὶ τῶν ἱερῶν κοινωνεῦντες) includes the consecrated properties, which clearly have a patrimonial meaning along with the more obvious sacral one. The goods listed in the initial dedication to Heracles comprise real estate (ll. 1–4): a precinct (τέμενος), guest houses (or rooms) within a garden (ξενῶνας τοὺς ἐν τῶι κάπωι) and other buildings (οἰκημάτια). The garden was a source of income because it was rented out to the freedman Libys and his children, who had to pay the rent in the month before that designated for the annual feast in honour of Heracles so that the money would be available for cult expenses

99 Cf. Bömer 1958–1963, 3, 475f.; 4, 941–950; Mikalson 2005, 156; Parker 2005, 15f.; Boedeker 2008, 231; Gherchanoc 2012, 74–76; slaves are sometimes present in Attic reliefs depicting family groups in the act of sacrificing: cf. Parker 2005, 37–42; cf. also Cohen 2011, 482f.

100 Both verbs, along with the correlate substantives διάκονος, διακονία, ὑπηρέτης and ὑπηρεσία, appear in reference to tasks accomplished by cult ministers especially in the context of sacrificial banquets: cf. Collins 1990, notably 156–168.

101 Fraser 1953, 44f.

102 Cf. Diomedon's foundation, ll. 115–119: τοῖς δὲ ἐπι[μ]ελομένοις ὅπως ἕκαστα συ̣[ντε]λῆται καθὰ διαγέγραπται [εἰ]ς δύναμιν εἶναι εὖ εἴη καὶ αὐ[τ]οῖς καὶ τοῖς ἐγγόνοις αὐτῶν; Poseidonios' foundation, ll. 51f.: τοῖς δὲ ταῦτα διαφυλάσσουσιν καὶ ποιοῦσιν ἄμεινον γίνοιτο ὑπὸ θεὸν καὶ ἄνθρωπον: this is Carbon's reading in Carbon/Pirenne-Delforge 2013, based on the stone; previous editors have changed the phrasing ὑπὸ θεὸν καὶ ἄνθρωπον to the more common genitive plural ὑπὸ θεῶν καὶ ἀνθρώπων: cf. Laum 1914 (see above, fn. 20); *LSAM* 72; for the phrasing actually present on the stone and for other kinds of inscriptions which include blessing formulas, cf. Carbon in Carbon/Pirenne-Delforge 2013, 114 (notes on ll. 51f.).

(ll. 11–17).[103] The next reference to funds occurs at ll. 69–80, where the *temenos*, the *xenones*, and probably the garden[104] are indicated as sources of income. The mention of the latter two is consistent with the initial regulation concerning the rent of the garden, where the *xenones* were located, whereas the addition of the *temenos* needs some clarification regarding the chronology of the text. In fact, based on palaeographic and linguistic considerations, the *editor princeps*, L. Ross, and later editors agree that the inscription was engraved in three stages.[105] Their chronology (end of the 4th century B.C. to about 280 B.C.) is established by Rudolf Herzog and accepted in all later editions, including the most recent one in *Inscriptiones Graecae* (2010).[106] It is significant that the regulation under discussion is placed right at the beginning of the third stage, probably due to the need for further clarification about the funds and their use. In this regard, a confirmation may come from the fact that the upkeep of the sanctuary had already been mentioned at ll. 47–51 (first stage), but without specifying the source of the πόθοδος used for the purpose. Accordingly, two possibilities are open: either the *temenos* was added to the sources of income in this third stage, or the *temenos* was part of the funds right from the outset, even though this was not explicitly stated. In any case, there may have been a financial rearrangement in the third stage. In fact, at the beginning the income from the garden is likely to have been directed to cult expenses concerning the annual feast in honour of Heracles (ll. 15–17: ἐς τὰν θυ[σίαν Πεταγειτ]ṿύọυ ἑκκαιδεκάται ḳ[αὶ ἑπτακαιδεκάται),[107] rather than for the maintenance of the sanctuary as stated in ll. 69–80; moreover, in these lines it is specified that the surplus, namely, the money not spent on maintenance, along with the ἐξαιρήματα ("amounts set aside for specific purposes")[108] had to be divided into parts intended for each sacrifice: this suggests that other sources of income were used beyond the garden and the *temenos*. In all likelihood, such additional funds were the plots of land (τεμένη) which are mentioned immediately after (ll. 81f.), within a set of prohibitions concerning the use of the spaces belonging to the sanctuary (ll. 80–86). The plural clearly distinguishes the *temene* from the *temenos* of l. 75, which must be identified with the precinct consecrated to Heracles at the beginning of the text. The prohibition against using these *temene* for agricultural purposes indirectly confirms

103 For the Coan calendar, in which the month of *Theudaisios* preceded that of *Petageitnyos*, see Paul 2013, 382.

104 ἀρ̣γύριον ἀπὸ τῶν προσόδ̣ων [τῶν ἀεὶ πιπτο]ṳσ̣ῶν ἀπὸ τοῦ τεμένευ̣ς [καὶ τοῦ κήπου] κα̣ὶ τῶν ξενώνων (ll. 74–76); the term *kepos* has been restored, but the mention of the garden is very likely, as it is in conjunction with the *xenones* which were located inside it (cf. l. 3).

105 Ross 1945, 45–54, no. 311; cf. Paton/Hicks 1891, no. 36; *Syll.*² 734; *Syll.*³ 1106.

106 Herzog 1928, 28–32, no. 10; cf. Segre 1993, ED 149.

107 The lines immediately following (17–23) are very damaged, but are likely to contain other duties of the freedmen referring to preparation for ritual performances (cf. above in this paper): the eventual expenses, therefore, must have been included in the specifications concerning the cult.

108 Cf. Klaffenbach, 1930, 214.

that they were actually suitable for farming. Besides indicating the sacred precinct as the divine dwelling and the place where the cult activities were performed, the term *temenos* can also refer to some agricultural land consecrated to a given deity as the financial support for its cult.[109] In the text under examination the prohibition against farming the *temene* is addressed to "those who share the sacred things", namely the members of the cult community, probably in order not to decrease the economic value of these plots, which are likely to have been rented out,[110] as was a common practice with land belonging to sanctuaries.[111] For the same reason, probably, "those who share the sacred things" were forbidden from dwelling in the *xenones* and in the *oikia* located within the *temenos* (ll. 82–84: μηδ᾽ ἐν τοῖς ξε̣[νῶσι] ἐνοικεῖν μηδ᾽ ἐν τῆι οἰκίαι τῆι ἐπὶ τ̣[οῦ τε]μένευς) as these buildings were evidently fit for residential use and could be rented out for this purpose. The indication concerning the location of the *oikia* within the *temenos* clarifies the way in which the *temenos* itself was a source of income (cf. l. 75) and allows this *temenos* to be further distinguished from the *temene* mentioned at l. 82: both were probably rented out, but the former was fit for residential use, the latter, probably, for agricultural use only. A rent contract from Rhamnous provides a close parallel for the rent of a *temenos* including an *oikia* where the tenant had to dwell (ll. 21f.: ο̣ἰ̣κ̣ήσει̣ [τ]ὴ̣ν̣ ο[ἰ]κίαν ὁ μισ[θω]σάμενος <τὴ>ν ἐπὶ τῶι τεμ̣ένει τούτωι).[112] The *temenos* belonged to the goddess Nemesis and was rented out by a group of demesmen for agricultural purposes.[113] Based on this instance, the possibility that the *temenos* of Diomedon's foundation was also fit for farming or something else cannot be completely ruled out, even though it was an actual cult place.[114]

The *oikia* itself turns out to be decisive in confirming that the *temenos* of ll. 75 and 83f. is to be identified with that consecrated to Heracles at the beginning of the text. This "house" is very likely the same *oikia* mentioned in ll. 56–59 as the place where the statues and the votive offerings were to be kept, since the inscription does not mention other *oikiai* apart from the οἰκία ἀνδρεία and the οἰκία γυναικεία (ll. 104–111), which are nevertheless easily recognisable because of the two adjectives matched

---

109 Cf. Parker 1983, 160–168; Hellmann 1992, 170–172; Cf. Horster 2004, 24f.; Lambrinoudakis et al. 2005, 308–310; for the semantic development of the term *temenos*, see Casevitz 1984, 84–87.

110 This convincing hypothesis goes back to Dittenberger: *Syll.*² 734, n. 30; *Syll.*³ 1106, n. 27.

111 See Isager/Skydsgaard 1992, 181–190; Horster 2004, 139–191; as an explanatory example of agricultural *temenos*, see the rent contract of the *temenos* (l. 36) of Zeus *Temenites* in Arkesine (Amorgos; 2nd century B.C.): see Brunet in Brunet/Rougemont/Rousset 1998, 222–231, no. 52 (text, translation, and commentary).

112 Petrakos 1999, no. 180.

113 See the commentary by Pernin 2007, 61–62.

114 The possibly multi-purpose nature of a *temenos* can be illustrated by a Coan inscription concerning the sale of the priesthood of Zeus *Alseios*: see below; sanctuary buildings could also be rented out for banquets performed by cult associations or family groups: cf. Leypold 2008, 197–199; the author also suggests that the *oikemata* of Heracles *Diomedonteios*' sanctuary were dining rooms.

in this case with the substantive οἰκία. As the place dedicated to keeping ἀγάλματα and ἀναθήματα, the *oikia* can be a *thesauros*, but, besides this or as an alternative to this function, it must be considered a shrine, since the ἀγάλματα were used in the rituals performed by the cult community and were therefore actual divine *simulacra*.[115] Accordingly, the *temenos* where the *oikia* lay is to be interpreted as a sacred precinct. The text provides confirmation, moreover, that the *oikia* and, consequently, the *temenos* itself must have already existed with the same function as here in the original foundation statute. Indeed, the regulation concerning the keeping of the statues and votive offerings occurs right at the beginning of the second stage of the text and is presented as ratification of a praxis already in force: "the statues and the votive offerings should be kept in the house on the spot as it is now" (ll. 56–59: τὰ δὲ ἀγάλματα καὶ τὰ ἀναθήματα ἔστω ἐν τᾶι οἰκία[ι] ḳατὰ χώραν ὥσπερ καὶ νῦṿ ἔχει). Up to that time, therefore, the "house" had been dedicated to this function, but had not been mentioned before, probably because its existence was implied in the dedication of the *temenos* at the very beginning of the inscription.

These conclusions have consequences for understanding the value of the sanctuary founded by *Diomedon* as family cult place and assets as well: based on this, the sharing of the *hiera* among the community members gains a patrimonial meaning too.

The practice of using cult places as financial support for maintenance and cult expenses is not rare, especially when sanctuaries belonging to religious associations are involved.[116] Significant evidence is provided by a well-known 4th century B.C. inscription from Athens that refers to the rent contract of the private sanctuary of Egretes, a hero worshipped by a cult association of *orgeones*.[117] The tenant Diognetos was allowed to use the buildings belonging to the cult place, provided that he made the "*oikia* where the *hieron* lies" available (ll. 26–28: παρέχειν Διόγνητον τὴν οἰκίαν, οὗ τὸ ἱερόν ἐστι, ἀνεωιγμένην) when the *orgeones* celebrated the annual feast in honour of their patron hero. The term ἱερόν raises interpretative problems, because here it

115 The hypothesis that this *oikia* was a *thesauros* goes back to Poland 1909, 460f., who also excluded the possibility that it could be rented out; that it functioned as a temple is argued by Svenson-Evers 1997, 142; the ritual function of the ἀγάλματα emerges from ll. 95–97: the statues, probably those of the other deities worshipped by the cult community along with Heracles, were arranged around the *kline* prepared for the god on the occasion of the *xenismos*; on the same occasion, the statues were probably decorated with the gold crowns mentioned in the inventory of ll. 120–130: στεφανίσκους πέντε τοῖς ἀγάλμασιν χρυσοῦς (ll. 124f.); from this, one could also infer that the statues were five in number; for the role of the divine *simulacra* in the *theoxenia* rituals, see Bettinetti 2001, 211–231 (223–227, in particular, for the *xenismos* in Diomedon's foundation); cf. also Campanelli 2011, 658; for criticism of the distinction between cult statue and votive statue, see Donohue 1997.

116 Cf. Svenson-Evers 1997, 139; this aspect is well illustrated by the documentation concerning the Attic *orgeones*: Horster 2004, 162–164; Papazarkadas 2011, 191–197.

117 *IG* II^2 2499 (307/306 B.C.).

cannot indicate the sanctuary as a whole, as it does in other points of the text, where this is undoubtedly the right meaning (ll. 2f.: [Ο]ἱ ὀργεῶνες ἐμίσθωσαν τὸ ἱερὸν το[ῦ] Ἐγρέτου; ll. 5–7: χρῆσθαι τῶι ἱερῶι καὶ ταῖς οἰκία<ι>ς ταῖς ἐνωικοδομημέναις ὡς ἱερῶι). Among the proposed interpretations are "cella in qua simulacrum herois est"[118] and "object sacré".[119] If ἱερόν indicates a divine symbol or the internal room where the divine image was housed, the *oikia* must be a shrine similar to the *oikia* of Diomedon's foundation. Evidently, this function did not conflict with the use of the sanctuary as a source of income. Indeed, the *orgeones* retained the right to use their sanctuary for the annual religious meeting in honour of Egretes. On that occasion, besides the shrine, they also had at their disposal other spaces of the sanctuary, as is suggested by the clauses concerning the obligations of the tenant (cf. ll. 24–30). The furniture and facilities he had to provide for the annual feast included dinner-couches and tables for two dining rooms, a kitchen and a temporary covered structure (l. 28: στέγη),[120] which was evidently set up somewhere within the open space of the sanctuary.

A similar management can be supposed for the *oikia* and other rented out structures of Diomedon's foundation. The *xenones*, in particular, were probably inhabited by Libys and his offspring, who paid rent for the garden where these buildings stood. In fact, literary and epigraphic evidence[121] suggests that the term ξενών indicates rooms or free-standing buildings suitable for both convivial and residential use. Furthermore, the duties of Libys and his offspring, which included providing furniture for religious celebrations, were similar to those of the tenant Diognetos. It is possible, therefore, that the *xenones* were made available for the annual feast in honour of Heracles to allow the cult community to use them for the banquets mandated by the statute of the foundation.[122] The text offers further evidence that buildings usually dedicated to residential use could be turned into meeting places for the cult community. This happened on the exceptional occasion when one of Diomedon's descendants in the male line was allowed to celebrate his wedding in the sanctuary

---

118 *Syll.*² 937, no. 4; *Syll.*³ 1097, n. 3; cf. Kloppenborg/Ascough 2011, 48–52, no. 7.

119 Hellmann 1999, 104.

120 This is the interpretation of Hellmann 1999, 104.

121 See the sources collected by Campanelli 2011, 661f.; a parallel for a *xenon* attached to a garden and used as a dwelling house is provided by the ξενὼν πρὸς τῷ κήπῳ mentioned in the testament of Aristotle (D.L. V 14) as the property bequeathed to his concubine Herpyllis should she decide to reside in Chalcis (ἐὰν μὲν ἐν Χαλκίδι βούληται οἰκεῖν); as an alternative, she could choose to dwell in the paternal house of the philosopher in Stagira.

122 The term which might refer to a banquet is συν]αιγλία (ll. 90f.), but convivial activities can also be implied in the terms διαγ̣[ομήν] (l. 91) and συνα]γωγή (ll. 93f.; cf. l. 141: συνάγειν): see Campanelli 2011, 659f.; cf. also ἀρ̣τ̣οφαγ[ίαν (ll. 37f.); in any case, the *thysia*, which was performed on the 16th of the month of *Petageitnyos* (cf. ll. 15–17, 59f.), is a kind of sacrifice which generally entailed a banquet: cf. e.g. the overview of Van Straten 2005; a reference to the portions of sacrificial meat distributed among the participants can be seen in the term μερίδας of l. 158: cf. the same term in Poseidonios' foundation (ll. 42f.).

immediately after the days dedicated to the feast of Heracles (the 16th and 17th of the month of *Petageitnyos*).[123] In such cases, the holders of the οἰκία ἀνδρεία and the οἰκία γυναικεία were required to make the two houses available, storing the σκεύη in some other building (*oikemata*: ll. 104–108). The use of the participle ἐκτ[η]μένοι to indicate the holders of the two *oikiai* suggests that their juridical status was different from the shared ownership that characterized the rest of the real estate mentioned throughout the text. As Dittenberger argues, it is probable that the two houses had been inherited by some of Diomedon's descendants in their private capacity.[124] The residential function of these buildings is further suggested by the term σκεύη, which is likely to refer to domestic furniture.[125]

The adjectives ἀνδρεία and γυναικεία point to a gender separation, probably part of the logistic rearrangement necessary on the occasion of a wedding. In fact, when a wedding celebration was scheduled immediately after the annual feast of Heracles, the main Heraclean rituals, namely the sacrifice (*thysia*) and the entertainment of the god at a banquet (*xenismos*), had to be performed in the οἰκία ἀνδρεία (ll. 108–111) and the *xenismos* had to remain fitted out while the wedding was celebrated (95–98), as if the god were participating in the nuptials by acting as the protector of Diomedon's family.[126] This suggests that the women were excluded from Heraclean rituals,[127] but were permitted to participate in the concomitant wedding celebration within a building reserved for them, the οἰκία γυναικεία.[128]

The realty of religious associations not infrequently included *oikoi* and *oikiai* which attest to a functional flexibility similar to that observed in the case of the three *oikiai* connected, albeit in different ways, to the sanctuary founded by Diomedon.[129] An Athenian inscription, on the other hand, provides evidence for a case of conversion of a private house into divine property: in accordance with an oracle, a certain

---

123 For an interpretation of the problematic passage concerning the wedding celebrations in the sanctuary (ll. 86–111), see Campanelli 2011, 659–661.

124 *Syll.*² 734, n. 36; *Syll.*³ 1106, n. 34.

125 Cf. the already-mentioned passage of Aristotle's testament (above, fn. 121) where the term σκεύη indicates the furniture needed to equip the house chosen by Herpyllis as her dwelling; the testament of Theophrastus is even more explicit because it contains the expression οἰκηματικὰ σκεύη, which indicates the pieces of domestic furniture bequeathed to the freedman Pompylos; in Ath. 170e the τραπεζοποιός is the minister charged with washing the tableware, which is indicated by the term σκεύη; see also Andrianou 2006, 251f.

126 This convincing observation goes back to Paton/Hicks 1891, 76; cf. also Campanelli 2011, 670.

127 For other cases of female exclusion from Heraclean rituals, see Graf 1985, 298–316 (especially 308 with fn. 105); Cole 1992, 105–107.

128 For this convincing hypothesis, see Jameson 1994, 42.

129 Cf. Poland 1909, 459–463, 485–487; for (*hierai*) *oikiai* functioning as shrines, see Lauter 1985, 159–177 (with particular reference to gentilicial groups); for *oikai* belonging to Attic phratries and *gene*, cf. Papazarkadas 2011, especially 164–166, 174, 177f.; an *oikia*, whose function is not specified, belonged to the *temenos* of Zeus *Alseios* on Cos: *IG* XII, 4, 1, 238, ll. 21f. (on this inscription see below).

Demon dedicated his *oikia* with adjacent garden to Asclepius and became his priest. The inscription does not provide clues to the intended use of this *oikia*, nor does it indicate whether the owner continued to dwell therein, but it is likely to have been rented out to contribute to cult expenses.[130]

The foundation of Charmylos offers further opportunity for analysis of the values of the spaces, as far as it is possible to draw inferences from the inscription on a boundary stone, which is the only preserved document pertaining to this foundation. But the very nature of this document and the archaeological context where it was found make it possible to gather new elements, which turn out to be crucial for the purpose of this discussion.

Fig. 5: The funerary crypt of the *Charmyleion* and remains of the ground floor embedded into the walls of the Church of the Stavros, Pyli, Cos (photos S. Campanelli).

The inscription was found among the ruins of an ancient building in a locality currently called Pyli. The remains still visible today are an underground vaulted chamber with twelve funerary *loculi* along its sides and part of the walls belonging to the ground floor, which have been embedded into the construction of a small medieval church dedicated to the Stavros (fig. 5). Based on these remains and on the architectural fragments found *in situ*, in 1934 Schazmann proposed a reconstruction of the building as a temple-tomb made up of three levels: the funerary crypt, a ground floor elevated on a high podium, and an upper storey shaped like an ionic shrine *in antis* (fig. 6).[131] The monument has, understandably, been connected with the inscrip-

130 *IG* II^2 4969 (4th century B.C.?); cf. Papazarkadas 2011, 43f.

131 Schazmann 1934; the *Charmyleion* is often included in studies on funerary architecture of the Hellenistic age as one of the best preserved examples of private *heroon*: cf. Scholl 1994, 261–266; Kader 1995, 201f.; Berns 2003, 21f.; Fedak 2006, 84. Roux 1987, 111–113, questions the existence of the second storey because of the lack of evidence of stairs leading up; he has also questions the division of the ground floor into two twin-rooms, considering that the dividing wall would have covered the holes made in the floor in order to pour libations into the underlying funerary chamber.

tion found *in situ* and currently built into the façade of the Stavros church. Both have been dated between the end of the 4th and the beginning of the 3rd century B.C. The structure of the monument as reconstructed by Schazmann suggests a twofold cult—divine and funerary—practised by the family community: the Twelve Gods and the heroized founder Charmylos are explicitly mentioned in the text, but it is likely that the statute of the foundation also established a cult for the deceased family members buried in the crypt, putting them under the protection of the deities chosen as the family patrons.

This concurrence of archaeological and epigraphic data provides a rare opportunity to compare the architectural terminology and the building to which the inscription actually refers in order to understand the structure of the cult place and the meaning of its various components. According to its function, the boundary stone marked the properties consecrated to the Twelve Gods and Charmylos: "land", "*oikia* on the land", "gardens", "*oikiai* in the gardens" (ἱερὰ ἁ γᾶ καὶ ἁ οἰκία ἁ ἐπὶ τᾶι γᾶι καὶ τοὶ κᾶποι καὶ ταὶ οἰκίαι ταὶ ἐπὶ τῶν κάπων Θεῶν Δυώδεκα καὶ Χαρμύλου ἥρω *vac.* τῶν Χαρμυλέων).

The *oikia* mentioned first and in the singular seems to distinguish itself from the plurality of the *oikiai* lying in the gardens. It might be identified, therefore, with the temple-tomb partially preserved *in situ*, as is also suggested by the parallels with both Diomedon's foundation and the rent contract of Egretes' sanctuary, where the term *oikia* is very likely to refer to shrines. The latter text, in particular, also provides evidence for the concomitant presence of a single *oikia*-temple and a number of other *oikiai* also lying within the sanctuary. Similar to the Athenian rent contract, which allowed the tenant to use these *oikiai* (albeit taking into account the sanctity of the place; cf. l. 7: ὡς ἱερῶι), it is possible that in Charmylos' inscription the *oikiai* and the gardens represented the funds on which the foundation was based.[132] This would also match the rent of the garden in the foundation of Diomedon, obviously including the buildings which were overlooking it, the *xenones*. An even closer comparison for *oikiai* meant as a source of income is provided by another foundation text from Hellenistic Cos,[133] which is not specifically dealt with in this paper because the city in this case is involved as both the authority in charge and the beneficiary of the foundation. A certain Phanomachos financed a civic cult of Zeus and the *Damos* along with the building of their sanctuary through the income from some land and a number of *oikiai* and other structures located there. The formulation adopted by the text offers almost a literal parallel for Charmylos' inscription:

132 For the rent of the *oikiai* belonging to Apollo on Delos and for their various functions cf. Hellmann 1992, 291–294; in general, rent contracts of agricultural properties including one or more *oikiai* are well documented: cf. Chandezon 1998, 45f. (rent contracts of farmsteads from Mylasa); Pernin 2007, 51f.; for the κῆπος as a vegetable garden and for its economic use cf. Vatin 1974; Hellmann 1992, 207–210.
133 *IG* XII 4, 1, 79 (early 2nd century B.C.).

ἀνιεροῖ τῶι τε Διὶ καὶ τῶι Δάμωι καὶ δίδωτι τὰς <γ>ᾶς τὰς αὐτοῦ τάς τε ἐν Ἀκόρναι καὶ τὰς οἰκίας τὰς ἐπὶ ταῖς γαῖς καὶ τἆλλα τὰ <ἐ>ν ταῖς γα[ῖς] πάντα, αἷς γείτονες Ἀριστοκλῆς Χαρμίππου, Ἁγησίας Ἁγησία κατὰ φύσιν δὲ Γενοκλεῦς, Θρασύανδρος Ἁγησία⟦ς⟧, καὶ τὰν γᾶν τὰν ἐν Αἰγήλωι καὶ τὰν οἰκίαν τὰν ἐπὶ τᾶι γᾶι καὶ τἆλλα τὰ ἐν τᾶι γᾶι πάντα καὶ τὰ προσόν̣τα τᾶι γᾶι ταύται π[ά]ν̣τ̣[α] ἇι γείτονες [Θε]ύδωρος Χαιρεδ̣[άμο]υ̣, Δ̣ιομέδω̣[ν Πυ]ρρίχου, . . - - - - - - - - - - - - (ll. 7–19).

[Phanomachos] consecrates and gives to Zeus and the *Damos* his plots in Akorna and the houses on the plots and everything else in the plots, which have as the neighbours Aristokles the son of Charmippos, Hagesias the son of Hagesias, natural son of Genokles, Thrasyandros the son of Hagesias, and the land in Aigelos and the house on the land and everything else in the land and everything that belongs to this land, which has as the neighbours Theudoros the son of Chairedamos, Diomedon the son of Pyrrichos, . . - - - - - - - - - - - -

There is, however, a substantial difference. The land and buildings consecrated by Phanomachos must have been located in rural areas different from the place where the inscription was set up, as can be inferred from the identification of the allocated properties by means of toponyms and names of the neighbouring owners. The cult activities, moreover, are likely to have taken place in the city, since a cult such as that of Zeus coupled with the personified *Damos* of the Coans must have had a strong civic value.[134] This would be consistent with the findspot of the inscription in the city of Kos, where the sanctuary of Zeus and the *Damos* was probably also built.[135] Accordingly, the landed properties listed in the text had an exclusively financial value, unlike Charmylos' land and gardens, which must have lain in the same place where both the inscription delimiting them was displayed and the cult place was located. If this hypothesis is viable, a rural sanctuary can be envisaged, with the temple-tomb and other buildings (*oikiai*) surrounded by landed properties that comprised the financial basis of the foundation.[136] This would also fit the location of the *Charmyleion*, which lies in an area that even now is scarcely urbanized.

---

134 Cf. Paul 2013, 43, 157f., 283; she is incorrect, however, in considering the plots listed in ll. 7–19 as the place where the sanctuary was being built (157): first, the consecration of already-existing buildings along with the land can have no other aim except allocating them as sources of income (for the rent of rural *oikiai* lying on landed properties see above, fn. 132); secondly, in ll. 35f. (πλὰν τῶν γραφομένων χάριν τᾶς κατασκευᾶς τοῦ ἱεροῦ) the participle τῶν γραφομένων implies χρημάτων (cf. *IG*, n. *ad loc.*) and refers, therefore, to the funds and not to the place where the sanctuary was being erected; finally, the verbs ἀνιερόω and δίδωμι, which introduce the list of the properties, are common in foundation texts with reference to the funds allocated by the founders for financing the established activities (ἀνιερόω, in particular, when the foundations have cult purposes): cf. Laum 1914, 1, 120–122.
135 Cfr. *IG* XII 4, 1, 79: "inventum in fundamentis curiae urbanae"; the urban location of the cult would be even more probable if the restoration proposed for l. 29, referring to the route of the procession, was correct: ἐκ τ[οῦ πρυτανείου ἐς] τὸ ἱερόν, even though processions starting from the city and destined for extra-urban sanctuaries are well known: see in general True et al. 2004.
136 One can wonder whether a fragment from Χαρμύλι (*IG* XII 4, 1, 392; Hellenistic period?), as the area around the Stavros church in Pyli is called by the locals (cf. Ross 1845, 45), refers to the *Charmyleion*: in fact, the words γεωπόνου (l. 1) and δένδρον (l. 2) fit a sanctuary which included culti-

The *horos* explicitly claims the sacredness of the place, including buildings and land that were likely used as a source of income. The marking out of the estate that became divine property through the foundation is known from other documents which show, like the Coan one, the definition of the borders as a procedure closely connected with the consecration itself, since the boundary stones made *hieros* the space they physically separated from the surrounding area.[137] A *horos* records one of the most ancient foundations epigraphically known, that established by a certain Archinos on Thera. The inscription on the boundary stone states its function of demarcating a plot consecrated to the Mother of the Gods (ll. 1f.: οὖροι γᾶς Θεῶν Ματρί) and concisely regulates the sacrifices to be performed twice a year, possibly by using the land revenues which may have included the harvest itself (cf. ll. 14f.: ἐπάργματα ὧν αἱ ὧραι φέρουσιν).[138] A similar procedure is attested by a literary source, the well-known passage of Xenophon's *Anabasis* concerning the foundation in honour of the Ephesian Artemis, which the author himself established in the Peloponnesian locality of Scillous, where he bought a plot in order to build a sanctuary for the goddess and fund the related cult activities. According to the text, a stele placed near the temple recorded the consecration of the estate to Artemis (ἱερὸς ὁ χῶρος τῆς Ἀρτέμιδος) and included concise regulations concerning the use of the land revenues for cult and maintenance expenses.[139] Finally, a very telling inscription about this point is the "Testament of Epikrates" from Lydia, which might fall into the developments of family cult foundations in the Imperial age.[140] The text deals with the funerary cult established by the founder in honour of his deceased and heroized son Diophantos. For this purpose, Epikrates allocated a number of plots which are carefully listed and identified by means of landscape features, toponyms, and names of neighbouring owners. The formula used to introduce each plot highlights the simultaneous process of marking out the boundaries and consecrating the land (ll. 3f.; 11f.; 14f.: ὁμοίως ἀφώρισα καὶ συνκαθωσίωσα τῷ μνημείῳ). The definition of the boundaries involves the physical action of arranging boundary stones on the ground, as the text seems to make explicit through the expression "up to the *horoi* I have marked" (l. 31: μέχρις ὧν σεσημείωμαι ὅρων). The sacred status of the demarcated land seems to stem from its connection to the funerary monument, which can be seen as a *locus religiosus*

---

vated land, as the *Charmyleion* must have been; the letters κοψΙ[- - - - - - - - - may refer to a form of the verb κόπτω, which appears in inscriptions concerning the protection of the sanctuaries (τῶν δένδρων μη]θὲν κόψη[ι, *IG*); cutting down trees was generally forbidden or only permitted under certain conditions: cf. Dillon 1997, 115–122; Horster 2004, 110–120.

137 Cf. Horster 2004, 23–33; Naerebout 2009, 201–208; Brulé 2012, 83–89.

138 *IG* XII 3, 436 (4th century B.C.); cf. Mannzmann 1962, 127f.

139 *An.* 5.3.4–13; due to its documentary value, this passage has often drawn the attention of scholars, including Purvis 2003, 65–120, and Brulé 2012, 98–109.

140 Herrmann/Polatkan 1969, 7–36, no. 1; see Campanelli 2012b for the possibility of including this inscription in the framework of family cult foundations.

since it was the dwelling of the hero Diophantos. This is all the more significant if one considers that the tomb, also intended for the future burial of Epikrates and other members of his family, is likely to have lain on the landed properties the founder had marked out and consecrated: a spatial arrangement already adopted by Charmylos some centuries before.[141]

Consecration was a security measure to which the founders occasionally resorted in order to protect the foundation's capital, even when the main purpose of a foundation was not cultic in nature.[142] This role of consecration emerges all the more clearly in such cases as that of the *Charmyleion*, where a physical overlap occurs between the ritual space and the real estate (land, gardens and *oikiai*) that likely represented the foundation's capital. In this respect, Charmylos' boundary stone seems to make explicit what is probably implied in the expression τοὶ τῶν ἱερῶν κοινωνεῦντες of Diomedon's foundation, where, again, the sharing among the community members can be extended to the physical space of the sanctuary meant as a twofold inheritance, based on the overlap between ritual space and family assets.[143] In the light of the reconstruction proposed above for the sanctuary founded by Charmylos, the one founded by Diomedon can be similarly viewed as a rural cult place lying on the family's landed estate.[144] The *temene*, indeed, might have lain in the immediate surroundings of the sanctuary since no identifying topographical references are given. The localization in the countryside might also be confirmed by the findspot of the inscription, which comes from the garden of a private house located in a suburb of Kos city, where it had been built into some modern stairs, according to the only information provided by the *editor princeps*, L. Ross.[145]

Based on the architectural terminology recorded in the inscription, the sanctuary founded by Diomedon can be envisaged as similar to but perhaps more complex than the *Charmyleion*, even though it is not possible to verify whether the former included a burial place for family members, since the preserved text only provides evidence for the cult of the ancestors, but does not allow for understanding of what it entailed.

141 Cf. Campanelli 2012b, 75–83, where a close connection between Diophantos' heroization and land ownership is also argued.

142 Cf. Laum 1914, 1, 169.

143 This patrimonial concept of the "sacred things" can be added to the various meanings of the neuter plural τὰ ἱερά which emerge from the Coan documentation according to the recent analysis of Paul 2013, 340–344 (cf. also 196–203): among the uses of the term pinpointed by the author, the one which comes closest to the interpretation proposed above is "affaires sacrées", but in the foundations of Diomedon and Charmylos this meaning seems to acquire a more material nuance, if it includes the sacred space meant as family assets to be managed in common by the cult community.

144 Cf. Campanelli 2011 for criticism of the hypothesis of De Matteis 2004, 191–196; 207; 219; 231f., who identifies the sanctuary founded by Diomedon with the sanctuary of Heracles located in the Harbour Quarter of Kos city, namely, a public and strongly urbanized area; De Matteis' hypothesis is also rejected by Paul 2013, 111–113.

145 Cf. above, fn. 105.

The sanctuary of Heracles *Diomedonteios* comprised a *temenos* including an *oikia*-temple, a garden with *xenones* overlooking it, and other buildings (*oikematia* and *oikemata*), some of which might have had service functions[146] as is evidently the case for the *oikemata* in which domestic furnishings were stored when the οἰκία ἀνδρεία and the οἰκία γυναικεία had to be made available for ritual and wedding celebrations. The latter buildings, as argued above, were probably private property, but their adjacency to the sanctuary is likely, because on these particular occasions the furniture needed for Heraclean *xenismos* (dinner couch and related coverings, table, god statues and so on) could be transferred to the οἰκία ἀνδρεία and, at the same time, the σκεύη which usually stood inside the two houses were moved to *oikemata* certainly belonging to the sanctuary itself, as buildings named like this are mentioned in other points of the text (ll. 44; 48). Moreover, it can be supposed that the original sanctuary cluster expanded over time, since buildings not previously mentioned first appear in the prohibitions of ll. 80–86, which belong to the third stage of the text. Besides being forbidden to farm the *temene* and dwell in the *oikia* of the *temenos* and in the *xenones*, the members of the cult community were not permitted to use the building located within the *hieron* (ll. 84f.: τῆ[ι λέσ]χηι τῆι ἐν τῶι ἱερῶι)[147] and the *peripatos* as an ἀποθήκη ("storehouse" or "refuge"). The only exception to these regulations was in the case of war. The *hieron* is to be understood here as a sacred precinct, because at least one building, the *lesche* (?), lay within it; on the other hand, it cannot be the same as the *temenos*, because the two terms both appear in this passage and the prohibitions which concern each of them are different, and must therefore refer to different spaces. The only possibility is that the *hieron* was a second sacred precinct attached to the sanctuary in addition to the *temenos*.[148] There is evidence, in fact, for the concomitant presence of the two terms *temenos* and *hieron* with reference to the same sanctuary. In some cases this might have a conceptual explanation, since *temenos* seems to indicate the sanctuary as a physical space delimited by an enclosure, whereas *hieron* refers to the same space but in a ritual sense. This is consistent with the literal meaning generally attributed to the term *temenos* as a piece of land "cut off" and "set aside" to become the property of a god independently of the use for which it was intended.[149] The term *temenos*, therefore, can also indicate the area of the sanctuary providing income directed to cult expenses, whereas *hieron* is the area where the deity dwelt and the cult was practised. The functional distinction between *temenos* and *hieron* can also be based on cultic need when the two terms are used to distinguish areas reserved for different cults or cult activities within the same

---

146 This interpretation is also supported by the generic meaning of the term οἴκημα, which could indicate various kinds of buildings fit for different functions: cf. Hellmann 1992, 288–290.

147 As an alternative, αὐλ]ῆι was proposed by Paton/Hicks 1891 (cf. above fn. 105).

148 For this hypothesis see Campanelli 2012, 665–667.

149 Cf. the bibliographical references quoted above, fn. 109.

sanctuary or to define parts of it characterized by a higher degree of sacredness.[150] This might have happened especially when a sanctuary was affected by expansion over time.[151] A distinction between *temenos* and *hieron* for cultic reasons can also be hypothesized for the sanctuary of Heracles *Diomedonteios*, considering that the *temenos* was an actual cult place and therefore cannot be seen as a mere piece of land whose revenues were being used for cult expenses. Accordingly, it can be supposed that the two sacred spaces were needed to physically separate cult areas reserved for some of the deities belonging to the *pantheon* created by Diomedon; as an alternative, the *hieron* can be thought to have become, perhaps at a later stage, the area dedicated to the sacred meals, if the construction lying inside it was actually a *lesche*, a building which generally appears in sanctuary contexts as a meeting place that can also be used for banquets.[152] The existence of areas intended for different uses within the same sacred precinct is well illustrated by the Coan inscription concerning the sale of the priesthood of Zeus *Alseios*.[153] The purchaser was granted the right to enjoy the *temenos* (ll. 18–20: καρπεύσεται δὲ καὶ τὸ τέμενος καθότι καὶ ὁ πρότερον ἱερεὺς ἐκαρπεύετο),[154] but the text specifies that a particular area was reserved for rituals in honour of a goddess who can be identified as Athena *Alseia* (ll. 20–24).[155] This provides evidence not only for the economic use of (part of) a *temenos*, but also for its internal arrangement in distinct areas (albeit not always architecturally separated) as a function of cultic need.

Temenos and *hieron* would appear together again in Pythion's foundation, if the restoration proposed in the *IG* edition for the first preserved line of the text is correct: ⟦*rasura*?⟧ τὸ τέ[μενος τόδε ἔστω] *vacat* ἱερὸν Ἀρτέμιτο[ς – – –]ας καὶ Διὸς Ἱκ[ε]σίου καὶ Θεῶν πατρώιων (ll. 1–3). This reading raises problems because the *temenos* is no longer mentioned in the continuation of the text, where the term *hieron* appears in all

150 For the various possible relationships between *temenos* and *hieron*, especially when the two terms appear together in the same inscription, see Casevitz 1984, 85–87; Le Roy 1986, 285f.; Cole 2004, 59–63; Patera 2010, 538–547; one of the inscriptions most frequently mentioned when discussing the semantic and functional relationships between *temenos* and *hieron* is *IG* I^3 84 (418/417 B.C.); it refers to the Athenian sanctuary of Neleus, Basile and Kodros, which consisted of two separate but adjacent spaces: the *hieron*, where the cult was practised, and the *temenos*, rented out for farming: cf. Wycherley 1960, especially 62.

151 For the development of sanctuaries over time, cf. Bergquist 1967, especially 57–61; Hellmann 2006, 193–196.

152 Cf. Leypold 2008, 13; 174f., 200f.

153 *IG* XII 4, 1, 328 (1st century B.C.).

154 For the right to enjoy landed properties and for the use of the verb καρπεύω in this context, see Velissaropoulos-Karakostas 2011, 2, 70–73; note that this verb appears in Poseidonios' foundation with reference to the eldest son of the founder, who was granted the right to enjoy the properties allocated for cult purposes (ll. 18f.; 28); the related substantive καρπεία is present in Epikteta's foundation (l. 72).

155 Cf. Paul 2013, 49f., 58.

passages involving the sanctuary: its care (l. 6), the supply for the joint sacrifice performed therein (l. 8), the wish for the cult's everlasting prosperity (l. 13), and finally, the sharing of the sanctuary among all of the founder's children (l. 15). This is probably the reason behind Fraser's proposed restoration:[156] [ἱερὸν ἔστω τόδε] τὸ τέ[μενος καὶ τὸ] *vac.* 5 ἱερὸν Ἀρτέμιτο[ς . . . .]ας καὶ Διὸς Ἰκ[ε]σίου καὶ Θεῶν Πατρωίων, where, nevertheless, the tautology ἱερὸν ἔστω ...τὸ ἱερόν is scarcely convincing.

On the assumption that both a *temenos* and a *hieron* are involved in this foundation, one may hypothesize that *hieron* has the restricted meaning of "temple", lying within the sacred precinct represented by the *temenos*:[157] in this case, however, it would be difficult to understand why all of the above-mentioned regulations only refer to the temple, without involving the sacred precinct as a whole, since the latter, with its altars and other possible structures, generally played an essential role in ancient Greek cult practices.[158] A preferable alternative might be to suppose that both terms refer to the sanctuary space, but with a conceptual differentiation between the two. Thus, *temenos* might indicate the extent of the sanctuary in a physical sense, and *hieron* might refer to the same space in a ritual sense, or to the part more specifically reserved for cult performances. This seems to be the case in an inscription from the *Amphiareion* at Oropos that forbids anyone to carry the meat of the victims sacrificed within the *hieron* outside the *temenos*:[159] it is clear that there is a conceptual distinction between the ritual space of the sanctuary (*hieron*), where the sacrifices were performed, and its total extent (*temenos*), whose perimeter was the limit beyond which the consumption of the sacrificial meat was forbidden.[160] A similar conception of the spaces might be seen in the aforementioned *temenos* of Zeus *Alseios* on Cos. This interpretation might explain the apparent inconsistency of the *IG* reading of ll. 1–3 of Pythion's inscription, where there is a mention of a *temenos* that is no longer involved in the rest of the text. Its content, in fact, seems to be consistent with the *hieron* meant as the ritual space of the sanctuary: the regulations engraved on the stone specifically concern the religious sphere; as for the rest, the text refers to other documents where the "profane things"[161] were regulated as well. Moreover, it should be noted that the last lines of the inscription contain requirements of ritual purity to gain access to

---

156 Fraser 1953, 36, 37f. (notes on these lines).

157 For *hieron* meant as a temple cf. Hellmann 1992, 170; Cole 2004, 40; Patera 2010, 546f.; cf. also above for the meaning of *cella*, which has been attributed to the *hieron* located within the *oikia* of Egretes' sanctuary.

158 See in general Sinn in Sinn et al. 2005, notably 1–4; for the relatively late appearance of the temple in cult places, cf. Bergquist 1967, 132; Burkert 1996.

159 *I.Oropos* 277, ll. 29–32.

160 Cf. Le Roy 1986, 285f.; in general, for the prohibition against taking away the sacrificial meat and the obligation to consume it "on the spot", that is, within the sanctuary, see Paul 2013, 358–364 (with a review of the scholarship on the subject).

161 Cf. above, fn. 93.

the sacred place (ll. 15–17): regulations of this kind are typical of the inscriptions set up at the entrance of sanctuaries in order to communicate the conditions needed to enter.[162] The same function, accordingly, has to be attributed to the Coan one (cf. in particular l. 15: ἁγνὸν εἰσπορεύεσθαι). On the other hand, the mention of the *temenos* in l. 1 becomes understandable when one considers that ll. 1–3 contain the dedication of the sanctuary and therefore necessarily concern the total extent of the consecrated area (the *temenos*), possibly including the part used as a source of income. What might be inconsistent with this interpretation is the regulation which states that the *hieron* belongs to all of the founder's children, who would be expected to have owned in common all of the properties of the sacred place, that is, the *temenos* as a whole. It can be supposed, however, that the children are regarded here as the cult community of "those who sacrifice together" (cf. l. 7: συνθύοντες) within the *hieron*, and, in this way, a ritual sense can be attributed to the "ownership" of the sanctuary as well.

The definition of the sacred space with its meanings and logistic arrangement is more difficult to grasp in the case of the foundation of Poseidonios. As for the funds, the founder allocated an ἀγρός, identified through a toponym and the names of the neighbouring owners, an αὐλή, a κῆπος, "what lies around the μνημεῖον," and a half of the "rights of tillage"[163] relating to a plot identified by another toponym (ll. 15–18).[164] The enjoyment of these properties was assigned to the eldest son of Poseidonios, provided that he paid the cult community the amount decreed for the celebration of the two-day religious feast every year (ll. 18–22). The properties were subject to a bond which can be interpreted as something looser than an actual mortgage, despite the terminology used in the text (l. 12: ὑπέθηκεν; l. 25: τῆς ὑποθήκης; l. 28: τὰ ὑποκείμενα; l. 49: τὴν ὑποθήκην).[165] This *hypotheke*, therefore, can be defined as

---

162 Cf. Lupu 2005, 14–21, 35 (reference to Pythion's foundation).

163 The term ἐνηρόσιον was translated by Laum 1914 (cf. above, fn. 20) as "Ackerland" at l. 18 and as "Pacht" at ll. 30f.; for this last interpretation cf. 1, 154; Carbon in Carbon/Pirenne-Delforge 2013, 109f. (notes on l. 18) interprets ἐνηρόσιον as the cash income from an arable plot which was rented out or farmed: the scholar suggests that this land was sacred based on the Delian evidence for the term used with reference to ἱερὰ τεμένη and ἱερὰ χωρία; Carbon's suggestion that Poseidonios allocated only half of these revenues for the foundation because the rest belonged to a sanctuary might be supported by two inscriptions from Mylasa (*I.Mylasa* 205 and 206), which attest the sale of half of a property to Zeus of the *Otorkondeis*: cf. Dignas 2002, 104; a comparable case, albeit concerning a public garden, is represented by *IG* VII 43 (= Laum 1914, 2, no. 21; Aigosthena, 3rd century B.C.): funding of a religious celebration and a contest through the revenues from half of a garden which Arete Aristandrou purchased from the *Aigosthenitai*, and where she had a *temenos* built (ll. 4–7: τοῦ κήπου τὸ ἥμυσυ, ἀγοράσασα παρὰ τῶν Αἰγοσθενιτῶν δραχμῶν χιλίων, τὸ πρὸς θάλασσαν, καὶ ποεῖ τέμενος Ποσειδώνιον); on this inscription cf. Vatin 1974, 354.

164 On the toponyms attested in Poseidonios' foundation see Carbon in Carbon/Pirenne-Delforge 2013, 108–110 (notes on ll. 15–18).

165 See Velissaropoulos-Karakostas 2011, 2, 141–189 for the *hypotheke* and its terminology, which does not sharply differentiate between pledge and actual mortgage: both can be indicated by the verb

a pledge consisting of a "trust fund" created by Poseidonios to the perpetual benefit of his family.[166] The system is similar to that adopted in Epikteta's foundation, where, unlike in Poseidonios' inscription, the monetary value of the pledged land is explicitly stated as three thousand drachmas that comprised the foundation capital as a whole (ll. 29–34); the testament granted Epikteta's heirs some margin to use this land freely, since they were allowed to transfer the *hypotheke* to another landed property of equal value (ll. 75–79). The *hypotheke*, however, was *de facto* fictitious: in the event that the heiress Epiteleia (or later heirs) failed to provide the amount decreed for the annual religious feast (210 drachmas) every year, the cult community was permitted to recoup the funds by directly using the land's income, but only up to the amount of 210 drachmas, whereas the actual property right remained untouched (ll. 70–75).[167] In contrast, in the case of failure by Poseidonios' firstborn, the landed inheritance passed to the shared ownership of the cult community, and the annually appointed *epimenioi* had to provide for renting out τὰ ὑποκείμενα, namely, the properties mentioned before, and the *temenos* (ll. 27–30). In this sense, the security system emerging from Poseidonios' inscription seems to be closer to an actual mortgage.

Whether in an exceptional or a regular situation, the *epimenioi* were entrusted with financial duties, including the management of the income to provide for the annual celebration (31–33) and the sale of the fleeces of the sacrificial victims; the income surplus was spent on *anathemata* (l. 47f.). The *epimenioi* were also required to account in detail for the money spent every year (ll. 45–47). In the latest edition of the text by Carbon the reading based on the stone (ΔΕΙΠΝΟΥ) is re-established at this point,[168] rejecting thereby the previous reading (πρὸ τοῦ δήμου), which entailed a participation of the people's assembly of Halicarnassus in supervising the cult group's accounting.[169] Thus, the *epimenioi* had to give their account "on the second day (*scil.* of the annual feast) before the dinner" (πρὸ τοῦ δείπνου), not "before the people's assembly".[170]

ὑποτίθημι and the correlated substantives; on the *hypotheke* as a security measure aimed at protecting the foundation capital see Laum 1914, 1, 170–174.

166 Carbon in Carbon/Pirenne-Delforge 2013, 108 (notes on l. 12).

167 Cf. Wittenburg 1990, 115.

168 Carbon in Carbon/Pirenne-Delforge 2013, 104 (*apparatus*); cf. 74, fn. 32.

169 Cf. e.g. the edition of Sokolowski (*LSAM* 72), where the amendment which had been made at this point is neither indicated through punctuation nor notified in the *apparatus*.

170 Cf. the statute of Epikteta's foundation, which decrees the annual convocation of the members' assembly on the second day of the feast (ll. 203f.). The reading πρὸ τοῦ δείπνου weakens the thesis of A. Wittenburg 1998, especially 453, who considers the involvement of a civic institution in Poseidonios' foundation as an indication that family associations, particularly those of Poseidonios and Epikteta, aimed at breaking the boundaries between the private and public sphere in order to reaffirm their prestige of local established aristocracy in the changed socio-political conditions of the Hellenistic age; however, the author's remarks on the adoption of civic procedures and language by these family cult foundations remain valid: on this point see also Stavrianopoulou 2006, especially 300 and 302.

Unlike the texts seen so far, the inscription of Poseidonios does not refer explicitly to the sacred status of land and buildings: indeed, the verb used to indicate the allocation of properties for cult purposes is ὑποτίθημι, which refers to the bond to which they were subject, rather than to their sacredness, in contrast with the verb ἀνατίθημι and the adjective ἱερός in the Coan family cult foundations. The sacred status of the land ἐμ Μελαιναῖς, allocated as a pledge by Epikteta, is all the more doubtful since her heirs had the option of transferring the *hypotheke* to another property. The properties of the gods were theoretically inalienable and unexchangeable, but quite a high degree of flexibility in the actual management of the sacred land must be taken into account.[171] Sacred land, for example, could be mortgaged by cities,[172] and there are cases in which assets belonging to deities worshipped by cult groups were alienated;[173] moreover, the special status of land belonging to gods could be disputed. This is why claims to sacredness had to be bolstered by regulations and administrative procedures ensuring the correct use of the sacred properties.[174] As far as Poseidonios' foundation is concerned, the sacredness can be implied at least in the terms *temenos* and perhaps ἐνηρόσιον, as the latter seems to appear exclusively in connection with ἱερὰ τεμένη and ἱερὰ χωρία.[175]

As for the places intended for cult activities, it is difficult to determine how they were conceived and arranged. The αὐλή, the κῆπος and the μνημεῖον of l. 17 probably formed one and the same complex, as they are mentioned in close sequence and without any clarification of their location, unlike the plots mentioned immediately before and after this line. However, the phrasing of the text does not completely rule out the possibility that the *mnemeion* and its attached properties lay in the same place as the *agros* ἐν Ἀστυπ̣α̣λαίαι. While a funerary building attached to other structures and surrounded by land fit for agricultural use (τὰ περὶ τὸ μνημεῖον) is perfectly plausible,[176] the picture is complicated by the *temenos* being mentioned for the first time at l. 29, where its shared ownership is decreed in the event that the eldest son of Poseidonios fails to fulfil his obligations (τὸ δὲ τέμενος εἶναι [κο]ι̣νόν). The *temenos* is

171 Cf. Isager/Skydsgaard 1992, 182; Horster 2004, 12f.; Horster 2010, 444; cf. above, fn. 18.

172 Cf. Dignas 2002, 27f.; Horster 2004, 47–49; Velissaropoulos-Karakostas 2011, 2, 34f.

173 See Papazarkadas 2011, 197–204, for the alienation of properties belonging to Attic orgeonic groups.

174 In this respect, the most debated and emblematic case is the dispute over the *Hiera Orgas*, the sacred land of Demeter and Kore lying on the boundary between Athens and Megara: see now the overview of Papazarkadas 2011, 244–259.

175 Cf. above, fn. 163.

176 Cf. Kubińska 1968, 142–147 (funerary gardens and their possible use as a source of income); 129–132, 163 (land surrounding the funerary monument); 148–150 (structures around the funerary monument); Ritti 2004, 470f., 509f. (with particular reference to Hierapolis of Phrygia in the Imperial period); Vatin 1974, 355 (with reference to the *kepos* of Poseidonios' inscription); Carroll-Spillecke 1989, 38, 58.

Fig. 6: Reconstruction of the *Charmyleion* by P. Schazmann (taken from Schazmann 1934, 117, Abb. 5).

mentioned separately from the other goods liable to become the property of the cult community in the same situation (l. 28: εἶναι τὰ ὑποκείμενα κ[οι]νά). The latter are likely to be the same as the ἀγρός and the ἐνηρόσιον of ll. 15f. and 18 respectively, but it is not clear whether the funerary monument and its related properties are implied in the expression τὰ ὑποκείμενα as well. If this is the case, the *temenos* has to be considered as an additional space, but could not be a mere piece of land used as a source of income: indeed, although it was rented out and was therefore fit for residential or agricultural use, it was also deemed appropriate for displaying the stele on which the text under examination was inscribed (ll. 49–51), and thus had to be an actual sacred precinct intended for worship. In fact, when the *temenos* was an agricultural plot lying elsewhere from the sanctuary of the deity who owned it, it was more commonly equipped with boundary stones which demarcated its divine ownership; on the other hand, rent contracts of sacred land and regulations concerning its use were usually displayed in the sanctuaries.[177] As a parallel, the engraved base bearing the

177 Cf. Horster 2010, 441f., 454f.; a good example of boundary stones demarcating sacred land not physically attached to a sanctuary is provided by a series of *horoi* from Aegina: cf. Polinskaya 2009.

"Testament of Epikteta" can be considered: it was obviously set up in the cult place (ll. 273–276) and not in the landed property which Epikteta had allocated as a pledge to ensure the annual payment of the amount decreed for cult expenses. Similarly, the block inscribed with the foundation of Diomedon is likely to have been placed in the sacred precinct, especially if the restoration of the deictic τόδε at l. 1 is correct: Δ̣[ιομέδων ἀνέθηκ]ε τὸ τ̣έμ̣ενος [τόδε].

If the *temenos* and the *mnemeion* were two different structures, the former could have been dedicated to the cult of the gods (Apollo of Telmessos, Zeus *Patroos*, and the Mother of the Gods) and the latter to the burial of Poseidonios' parents, Poseidonios himself along with his wife and, probably, other family members, as was the case in the Coan *Charmyleion*. A distinction between two sacred areas based on their different cultic functions would find correspondence in the sanctuary completed by Epikteta, which consisted of a *Mouseion*, dedicated to the cult of the Muses, and a *temenos*, where the funerary monuments of the heroized family members stood (cf. ll. 35f.; 41–51).[178] The two areas must have been distinct but adjacent since the inscribed base that supported the lost statues of Epikteta and her two sons (fig. 7) was placed in the *Mouseion* instead of the *temenos*, which could be envisaged as the most suitable location because it was specifically dedicated to the commemoration of the deceased family members. Epikteta's inscription, however, attests that the *temenos* was meant also as a funerary precinct: because of this, the possibility cannot be ruled out that in Poseidonios' inscription, too, the *temenos* was a funerary precinct including the *mnemeion* and the attached properties.[179] In this case, the arrangement would be similar to that of the *Charmyleion*, where one building, surrounded by other structures, gardens, and land, was probably used for both the funerary and the divine cult. In contrast, the ἀγρός and the ἐνηρόσιον of ll. 15f. and 18 respectively were located elsewhere, as is shown by the topographical clarification concerning them. It cannot be ruled out, however, that the *mnemeion* was also ἐν Ἀστυπ̣α̣λαίαι, like the ἀγρός: if this was the case, the funerary monument and the *temenos* have to be considered as two different structures, even though the comparison to the other family sanctuaries/tombs from Cos and Thera speaks in favour of an adjacency of (if not overlap between) the areas dedicated to the divine and the funerary cults respectively.

The absence of explicit reference to the structures used for the worship of the gods might be surprising, because they played an essential role in the cult as the recipients of the sacrifices (ll. 36–38). Setting aside the Theran *Mouseion*, whose cultic function is self-evident, the architectural terminology of Diomedon's inscription also provides clues to structures connected in some way to the divine cult (*temenos*, *oikia*,

178 Cf. Wittenburg, 139–143.
179 Carbon in Carbon/Pirenne-Delforge 2013, 109 (notes on l. 17) only touches on this possibility, but the comparison with the *temenos* and the *kepos* of Diomedon's foundation is not completely satisfactory because these were not specifically funerary in nature.

Fig. 7: Reconstruction of the base bearing the "Testament of Epikteta" with the statues originally arranged on its top (taken from Wittenburg 1990, Tav. 2).

hieron), and the same goes for the inscriptions of Pythion (*hieron*) and Charmylos (*oikia*). As for Poseidonios' inscription, some clues in this respect might come from the term αὐλή, which is attested in connection with sanctuaries as a courtyard where statues, stelae and votive offerings could be displayed.[180] If this was the case, the αὐλή of Poseidonios' inscription might also be the place where the votive offerings mentioned at l. 48 were arranged. A good example of an αὐλή used for this and other purposes is the so-called "Établissement des Poseidoniastes de Bérytos" on Delos, since it is the cult and meeting place of a private religious association, albeit ethnic in nature. In an honorary inscription to one of its benefactors, the Roman banker Marcus Minatius Sextus, the association of the "Berytian Merchants, Shippers and warehouse Workers" decreed the location of his statue in "a place chosen by him in the courtyard [...] or in whatever other place he may decide, except a place in the temples and the porticos".[181] The architectural complex, which is relatively well pre-

---

180 See Pippidi 1966, 90–93; cf. e.g. *IG* II^2 1329: the honorary decree for Chaireas Dionysiou, secretary of the Attic *orgeones* of the Mother of the Gods, was to be displayed in the courtyard of the sanctuary (ll. 28f.: ἀναγράψαι δὲ τόδε τὸ ψήφισμα εἰς στήλην λιθίνην καὶ στῆσαι ἐν τεῖ αὐλεῖ τοῦ ἱεροῦ).

181 *I.Délos* 1520 (after 153/152 B.C.), ll. 23–25; translation (slightly modified) from Ascough/Harland/Kloppenborg 2012, 136–139, no. 224.

served, includes three courtyards.[182] The *aule* mentioned in the inscription is probably to be identified with the courtyard indicated by the letter X on the plan drawn by Picard (fig. 8), because it was a cultic and representative place ("cour d'honneur"), as has been inferred from the findings *in situ* (altars and an exedra intended for displaying honorary statues) and from its proximity to the most sacred area of the complex with rooms serving as shrines.[183]

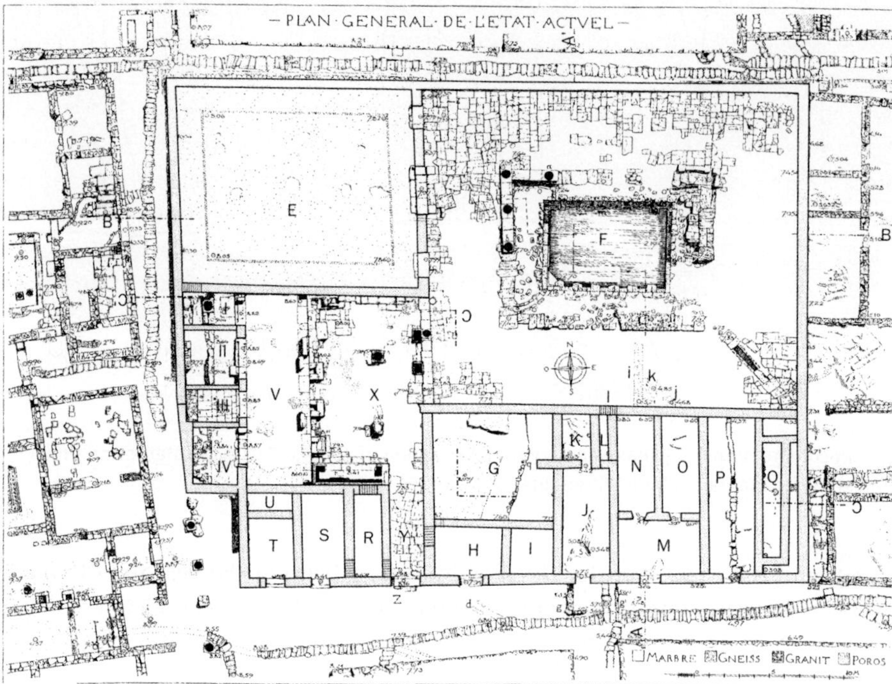

Fig. 8: Plan of the "Établissement des Poseidoniastes de Bérytos", Delos (taken from Trümper 2002, 267, Abb. 1)

By comparison, the αὐλή seems to be more rarely attested in the terminology specifically concerned with funerary architecture,[184] even though monumental tombs

182 Cf. Picard 1921, 77–112; for a concise description of the structure of this building as a whole, see Trümper 2002, 266 with Abb. 1 (plan based on Picard 1921, pl. I).

183 Cf. Bruneau 1970, 623–626; cf. Hellmann 1992, 60f.; for a description of the courtyard X and the findings *in situ*, see Picard 1921, 21–33.

184 The term is not listed among the Greek words analyzed by Kubińska 1968 (see Index, 175–178) in her work on the architectural terminology referring to the funerary monuments of Asia Minor: the only (but rare) terms which come close to αὐλή are περίαυλον and περιαύλιον, which seem to refer to the land surrounding the funerary monument (132, 163); see Sartre in Lewis/Sartre-Fauriat/Sartre

of the Hellenistic age with various buildings or rooms organized around (peristyle) courts are archaeologically known.[185] On the other hand, we cannot ignore the fact that the term is frequently used in connection with farmsteads, all the more so as this use is attested in Caria. Even though αὐλή literally indicates the courtyard of the farmstead, it seems also to have been used to refer to the estate as a whole as a *pars pro toto*.[186] With regard to the inscription being examined, it is interesting to note that in some cases funerary monuments have been discovered within or in close proximity to Anatolian farmsteads dating from the Classical age onwards. These burials must have belonged to the owners of the properties concerned.[187] The inclusion of Poseidonios' inscription in a similar framework might entail a location of the *mnemeion*-complex in the countryside. Even though an urban localization of properties fitted for agricultural use cannot always be ruled out,[188] the presence of the funerary monument point to a rural structure. The findspot of the stele, nevertheless, does not help in this respect, because at its discovery it was broken in several fragments (later recomposed), which were built into the house of Hadji Captan, not far west of the Maussolleion terrace, that is, within the city walls of Halicarnassus.[189] That this was the original location of the stone, however, is unlikely, because the whole action of Poseidonios is the result of a private initiative, without any involvement of the city, which was the only authority entitled to bestow the honour of burial within the urban centre on prominent citizens and their families.[190] If it was located outside the city,

---

1996, 87, no. 52, for an inscription from Hauran (Boṣrā al-Ḥarīri), which records the building of a μνημῖον and an αὐλή in front of it: the latter has been interpreted as a "résidence à cour"; a similar interpretation can be found in Sartre-Fauriat 2001, 19f., with regard to other inscriptions from Hauran showing the construction of tombs near *aulai*: the author does not rule out the possibility that αὐλή indicates the precinct of the tomb (possibly including a funerary garden), but points out that αὐλή as a rural court-residence is more frequent; based on this, she hypothesizes that the tombs concerned were built in close proximity to country dwellings or within their courtyards; note, however, that the context of Hauran is obviously different from that of the Halicarnassus inscription.

185 For instance, the *Heroon* of Kalydon and the *Heroon* I of Milet: see Kader 1995, 205–211 with Abb. 3.4 and 3.6.

186 Cf. Hellmann 1992, 59f.; Schuler 1998, 59–61; Chandezon 1998, 43–46.

187 Cf. Geppert 1996; Hailer 2000; Lohmann 1999, notably 454–456, 464; Chandezon 2003, 204, 213; Carstens 2004, 50–52; isolated funerary *periboloi* of the Classical period found in the Attic countryside have been connected to family farmsteads or landed properties: cf. Marchiandi 2011, 107–109 (with interesting reference to literary sources which highlight the praxis of placing the tomb on one's landed property), 158–160; a similar feature has been observed in the north-Aegean area (e.g. present-day Bulgaria), where imposing mound burials located in isolated sites have been discovered: Archibald 2013, 297–300; for the location of these tombs see 151, fig. 4.7; 153, fig. 4.8.

188 Cf. Schuler 1998, 60f. with fn. 18.

189 Cf. Carbon in Carbon/Pirenne-Delforge 2013, 101.

190 For the same reason Carbon in Carbon/Pirenne-Delforge 2013, 109, admits the possibility that Poseidonios' stele is a *pierre errante*; on the burial within the city in the Hellenistic age, see Gauthier 1985, 60–66; Chiricat 2005; Galli 2007–2008; Ehrhardt 2008.

the cult complex established by Poseidonios would join the *Charmyleion* and probably the sanctuary of Heracles *Diomedonteios* in the type of family sanctuaries/tombs erected on the estate belonging to the family.

Since Epikteta's foundation provides a close comparison for land allocated as a pledge (but also for the arrangement of the cult complex established by Poseidonios), some of its features have already been mentioned, but more is needed to complete the picture. By all appearances, the funds on which the foundation was based came from Epikteta's personal possessions: the plots allocated as a pledge on the total amount of the foundation funds (3,000 drachmas) are defined as αὐτόκτητα χωρία (l. 32), namely, properties personally held by Epikteta, even though the way in which they were acquired remained unspecified. These have to be distinguished from her other means to which the text refers, τὰ ἄλλα μοι ὑπάρχοντα (ll. 37f.), probably connected with Epikteta's paternal inheritance, including her dowry.[191] Since Epikteta is named as the executor of the last will of her husband and then of her second deceased son (cf. ll. 7–27),[192] the issue remains open as to whether the directions they gave her on the point of death included endowments aimed at bringing the project to completion.[193] As far as it is possible to understand from the mere textual data, the property allocated for funding the cult institution seems to fall into the category of the ματρῷια.[194] In accordance with the legitimate inheritance rights, Epikteta's assets were bequeathed to her daughter Epiteleia and subsequently to her descendants (ll. 30–41). All of the related obligations rested on the heirs, including the annual remittance of the income intended for the religious meeting of the cult community (210 drachmas),[195] but the management of this amount was in the hands of the κοινὸν τοῦ ἀνδρείου τῶν συγγενῶν. For this purpose, a complex administrative system, hierarchical in nature, was enacted in the foundation statute which immediately follows Epikteta's will (ll. 109–288). The community officials were appointed annually by the members' assembly (σύλλογος), which was the highest deliberative authority of the association and probably consisted of the twenty-five men listed in ll. 81–93 of the testament, but their sons were also almost certainly included in the number of councillors, because the text specifies that they were to attain membership in the cult community with full rights at the end of the *ephebia* (ll. 135–139; cf. ll. 96f.).[196] An administrator (ἀρτυτήρ) was entrusted with the collection of the annual income from

191 Cf. Wittenburg 1990, 73, 85–88; Stavrianopoulou 2006, 141, 152f.; see Velissaropoulos-Karakostas 2011, 2, 52–54, on the αὐτόκτητα possessions and on the distinction between them and those that women acquired by inheritance or dowry (with reference to the "Testament of Epikteta").

192 Cf. Wittenburg 1990, 84; Stavrianopoulou 2006, 297–300.

193 Cf. Wittenburg 1990, 77–79; Stavrianopoulou 2006, 142.

194 On the ματρῷια and πατρῷια goods cf. Velissaropoulos-Karakostas 2011, 2, 48–50.

195 Cf. Wittenburg 1990, 74f.

196 Cf. Wittenburg 1990, 112–114.

Epikteta's heirs and with distributing it as required (ll. 221–229).[197] He was subordinate to a supervisor (ἐπίσσοφος; cf. ll. 223f.), who was in charge of the accounting as a whole (ll. 213–215) along with many other duties connected with overseeing the observance of the testament and statute (ll. 205–207); his tasks also included the role of secretary, as he had to draw up the documents relating to the association (lists of the officials, payments of any fines incurred in violations against the decreed regulations, decisions made by the association: ll. 207–213).[198] All of the male members of the association had to serve in turn as cult ministers (*epimenioi*). They were obliged to fulfil the office once at their own expense in order of seniority, supplying wine, crowns, musical entertainment and perfumed oil (ll. 134–141), whereas the sacrificial victims were probably paid for by the association. After all of the community members had served as cult ministers at their own expense, the three *epimenioi* appointed every year received the necessary funds from the administrator (ll. 155–161).[199] Because of this free service obligation, the association had, at least in the early period of its life, an income surplus which was invested in the form of loans by specifically appointed officials (ἐγδανεισταί: ll. 146–155).[200] Several measures were taken in order to protect the foundation, including fines and penalties (such as temporary exclusion from the cult community) as well as the appointment of special committees called to intervene in any violations against the testament and the statute (241–254).[201]

A similar balance between inheritance rights and management rights can be seen with respect to the cult place, which was bequeathed to Epiteleia as the legitimate heiress, but was submitted to the control of the cult community in order to ensure its proper use in accordance with Epikteta's will (ll. 35–57). As mentioned, the cult place was made up of two components corresponding to the twofold cult practised by the *koinon*. There are no clues to the structure of the *Mouseion*, which might have consisted of a temple or an open-air precinct with an altar inside.[202] In both cases, however, dining structures must have been included, as the text states that the annual three-day meeting, during which at least one community meal was

---

197 The administrator was in general involved in the operations concerning cash management: cf. Wittenburg 1990, 108f.

198 The first supervisor was also in charge of recording the testament and statute on the stone and on wooden tablets; moreover, he had to appoint an archivist entrusted with the upkeep of the association's written documents (ll. 267–286); on the duties of the supervisor and the archivist see Wittenburg 1990, 103–107, 109–111.

199 Cf. Wittenburg 1990, 100–103; see above, fn. 55.

200 Cf. Wittenburg 1990, 109, 116.

201 Cf. Wittenburg 1990, 111f., 116–118.

202 Examples of sanctuaries shaped like this are mentioned by Hellmann 2006, 122–126: e.g. the *Mouseion* of Thespiae, which still in the 3rd century B.C. consisted of an altar, a long porch and a theatre; cf. also 146f. for the altar as the essential element of every sanctuary.

scheduled,[203] took place in the *Mouseion* (ll. 61f.; 118f.; 131–133); banquets could also be held in the open spaces of the sanctuaries or within temporary structures set up there.[204] Actually, Epikteta hopes in her testament for the future construction of a *stoa* in the *temenos* (ll. 48–50: μηδὲ ἐνοικοδομῆσαι ἐν τῶι τεμένει μηθέν, εἴ κα μή τις στοὰν οἰκοδομῆσαι προαιρεῖται; "nor [shall anybody have the right] to build anything in the *temenos*, unless somebody wants to build a *stoa*"); this could be for convivial use, but other functions cannot be ruled out.[205] When used for banquets within sanctuaries, a *stoa* mainly served as the connecting structure for dining rooms (often called *oikoi*) arranged in a row, also providing them with a monumental façade as well as a covered space in front; there are also cases in which the banquets were held in the *stoa* itself.[206]

The *temenos* was certainly a precinct enclosing the funerary monuments, here called ἡρῶια. This term lends itself to a wide range of applications including a whole funerary complex, a free-standing building, or sepulchral monuments such as niches, aedicules, or sarcophagi located in the middle of a precinct or built into its walls.[207] One of the latter must have been the case of the *heroa* under examination, but the text does not provide other clues to the shape of these funerary monuments. The arrangement of a burial within a *temenos* might recall some open-air hero cult places from the Geometric period onwards, arisen around the presumed tombs of mythical or historical personalities. Two examples of this are the *Heroon* of Pelops in the sanctuary at Olympia (5th century B.C.) and the *Heroon* of Opheltes in the sanctuary at Nemea (several phases starting from the 4th century B.C.).[208] A similar shape recurs in some Hellenistic *heroa* which were privately built for prominent individuals recently deceased, whether explicitly heroized or not.[209] A good example, which has indeed been mentioned as a comparison for the family *temenos* of Thera,[210] is the *heroon* of Gölbaşı Trysa (Lycia; ca. 360–350 B.C.) (figs. 9 and 10),[211] which probably

203 Cf. Wittenburg 1990, 133–137.

204 Cf. Leypold 2008, 193.

205 E.g. shelter for the participants in cult activities or display of votive offerings: on the multifunctional nature and various possible shapes of *stoai* in sanctuaries, cf. Hellmann 2006, 212–218.

206 Cf. Börker 1983; Hellmann 2006, 216, 222–225; epigraphic evidence for a *stoa* built against a row of dining rooms (*oikoi*) is provided by a rent contract from the *Herakleion* of Thasos, which entails building activities by the lessee (*IG* XII Suppl. 353, l. 13: [οἰκοδομήσει δὲ καὶ οἴκους πέντε] οὐκ ἐλάσσους ἑπτὰ κλινῶν καὶ παρὰ τούτους στωιήν): see Leypold 2008, 125–128.

207 Cf. Kubińska 1968, 26–31; Wittenburg 1990, 146f.; Weber 2004, 151f.

208 Cf. Hellmann 2006, 275–277 and figs. 220, no. 8 and 381; for other examples dating from the Geometric and Archaic periods, see Mazarakis Ainian in Damaskos et al. 2004, 133–140.

209 Modern archaeological language has generalised the use of the term *heroon* to indicate several kinds of monumental tombs independently of the actual heroization of the individuals buried therein: cf. Hellmann 2006, 275–287.

210 Cf. Wittenburg 1990, 139–141.

211 See Oberleitner 1994; for the sculptural decoration see Landskron 2011.

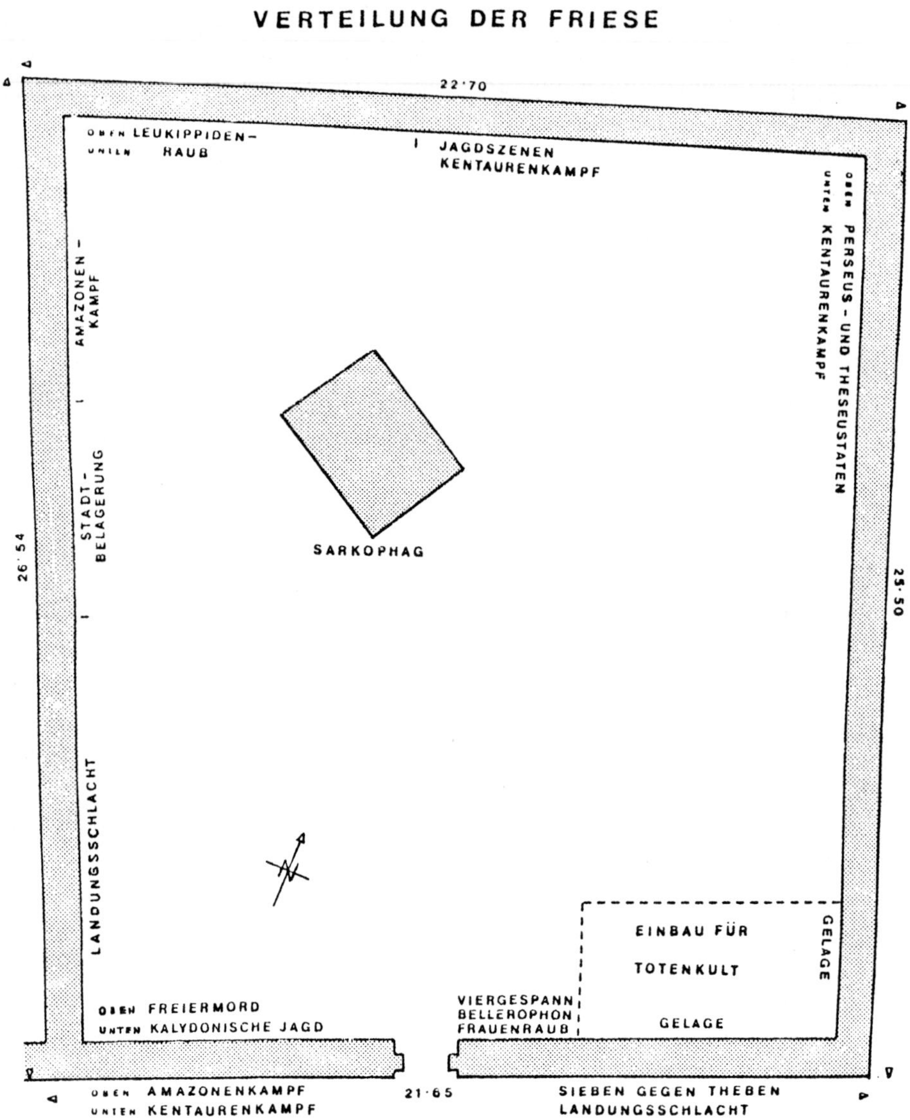

Fig. 9: Plan of the *Heroon* of Gölbaşı Trysa (Lycia) with indication of the scenes depicted in the peribolos' frieze (taken from Oberleitner 1994, 20, Abb. 24).

belonged to a local ruler and consisted of an open-air precinct enclosed by a *peribolos* lavishly embellished with friezes; the inner area included the burial, represented by a monumental sarcophagus in the middle of the *temenos*, a quadrangular base by the northern wall probably used as an altar, and traces of a wooden structure at the south-east corner which was likely used for banquets and seems to recall the *stoa* that Epikteta wanted for her funerary *temenos*. The so-called *periboloi* of Attica, dating to the 5th and 4th centuries B.C.,[212] can also provide a good comparison as they are open-air funerary monuments which gather the burials of a variable number of family members belonging in several cases to more than one generation and to various degrees of kinship.[213]

Fig. 10: Remains of the *Heroon* of Gölbaşı Trysa, Lycia (taken from Oberleitner 1994, 19, Abb. 23).

The sepulchral cluster is delimited by a pi-shaped enclosure, or at least by a wall as a monumental façade, on top of which stood several kinds of *semata* and inscriptions referring to the individuals buried within the precinct.[214] Even though the geographical and chronological context is different, the number of sculptures characteristic for Attic *periboloi* might recall the ἀνδριάντες, the ζῶια, and the ἀγάλματα

212 For the chronology of the *periboloi* and for the fenced-in family tombs predating the Classical period, see Marchiandi 2011, 19–34.
213 See Marchiandi 2011, notably 35–46.
214 See Marchiandi 2011, notably 47–78.

mentioned in Epikteta's inscription.[215] According to the finds *in situ*, funerary rituals took place in the internal space of the *periboloi*.[216]

Since the text does not mention landed properties surrounding the *Mouseion* and the funerary *temenos*, the sanctuary of Thera cannot be included with certainty in the pattern seen so far, which shows a family sanctuary/tomb resting on the family estate. Furthermore, unlike the cases from Cos and Halicarnassus, the cult place is not used as a source of income; therefore an actual economic value is not added to the sacral meaning of the place. Precisely because of this, however, a particular significance can be attributed to the regulations concerning the use of the spaces (ll. 41–57). The testament denies anyone the right to sell, mortgage, exchange, or alienate the *Mouseion* and the *temenos*, along with the statues located therein, or to carry away any object; moreover, nobody was allowed to build anything else in the *temenos* except for a *stoa*, nor did anybody have the right to use the *Mouseion* for personal reasons except for the weddings of Epiteleia's descendants. The *koinon* was vested with the responsibility and authority to prevent such violations (ll. 52f.: κύριον ἔστω τὸ κοινόν; cf. l. 57).

The adoption of measures aimed at ensuring the inalienable nature of the funds is a feature common to all of the foundations, including those intended for public enjoyment.[217] Similar prohibitions also occur in simple funerary inscriptions as a way to prevent misuse and usurpation of the burial place and attached properties, and the place where the tomb lay was sometimes explicitly called a *locus religiosus*;[218] however, norms of this kind are all the more significant in foundations such as those being examined, where a well-defined family group is called to share a cult space for joint ritual actions which are formally regulated and funded. The prevention of any free initiative within the sanctuary activates a dialectic between the individual and the group which is solved in favour of the group, but it has been shown that the cult place of Epikteta's family was not meant as a source of income and, accordingly, it was not an integral part of the funds. The only exceptions to the prohibition of using the sanctuary for purposes other than the annual community meeting were, significantly, the wedding celebrations of Epiteleia's descendants, crucial events for perpetuating the direct descent of Phoenix, father of Epiteleia and husband of Epikteta. The ultimate aim of the foundation, in fact, seems to have been the preservation of the male line, which is highlighted through both the official denomination of the cult community (which clearly values the male component: κοινὸν τοῦ ἀνδρείου τῶν συγγενῶν) and the bestowing of the priesthood on the son of Epiteleia and on his

---

215 The sculptural decoration of the *Mouseion* and the *temenos* is briefly dealt with by Wittenburg 1990, 144–147.

216 See Marchiandi 2011, notably 92–94.

217 Cf. Laum 1914, 1, 178–193.

218 Cf. Ritti 2004, 474f., 510–530 (with particular reference to Hierapolis of Phrygia in the Imperial age); cf. also Harter-Uibopuu 2010, notably 257–261.

descendants after him. By a twist of fate, this responsibility rested on the heads of the two surviving women of the family, one of whom was affiliated to Phoenix's family by marriage rather than by legitimate descent.[219]

The interpretation of the above prohibitions as a way to ensure the exclusively shared use of the sacred space sheds light on the analogous regulations in Diomedon's foundation (ll. 80–86), where "those who share the sacred things" are forbidden from farming the *temene*, from dwelling in some buildings which belonged to the sanctuary (the *xenones* and the *oikia* within the *temenos*), and from using the building in the *hieron* (*lesche*?) and the *peripatos* as a storehouse (or refuge). There may be an economic reason behind these prohibitions, since the cult place was also a source of income, but besides this, there may be a connection between the sharing of the "sacred things", which the denomination of the cult community clearly claims, and the joint use of the sacred spaces. The prohibitions aimed to prevent privatization of the sacred spaces by single members of the cult community for uses different from those decreed. It is hardly coincidental that here, too, an exception was made for the wedding celebrations of the founder's descendants, but this right was reserved for those who were in need at the moment of their marriage: a regulation which points towards family solidarity as a basic principle of the cult community (ll. 86–90).[220] The other permitted exceptions concern the usage of the sanctuary spaces in case of war, but the phrasing of the text does not clarify whether the permission only concerns the last item mentioned (ll. 84–86: μηδὲ ἀποθήκηι χρᾶσθαι τῆ̣[ι λέσ]χ̣ηι τῆι ἐν τῶι ἱερῶι μηδὲ{ν} τῶι περιπάτω̣[ι] ἂμ μὴ πόλεμος ἦ̃ι), or extends to the two previous ones (farming the *temene* and dwelling in the *xenones* and in the *oikia*). In any case, it is important to point out that in emergency situations the resources of the sanctuary could be converted to uses different from those for which they were conceived. Such uses can be compared with the "profane things" mentioned in Pythion's foundation, where the reference might apply to the sanctuary's goods, whose enjoyment was evidently subject to specific regulations. As the text states that the sanctuary had to be shared among all of the founder's descendants, the "profane things" might have also included possible use of the sacred spaces by individual members of the cult community for purposes not sacral, but similar to the agricultural and residential uses which the foundation of Diomedon forbids. This does not necessarily mean that any "profane" use was forbidden, but that eventual "profane" uses were regulated or only

219 Cf. van Bremen 1996, 214–216; Stavrianopoulou 2006, 142, 297–299, who also points out that Epikteta earned a legitimate position within Phoenix's family because of her total commitment to perpetuating the male line according to the wishes of Phoenix and Andragoras.

220 The same meaning is to be attributed to the fact that on the occasion of nuptials taking place in the sanctuary the *gere* to which the priest was normally entitled were given to the person celebrating his wedding, along with other portions of the sacrificial victims (ll. 100–104); cf. Campanelli 2011, notably 669f.

admitted in exceptional situations, such as war in Diomedon's foundation.[221] The latter contains another set of regulations concerning the management of the sacred spaces (ll. 43–47): besides the prohibitions of sale and mortgage, appropriating the *temenos* and the attached buildings to oneself (ἐξιδιάζεσθαι) is also forbidden. The verb ἐξιδιάζεσθαι is formed upon ἴδιος ("one's own", "private", "personal"), which is the opposite of κοινός,[222] and therefore calls attention, again, to sharing as the only acceptable way to properly enjoy the spaces.

The dialectic between *idion* and *koinon* as a common feature in the sphere of private associations can be illustrated through two examples, different in nature but both involving the management of community spaces. The first concerns the three decrees issued by the gentilicial group of the *Klytidai* on Chios in order to endorse the construction of an οἶκος τεμένιος ἱερός, also called κοινὸς οἶκος (4th century B.C.).[223] This "sacred house belonging to the *temenos*", or "shared house", was intended for keeping the ἱερὰ τὰ κοινά, otherwise indicated as τὰ πατρῷια ἱερά, that is, the ancestral sacred objects of the community which were previously preserved in the ἰδιωτικαὶ οἰκίαι, probably understood as single families which belonged to the *Klytidai* and had been the privileged holders of the *hiera* up to that time. The transfer of these to the shared house certainly aimed at cementing the group identity on a more egalitarian basis through a wider sharing of the sacred objects, which thereby became really *koina*. The process toward the *koinon* is fulfilled in the third decree, through the clause which sanctions the collective enjoyment (ll. 27f.: χρῆσθαι Κλυτίδας κοινῆι) of the house and the plot surrounding it, forbidding individual usage by any person or group (φατρία).[224]

The second example is a passage of Theophrastus' testament, which Laum includes in the number of the foundations he collects.[225] The text, in fact, shows many

---

221 For the relationship between the regulations concerning the usage of spaces in Diomedon's foundation and the "profane things" of Pythion's foundation, see Campanelli 2012a, especially 87–90, where reference is also made to sacred laws which forbid or limit individual uses of material resources belonging to sanctuaries.

222 Cf. Liddell/Scott/Jones 1968, s.v. ἐξιδιάζομαι and ἴδιος.

223 Reference is made to the edition of Graf 1985, 428f., no. 3; the considerations which follow are based on the interpretation of this inscription by Brulé 1998 (= Brulé 2007, 385–403).

224 For the interpretation of the word φατρία see Brulé 1998, 317f.

225 D.L. V 52, 9–54, 6 (= Laum 1914, 2, no. 15, not including the full passage); Dareste/Haussoullier/Reinach 1898, 108, compare Epikteta's foundation with Theophrastus' testament in view of both the inalienable nature of the private properties allocated as cult/meeting places and the designation by name of those having the right to membership in the respective communities; the authors argue, however, that the juridical status of the two were different, since the philosophical communities had no legal personality; in their view, this is demonstrated by the way in which Theophrastus' bequest was handed down to subsequent successors as can be reconstructed through the testaments of the two peripatetic philosophers, Strato and Lyco, who seem to have become, in succession, the exclusive holders of the property allocated by Theophrastus for the philosophical activities; cf. D.L. V 70 for similarities with the foundations examined here. The bequest of the philosopher was to the benefit of a well-defined group of κοινωνοῦντες whose names are listed. They were called to share the κῆπον καὶ τὸν περίπατον καὶ τὰς οἰκίας τὰς πρὸς τῷ κήπῳ πάσας, provided that they owned these spaces in common and enjoyed them in terms of familiarity and friendship (ὡς ἂν ἱερὸν κοινῇ κεκτημένοις καὶ τὰ πρὸς ἀλλήλους οἰκείως καὶ φιλικῶς χρωμένοις) for the established purpose, namely, joint philosophic practice. The estate, which was conceived as sacred (ὡς ἂν ἱερόν), was declared to be inalienable and nobody was permitted to appropriate it for himself. The prohibition of self-appropriation is expressed by the same verb used in Diomedon's foundation, ἐξιδιάζεσθαι, which contrasts with the following adverb κοινῇ, "in common". Both the affiliates of the philosophical community and the descendants of Diomedon were κοινωνοῦντες, and what they shared with their respective fellow members were not only activities, whether intellectual or ritual, but also the places where such activities were performed.

Albeit in different and not always explicit ways, the concept of *koinon* recurs in all of the family cult foundations seen so far, and goes beyond the self-evident sharing of ritual actions, also embracing spatial and patrimonial values. Both Diomedon's and Epikteta's foundations make this explicit through the designation of the respective cult communities. In both texts a number of prohibitions aimed at preventing any individual use of the cult place, which was reserved for collective enjoyment; however, there are some differences between the two foundations. In that of Diomedon the physical and patrimonial overlap between sacred funds and sacred spaces is phrased through the broad and multifaceted sense of *hiera*, "the sacred things" which were to be fully shared among the community members; in Epikteta's foundation, on the other hand, the situation is more elaborate, since a logistic and patrimonial distinction is made between the foundation assets and the cult place: indeed, the plots allocated as a pledge lay elsewhere from the sanctuary which, in turn, was not an integral part of the funding system; both were bequeathed to the legitimate heiress Epiteleia, but the management of the funds and the supervision over the cult place were shared among the community members, who enacted an administrative and deliberative *koinonia* by means of a complex organisational system.

A somewhat similar situation can be seen in Poseidonios' foundation, where the principle of primogeniture was respected in designating the founder's firstborn as the beneficiary of the properties allocated for funding the cult, but in this case, too, the community members were entrusted with management of the funds through the office of *epimenios*; they also appear as a deliberative body in the sanction formula of the *dogma* by which the foundation is ratified and enacted (ll. 22–23: ἔδοξεν Ποσειδωνίωι καὶ τοῖς ἐκγόνοις τοῖς ἐκ Ποσειδωνίου καὶ τοῖς εἰληφόσιν ἐξ αὐτῶν). Unlike Epikteta's

---

the passage of Lyco's testament concerning the bequest of the *peripatos* to a group of ten friends who intended to use it to carry on with their philosophical studies.

foundation and similar to that of Diomedon, the revenues came, at least partially, from the cult place, which thereby turns out to have an economic value too. Significantly, the term κοινόν appears in reference to the passage of the properties under the ownership of the cult community in the event that Poseidonios' eldest son failed to fulfil his obligations. In contexts such as this, the phrase εἶναι κοινόν (cf. ll. 28f.) indicates an internal transition to an extended ownership which occurs completely privately, in contrast with εἶναι δημόσιον, which refers to something that is or becomes public.[226] The latter formula appears for example in a civic decree from the island of Anaphe,[227] by which a place in the sanctuary of Apollo *Asgelatas* was granted to a private citizen who had requested it in order to build a public temple of Aphrodite (ἦμεν δαμόσιον: ll. 21; 28) at his own expense.[228]

The same difference between "public" and "in common" can be seen in the private status of the sanctuary founded by Pythion: the fact that this had to be held in common by all of his descendants (ll. 15f.: τὸ δὲ ἱερὸν ἔστω τῶν υἱῶν πάντων κοινόν) certainly fits the practice of a joint sacrifice therein (l. 7: συνθυόντων), but a shared ownership of the material resources of the sanctuary can be supposed, even though it is not possible to specify the conditions and requirements on which it was based, due to the document's brevity. The same goes for Charmylos' foundation, known from a *horos* which is by nature concise and does not specify the regulations concerning the use of the properties consecrated by the founder. However, the shared enjoyment of the sacred spaces by the family group named after Charmylos is very likely.

In conclusion, a brief reference can be made to the two fragments from Thera and Halicarnassus. The inscription from Thera[229] mentions χωρία (ll. 1; 3; 7), which may have served as sources of income, and probably refers to building or maintenance activities within the cult place (l. 2: ἐπισκευάξ[ηι]; l. 4: τὰ δέο]ντα ἐπιςκευ[ᾶς]; l. 5: σṭεγνόν). The remains of ll. 8f. concern prohibitions which, however, seem to be slightly different from those in the foundations of Diomedon and Epikteta, because the employed formulas are characteristic of the funerary sphere: μὴ ἐχέτ[ω δὲ ἐξουσίαν μηθεὶς θέμ]εν ἄλλον μηθένα̣ [ἐς τὸ τέμενος· ἔστω δὲ ὁ] θεὶς καὶ ὑπόδικος τῶ[ι δάμωι τῶι Θεραίων δρα]χ̣μᾶν χιλιᾶν, καὶ τὸ ἐπ̣[ιδέκατον ἔστω τοῦ κοινοῦ] ṭῶν συγγενῶν. The verb τίθημι, referring to a person (μηθένα̣), technically indicates the burial rights within a funerary monument ("to put someone", that is, "to bury"); fines against vio-

226 Cf. Migeotte 2006, especially 188–190, where he points out that the concept of *koinon* in Poseidonios' foundation has been misunderstood by Hegyi 1976, 85f., in her attempt to demonstrate the way in which sanctuaries originally in private ownership were liable to become public; in other contexts, however, the term *koinon* is used as a synonym of *demosion*: cf. e.g. Velissaropoulos-Karakostas 2011, 2, 36–38.

227 *IG* XII 3, 248 (ca. end of the 3rd century B.C.).

228 He was, however, allowed to use materials made available by the sanctuary: cf. ll. 9f.

229 Cf. above, fn. 86.

lations are also a common feature in these contexts.[230] Accordingly, it is likely that the foundation from Thera dealt with the funerary cult at a family tomb, but it is not possible to determine from the surviving text whether a cult of deities was also included.

The inscription from Halicarnassus[231] refers to an *oikia* which may have belonged to the *syggeneis* mentioned in the lines immediately preceding and following l. 4 (ἡ οἰκία ἔστω τῶν [συγγενῶν–]). It would have been interesting to know the nature of this "house", but from what remains of the text it is not possible to understand whether it was a mere source of income or had a cultic function similar to that of the *oikia* in Diomedon's and Charmylos' foundations. The reference to the polis in l. 7 (εμ πόλει πρὸς τοῖς [–]) certainly concerns the localization of properties allocated for funding the cult.[232] Whatever the nature of the cult was, it must have included an animal sacrifice, since the σπλάγχνα are mentioned at l. 9. The allocation of properties lying within the city might differentiate this foundation slightly from the previous ones, which seem to show a more rural context. Besides this urban real estate, however, landed properties must also have been involved, as the last preserved lines (11–16) deal with the rent conditions of an ἀγρός.

Finally, a fragment from Cos deserves to be mentioned because it might fall into the category of family cult foundations.[233] From what remains of the text, indeed, it is possible to reconstruct a testamentary formula similar to that in the "Testament of Epikteta": εἴη [μέν μοι ὑγιαίνοντι τὰ ἐμαυτοῦ διοικεῖν· ἐὰν δέ τι π]ẹρί με γίνηται τῶν ἀν[θρωπίνων, καταλείπω - - - - - - - -].[234] The other surviving words concern a consecration of an ἀγρός, possibly including an οἰκία, and something else which is lost: ἀνιερῶ τὸν ἀγρὸν τ[ὸν - - - - - - - καὶ τὴν οἰκί]ạν τὴν ἐπὶ τοῦ ἀγρο[ῦ - - - - - - - - - - - - - - ἀ]νιερῶ δὲ [καὶ. It is not possible to know, however, whether the family of the testator was the beneficiary of the bequest or the purpose for which this was intended, even though the consecration of land points to a cult foundation.[235]

---

230 Cf. Ritti 2004, 494f., 513f., 539–548 (with particular reference to Hierapolis in Phrygia in the Imperial period); the term *temenos* referring to the burial place was restored by Hiller von Gaertringen (cf. above, fn. 87) who was convinced that this inscription was connected with Epikteta's foundation.

231 Cf. above, fn. 88.

232 Cf. e.g. *TAM* II 1037 (= Laum 1914, 1, no. 138a): allocation of "three houses lying in front of the stadium" in order to fund annual libations for the deceased Pheidias and Arete (Olympos, Lycia; Imperial period).

233 Laum 1914, 2, 57, no. 48; for another similar fragment cf. Segre 1993, ED 219.

234 Cf. Epikteta's foundation, ll. 5–7: εἴη μέν μοι ὑγιαινούσαι καὶ σωιζομέναι τὰ ἴδια διοκέν, εἰ δέ τί κα γένηται περί με τῶν ἀνθρωπίνων, ἀπολείπω κτλ.

235 The same verb is used in the foundation of Phanomachos seen above.

5 Conclusions

During the Hellenistic age, the creation of family cults addressed either to deities or jointly to deities and deceased family members brought into being family cult communities which were responsible for both the continuity of the cult institution and the management of the funds intended for this purpose.

The fact that two fragments from Thera and Halicarnassus (and possibly a third fragment from Cos) can probably be added to the documentary cluster (all from the south-Aegean Doric area), suggests that the phenomenon, at least in these localities, might have been more widespread than the main dossier of the five relatively well-preserved inscriptions would lead us to believe.

The comparative analysis of the texts provides a consistent picture of a well-codified family institution. From a religious point of view, a trend toward the heroization of deceased family members is apparent, even though the chronology of the inscriptions and textual data do not allow us to accept the idea of a linear development toward the bestowal of the actual title of hero on the deceased, passing through an intermediate stage in which they were indirectly heroized by means of their cultic connection with deities or divine personifications. However, this old, going back to W. Kamps interpretation has the merit of identifying the placement of deceased family members under the aegis of a family's patron deities as a feature of family cult foundations in the Hellenistic age. Kamps is also right in considering the heroization, whether incipient or fulfilled, of recently deceased people as a religious expression of a new family solidarity more restricted than the clan, which recognized itself in the worship of remote shared forefathers. The application of this theory has allowed S. Sherwin-White to ascribe to the number of family cult foundations the *horos* from Cos, bearing traces of the foundation of Charmylos, who has been recognized as a recently deceased founder of a family cult instead of as a mythical eponymous hero. The essential data provided by this document and its archaeological context has contributed to further research. The result of later studies on family structure in different places and times lead us to reject the idea of a generalized passage from an extended to a nuclear family during the Hellenistic period. Nonetheless, Kamps' interpretation suggests the necessity of studying family cult foundations in the light of sociopolitical and historical factors that likely affected the structure and identity of the family. This has partially been attempted in more recent times, by interpreting family cult foundations as a strategy of aggregation to cope with external (sociopolitical upheavals) and internal (mortality, migration) dangers which could undermine the social recognition of the family or its survival.

The analysed texts show the intention for self-representation in the private *pantheon* created by the founders, where a common thread is the juxtaposition of gods closely connected with the household sphere and gods borrowed from the civic *pantheon* or from cults of sub-polis groups. The adherence to the religious traditions of the polis and its internal subdivisions reveals a will to display all of the social levels to which a family belonged, over and above membership in the household. The fatherland's religious customs were embedded in the family tradition in various ways: by establishing a privileged bond between the founder and a civic god, as is the case of the epithet *Diomedonteios* attributed to Heracles in Diomedon's foundation, or by dedicating a cult to deceased family members along with or under the aegis of civic gods, as is the case in Charmylos' foundation; especially significant is the choice of deities named *patrooi*, as in the foundations of Pythion and Poseidonios, where this multifaceted adjective reflects the merging of fatherland ancestry with household ancestry.

Some differences among the foundations can be detected in respect of the membership in the cult communities and the juridical means of handing down the funds allocated for the cult. The membership in the cult communities shows various degrees of inclusiveness, ranging from the generic entitlement of the founder's descendants in Diomedon's foundation to the designation by name of a number of extended family members outside the direct line of descent in Epikteta's foundation; Poseidonios' foundation can be perhaps placed at an intermediate point, because the descendants of the founder by both the male and female lines are included, but membership also seems to be extended to his sons-in-law; in Pythion's and Charmylos' foundations the entitled people are designated in even more general terms than in Diomedon's foundation, but equally underlining the concept of progeny. In general, however, the founder's direct descent down the male line turns out to be privileged according to the texts which allow us to verify this (foundations of Diomedon, Poseidonios and Epikteta).

From a cultic point of view, the priesthood is handed down from eldest son to eldest son in the foundations of Diomedon and Poseidonios, whereas in Epikteta's the situation is complicated by the absence of direct male offspring, and the principle of primogeniture is therefore reinstated by the appointment of the son of Epikteta's daughter and his descendants as priests of the Muses and heroes. Diomedon's foundation also decrees a sharp distinction between legitimate and illegitimate children, prohibiting the latter from access to priesthood and participation in the sacrifices to *Pasios* and the Moirai, which are reserved for legitimate male descendants.

From a financial point of view, the foundations of Poseidonios and Epikteta privilege the direct descendants of the founders over other community members in the transmission of the legacy according to the rights of inheritance; the remaining male members of the family associations are only entitled to manage the part of the income allocated for cult purposes.

The peculiarities of the individual foundations might be connected, at least in part, with individual family histories, and might mirror the strategies adopted by families to cope with events which threatened to undermine the continuity of the *oikos* and the proper inheritance of household assets.

Besides calling attention to the necessity of studying family cult foundations in the light of diachronic and synchronic developments of the family, these considerations also provide an appropriate framework for the main subject of this paper, namely, the conception of the physical spaces which seem to play an essential role in the foundation system. Two kinds of space emerge from the texts: the ritual space structured as a sanctuary, whether connected with a family tomb or not, and the real estate which provides income for funding the cult. The sacredness of both kinds of space is phrased through the dedication formulas in Diomedon's foundation and on the boundary stone associated with the foundation of Charmylos, where the adjective ἱερός is used with reference to all of the land and buildings included in the sanctuary. Sacredness seems to be implied in the terms *temenos* and *hieron*, which appear in all of the inscriptions studied here, except for the *horos* of Charmylos' sanctuary. The analysis of the texts confirms the wide range of meanings entailed in the basic nature of the *temenos* as a space "cut off" for entry into the sacred sphere. In the documents examined, its character ranges from that of consecrated land suitable for farming to that of sacred precinct housing a shrine or heroic tombs and used, accordingly, for cult activities. Both kinds of *temenos* are present in Diomedon's inscription, where they are indicated by the use of the plural *temene* and the singular *temenos*, respectively. When *temenos* and *hieron* appear in the same inscription (Diomedon's and probably Pythion's), the difference between the two can be supposed to be conceptual or functional in nature. The first possibility, which involves a difference in degree of sacredness, might be illustrated by Pythion's inscription, if a *temenos* is actually mentioned along with the *hieron*: the latter might be interpreted as the ritual space of the sanctuary and, accordingly, its most sacred part; *temenos*, on the other hand, might indicate the physical area of the cult place as a whole, encompassing the *hieron* itself. In the sanctuary founded by Diomedon, on the other hand, *temenos* and *hieron* might indicate two distinct but adjacent precincts which probably differ in their cultic function, given the plurality of deities worshipped by the cult community; the cultic use of the *temenos* is in any case assured by the *oikia* lying therein, which functioned as a shrine – the place where cult statues and votive offerings were kept. A cultic specialization can be attributed with certainty to the *temenos* of Epikteta's inscription, which was clearly for funerary purposes. It is more difficult, on the other hand, to pinpoint the nature of the *temenos* in Poseidonios' foundation and its topographical relationship with the family tomb (*mnemeion*), but it may be a sacred precinct intended for cult activities (e.g. the worship of the deities: Apollo of Telmessos, Zeus *Patroos*, the Mother of the Gods and the Moirai), because it was the place decreed for displaying the stele engraved with the foundation inscription.

As for the constructions lying within these sacred spaces, the most informative inscription is that of Diomedon, which also allows us to connect some of the buildings mentioned with the cult practices of the family community. As implied by the very name and confirmed by literary and epigraphic evidence, the *xenones* might have served as banquet halls, and the same function can perhaps be attributed to the *lesche* if indeed it is actually mentioned as the structure within the *hieron*. The *oikemata* can be interpreted as service buildings based on textual clues concerning the use of some of them as storehouses, at least in some situations. Two residential buildings, probably owned by members of Diomedon's family in their private capacity, were topographically attached to the sanctuary and bear names alluding to the ritual use for which they were fitted in exceptional cases (οἰκία ἀνδρεία and οἰκία γυναικεία). The interpretation of the *oikia* within the *temenos* as a shrine is strongly supported by the inscription on the *horos* of Charmylos' sanctuary: the *oikia*, which this text distinguishes from the other *oikiai* belonging to the cult place, is probably to be identified with the partially preserved tomb-temple whose remains also indicate the monumental shape which these private temples/tombs might have had.

The sanctuaries examined here show a flexibility in the use of the buildings which were fit for "profane" functions, too. This is why they could be rented out for residential use or other purposes. The freedmen who were tenants of the *kapos* belonging to the sanctuary of Heracles *Diomedonteios* likely lived in the *xenones* overlooking the garden itself, the same "guesthouses" or "guestrooms" where the annual community banquets probably took place; similarly, the *oikia*-temple was suitable for dwelling, as is indirectly confirmed by the prohibition against it. On the other hand, open spaces inside the cult places or immediately surrounding them could be rented out for agricultural use. This is certainly true for the *temene* and the garden belonging to the sanctuary of Heracles *Diomedonteios*, and for the number of plots and gardens marked out by the boundary stone of Charmylos' sanctuary, whereas the economic utilization of the *temenos* of Poseidonios' inscription is less clear, because the phrasing of the text does not allow us to understand whether it was the precinct enclosing the *mnemeion* or an independent cult area including buildings to be rented out for residential purposes; the land surrounding the family tomb, on the other hand, was certainly used for agricultural purposes: in this respect, the possibility that the αὐλή attached to the *mnemeion* itself refers to a farmstead cannot be ruled out. Possible economic uses of the sanctuary's spaces might also be implied in the "profane things" concisely mentioned in Pythion's inscription.

The fact that these cult places also represented a body of economic resources has important consequences for the conception of the sacred space, which can be understood, accordingly, as an actual *patrimonium* blending family religious traditions with family assets. The concept of *patrimonium* fits the interpretation of the phrase τοὶ τῶν ἱερῶν κοινωνεῦντες, used in Diomedon's inscription to designate the members of the cult community, as indicating those who share not only rituals but also sacred spaces with a financial value. The shared enjoyment of the sacred spaces among the members of the family communities recurs in various ways throughout the inscriptions: the sanctuary founded by Pythion was to be τῶν υἱῶν πάντων κοινόν; the *temenos* and other properties involved in Poseidonios' foundation were declared to be κοινά under particular circumstances; prohibitions against individual uses of the sacred spaces were included in Diomedon's and Epikteta's foundations. Since the *Mouseion* and the *temenos* of the latter were not an integral part of the funding system, these prohibitions specifically aimed at protecting the meeting place as the representation of the group identity.

The main form of protection of the shared spaces was their consecration. It established a distance between the cult community and the spaces which fell into divine ownership, without, of course, excluding their management and use by humans. The separation of the gods' properties from the surrounding territory entailed the arrangement of boundary stones on the ground. This well-documented procedure finds correspondence in the *horos* of the *Charmyleion*, which claimed the sacredness of all of the land plots and buildings it marked out independently of the specific use, whether ritual or financial, for which they were intended. By way of analogy, similar boundary stones may have demarcated the other family sanctuaries, tombs, and landed properties seen so far. But what is even more relevant is the location of the *Charmyleion* in the Coan countryside. Since the *horoi* were necessarily set up in the spaces they demarcated, the mention of the real estate (land, gardens and *oikiai*) together with the *oikia* on the boundary stone of the *Charmyleion* allows us to conclude that the landed properties of the *Charmyle(i)oi* lay in the immediate surroundings of the preserved tomb-temple. This arrangement, with a family cult place resting on a family estate, might represent the religious landscape typical of family sanctuaries/tombs established by foundations, especially so since this supposition finds correspondence in the epigraphic data, which suggest a topographical relationship between family cult places and landed properties, at least as far as Diomedon's and Poseidonios' foundations are concerned. Landed properties surrounded the family tomb of Poseidonios, even though its topographical connection with the *temenos*, which could be rented out, cannot be verified with certainty; on the other hand, undoubtedly attached to the *mnemeion* was the "courtyard", which would provide further evidence for farming activities connected with the cult place if the term αὐλή actually referred to a farmstead. The *temene* belonging to the sanctuary of Heracles *Diomedonteios* were also likely in its immediate surroundings, as the text does not give the topographical references usually seen in inscriptions such as rent contracts or foundation texts dealing with properties located elsewhere from the place where the inscriptions were displayed.

The above-mentioned concepts of family self-representation and self-preservation may provide an appropriate framework for interpreting the family sanctuary/tombs and attached properties which are inherent in this particular foundation system. Ultimately, family *monumenta* might have functioned as landmarks which were placed on the family estate in order to display the presence of landowners' families on given territories. But it is not only a matter of monumental visibility: family sanctuaries as well as family tombs, especially when they housed heroized deceased, also made manifest the act of consecration. In terms of family self-preservation, a foundation might have been intended to ensure the legitimate transmission of the inheritance under the aegis of the "ancestral" gods and family heroes. This is all the more likely if one considers that the cult places were themselves an integral part of family assets and were used in most cases as sources of income. However, further research is needed to verify whether family cult foundations can be interpreted as an actual patrimonial strategy from a juridical and financial point of view.

6 Acknowledgments

I am grateful to the SFB 933 "Materiale Textkulturen" of Heidelberg University for granting me this Fellowship, which has given me the opportunity to carry on the research for this paper. I would like to say thank you to Heidelberg University's "Seminar für Alte Geschichte und Epigraphik" for hosting me and supporting me in various ways during my stays in Heidelberg. Thank you, finally, to the participants in my conference and workshop, which took place in Heidelberg in the months of April and June 2013 respectively. The workshop, in particular, has been a fruitful space of discussion, which has given me the opportunity to reconsider aspects of my research and to find new causes for reflection, as well as some weak points which deserve further examination.

Abbreviations

The titles of the periodicals are abbreviated according to *L'Année Philologique*. The abbreviations for epigraphic *corpora* and collections conform to *SEG – Index* 36–45 (1986–1995) and *SEG* 55 (2005). Periodicals and epigraphic *corpora* and collections not included in the above lists are quoted in extenso in the bibliography below.

Bibliography

Aleshire, Sara B. (1994), "Towards a Definition of 'State Cult' for Ancient Athens", in: Robin Hägg (ed.), *Ancient Greek Cult Practice from Epigraphical Evidence*. Proceedings of the Second International Seminar on Ancient Greek Cult Athens, 22.–24. November 1991, Stockholm, 9–16.

Andrianou, Dimitra (2006), "Chairs, Beds, and Tables. Evidence for Furnished Interiors in Hellenistic Greece", in: *Hesperia* 75, 219–266.

Archibald, Zosia Halina (2013), *Ancient Economies of the Northern Aegean. Fifth to First Centuries BC*, Oxford.

Arnaoutoglou, Ilias N. (2012), "Cultural Transfer and Law in Hellenistic Lycia: The Case of Symmasis' Foundation", in: Bernard Legras (ed.), *Transferts culturels et droits dans le monde grec et hellénistique*. Actes du Colloque International Reims, 14.–17. Mai 2008 (Histoire ancienne et médiéval 110), Paris, 205–224.

Ascough, Richard S./Harland, Philip A./Kloppenborg, John S. (2012), *Associations in the Greco-Roman World. A Sourcebook*, Waco.

Baldassarra, Damiana (2010), "Le liste cultuali della Grecia nord-occidentale: Tipologie, protagonisti e fenomenologia rituale", in: Claudia Antonetti (ed.), *Lo spazio ionico e le comunità della Grecia nord-occidentale. Territorio, società, istituzioni*. Atti del Convegno Internazionale Venezia, 7.–9. Gennaio 2010, Pisa, 341–371.

Baldassarra, Damiana/Ruggeri, Anna (2010), "Intorno al sacrificio: *aozos* e *hierophoros*", in: Claudia Antonetti (ed.), *Lo spazio ionico e le comunità della Grecia nord-occidentale. Territorio, società, istituzioni*. Atti del Convegno Internazionale Venezia, 7.–9. Gennaio 2010, Pisa, 373–384.

Bérard, Victor (1891), "Inscriptions d'Asie Mineure", in: *BCH* 15 (2), 538–562.

Bergquist, Birgitta (1967), *The Archaic Greek Temenos. A Study of Structure and Function* (Skrifter utgivna av Svenska Institutet i Athen 4. 13), Lund.

Berns, Christof (2003), *Untersuchungen zu den Grabbauten der frühen Kaiserzeit in Kleinasien* (Asia Minor Studien 51), Bonn.

Bettinetti, Simona (2001), *La statua di culto nella pratica rituale greca* (Le rane 30), Bari.

Boedeker, Deborah (2008), "Family Matters: Domestic Religion in Classical Greece", in: John Bodel and Saul M. Olyan (eds.), *Household and Family Religion in Antiquity* (Ancient World, Comparative Histories), Oxford, 229–247.

Bömer, Franz (1958–1963), *Untersuchungen über die Religion der Sklaven in Griechenland und Rom* (Abhandlungen der Geistes- und Sozialwissenschaftlichen Klasse 1957/7, 1960/1, 1961/4, 1963/10), vols. 1–4, Mainz.

Börker, Christoph (1983), *Festbankett und griechische Architektur* (Xenia 4), Constance.

Bruck, Eberhard F. (1926), *Totenteil und Seelgerät im griechischen Recht* (Münchener Beiträge zur Papyrusforschung und antiken Rechtsgeschichte 9), Munich.

Bruck, Eberhard F. (1955), "Les facteurs moteurs de l'origine et du développement des fondations grecques et romaines", in: *RIDA* 3e série 2, 159–166.

Brulé, Pierre (1998), "La sainte maison commune des Klytides de Chios", in: *Ktema* 23, 307–324.

Brulé, Pierre (2006), "La parenté selon Zeus", in: Alain Bresson, Marie-Paule Masson, Stravros Perentidis and Jérôme Wilgaux (eds.), *Parenté et société dans le monde grec de l'Antiquité à l'Age moderne*. Colloque international Volos (Grèce), 19.–21. Juin 2003 (Études Ausonius 12), Bordeaux, 97–119.

Brulé, Pierre (2005), "'La cité est la somme des maisons'. Un commentaire religieux", in: Véronique Dasen and Marcel Piérart (eds.) (2005), Ἰδίᾳ καὶ δημοσίᾳ. *Les cadres "privés" et "publics" de la religion grecque antique* (Kernos, Supplément 15), Liège, 27–53.

Brulé, Pierre (2007), *La Grèce d'à côté. Réel et imaginaire en miroir en Grèce antique* (Collection "Histoire"), Rennes.

Brulé, Pierre (2012), *Comment percevoir le sanctuaire grec? Une analyse sensorielle du paysage sacré*, Paris.

Bruneau, Philippe (1970), *Recherches sur les cultes de Délos à l'époque hellénistique et à l'époque impériale* (Bibliothèque des Écoles Françaises d'Athènes et de Rome 217), Paris.

Brunet, Michèle/Rougemont, Georges/Rousset, Denis (1998), "Les contrats agraires dans la Grèce antique. Bilan historiographique illustré par quatre exemples", in: *Histoire et Sociétés Rurales* 9, 211–245.

Burkert, Walter (1996), "Greek Temple-Builders: Who, Where and Why?", in: Robin Hägg (ed.), *The Role of Religion in the Early Greek Polis*. Proceedings of the Third International Seminar on Ancient Greek Cult Athens, 16.–18. October 1992, Stockholm, 21–29.

Buraselis, Kostas (2000), *Kos between Hellenism and Rome. Studies on the Political, Institutional and Social History of Kos from ca. the Middle Second Century B.C. until Late Antiquity* (Transactions of the American Philosophical Society 90.4), Philadelphia.

Buxton, Richard (ed.) (2000), *Oxford Readings in Greek Religion*, Oxford.

Campanelli, Sara (2011), "Tra epigrafia e topografia. Su una recente ipotesi di identificazione del santuario di Eracle *Diomedonteios* a Cos", in: *Scienze dell'Antichità* 17, 643–685.

Campanelli, Sara (2012), "*Hiera kai bebala* in un'iscrizione di Cos. Lessico epigrafico e legislazione sacra", in: Gabriella Bevilacqua and Sara Campanelli (eds.), Ἀρετῆς ἕνεκεν καὶ σοφίας. *Un omaggio a Paola Lombardi*. Giornata di Studio Roma, 28 Ottobre 2010 (Opuscula Epigraphica 14), Rome, 83–95.

Campanelli, Sara (2012), *Eroizzazione e proprietà terriera nel 'Testamento di Epikrates'. Per una proposta di lettura delle fondazioni cultuali di carattere familiare*. http://www.unipa.it/dipartimenti/beniculturalistudiculturali/riviste/hormos/.content/documenti/5.-Sara-Campanelli-hormos-2012.pdf (last accessed: 29.09.2015).

Carbon, Jan-Mathieu/Pirenne-Delforge, Vinciane (2013), "Priests and Cult Personnel in Three Hellenistic Families", in: Marietta Horster and Anja Klöckner (eds.), *Cities and Priests. Cult Personnel in Asia Minor and the Aegean Islands from the Hellenistic to the Imperial Period* (Religionsgeschichtliche Versuche und Vorarbeiten 64), Berlin/Boston, 65–119.

Carlsson, Susanne (2010), *Hellenistic Democracies. Freedom, Independence and Political Procedure in Some East Greek City-States* (Historia-Einzelschriften 206), Stuttgart.

Carroll-Spillecke, Maureen (1989), ΚΗΠΟΣ. *Der antike griechische Garten* (Wohnen in der klassischen Polis 3), Munich.

Carstens, Anne Marie (2004), "Rock-cut Tombs in the Halikarnassos Peninsula", in: Jonas Eiring and Jørgen Mejer (eds.), *Proceedings of the Danish Institute at Athens*, vol. 4, Athens, 49–67.

Casabona, Jean (1966), *Recherches sur le vocabulaire des sacrifices en grec des origines à la fin de l'époque classique* (Publications des Annales de la Faculté des Lettres d'Aix-en-Provence n.s. 56), Aix-en-Provence.

Casevitz, Michel (1984), "Temples et sanctuaires: ce qu'apprend l'étude lessicale", in: Georges Roux (ed.), *Temples et sanctuaires*. Séminaire de recherche 1981–1983 (Travaux de la Maison de l'Orient 7), Lyon, 83–95.

Chandezon, Christophe (1998), "Paysage et économie rurale en Asie Mineure à l'époque hellénistique. À partir de quelques baux de Mylasa (IIe–Ie siècle avant J.-C.)", in: *Histoire et Sociétés Rurales* 9, 33–56.

Chandezon, Christophe (2003), "Les campagnes de l'Ouest de l'Asie Mineure à l'époque hellénistique", in: Francis Prost (ed.), *L'Orient méditerranéen de la mort d'Alexandre aux campagnes de Pompée. Cités et royaumes à l'époque hellénistique*. Actes du Colloque international Rennes, 4.–6. avril 2003 (Pallas 62), Rennes, 193–217.

Chantraine, Pierre (2009), *Dictionnaire étymologique de la langue grecque. Histoire des mots* (Librairie Klincksieck, Série linguistique 20), Paris.

Chiricat, Édouard (2005), "Funérailles publiques et enterrement au gymnase à l'époque hellénistique", in: Pierre Fröhlich and Christel Müller (eds.), *Citoyenneté et participation à la basse époque hellénistique*. Actes de la table ronde Paris, 22.–23. May 2004 (Hautes études du monde gréco-romain 35), Geneva, 207–223.

Cohen, Ada (2011), "Picturing Greek Families", in: Beryl Rawson (ed.), *Companion to Families in the Greek and Roman Worlds* (Blackwell Companions to the Ancient World), Malden (MA), 465–487.

Cole, Susan Guettel (1992), "*Gynaiki ou themis*: Gender Difference in the Greek *Leges Sacrae*", in: *Helios* 19, 104–122.

Cole, Susan Guettel (2004), *Landscapes, Gender, and Ritual Space. The Ancient Greek Experience* (Joan Palevsky Imprint in Classical Literature), Berkeley/Los Angeles/London.

Collins, John N. (1990), Diakonia. *Re-interpreting the Ancient Sources*, New York/Oxford.

Connor, Walter Robert (1988), "'Sacred' and 'Secular'. Ἱερὰ καὶ ὅσια and the Classical Athenian Concept of the State", in: *AncSoc* 19, 161–188.

Cox, Cheryl Anne (1998), *Household Interests. Property, Marriage Strategies, and Family Dynamics in Ancient Athens*, Princeton.

Cucuzza, Nicola (1997), "Artemis *Toxitis* a Coo. In margine alle guerre rodio-cretesi e ad Aglaos di Coo", in: *PP* 52, 13–31.

Damaskos, Dimitris et al. (2004), "Heroisierung", in: *Thesaurus Cultus et Rituum Antiquorum*, vol. 2, 129–158.
Dareste, Rodolphe/Haussoullier, Bernard/Reinach, Théodore (1898), *Recueil des Inscriptions Juridiques Grecques. Texte, Traduction, Commentaire*, vol. 2, Paris.
De Matteis, Lorella (2004), *Mosaici di Cos. Dagli scavi delle missioni italiane e tedesche (1900–1945)* (Monografie della Scuola Archeologica di Atene e delle Missioni Italiane in Oriente 17), Athens.
Dignas, Beate (2002), *Economy of the Sacred in Hellenistic and Roman Asia Minor* (Oxford Classical Monographs), Oxford/New York.
Dillon, Matthew P. J. (1997), "The Ecology of the Greek Sanctuary", in: *ZPE* 118, 113–127.
Donohue, Alice A. (1997), "The Greek Images of the Gods. Considerations on Terminology and Methodology", in: *Hephaistos* 15, 31–45.
Ehrhardt, Norbert (2008), "Der Bürger als Heros: Staatsbegräbnis und Öffentlichkeit in Kyzikos I", in: Engelbert Winter (ed.), *Vom Euphrat bis zum Bosporus. Kleinasien in der Antike. Festschrift für Elmar Schwertheim zum 65. Geburtstag*, vol. 1 (Asia Minor Studien 65), Bonn, 187–196.
Faraone, Christopher A. (2008), "Household Religion in Ancient Greece", in: John Bodel and Saul M. Olyan (eds.), *Household and Family Religion in Antiquity* (Ancient World, Comparative Histories), Oxford, 210–228.
Fedak, Janos (2006), "Tombs and Commemorative Monuments", in: Frederick E. Winter (ed.), *Studies in Hellenistic Architecture* (Phoenix, Supplementary Volume 42), Toronto/Buffalo/London, 71–95.
Ferrucci, Stefano (ed.) (2005), Iseo, *La successione di Kiron* (Studi e Testi di Storia Antica 15), Pisa.
Feyel, Christophe (2009), ΔΟΚΙΜΑΣΙΑ. *La place et le rôle de l'examen préliminaire dans les institutions des cités grecques* (Collection d'Etudes anciennes 36), Paris.
Fraser, Peter Marshall (1953), "An Inscription from Cos", in: *Bulletin de la Société Archéologique d'Alexandrie* 40, 35–62.
Gabrielsen, Vincent (2008), "The Public Banks of Hellenistic Cities", in: Koenraad Verboven, Katelijn Vandorpe and Véronique Chankowski (eds.), Pistoi dia tèn technèn. *Bankers, Loans and Archives in the Ancient World. Studies in honour of Raymond Bogaert* (Studia Hellenistica 44), Leuven, 115–130.
Galli, Marco (2007–2008), "*Sema*: sepolture eroiche e identità civiche nell'Oriente dell'Impero romano (II–I secolo a.C.)", in: Gilda Bartoloni and Gilda M. Benedettini (eds.), *Sepolti tra i vivi. Evidenza e interpretazione di contesti funerari in abitato*, Roma, 26.–29. Aprile 2006 (Scienze dell'Antichità 14), Rome, 567–590.
Gauthier, Philippe (1985), *Les cités grecques et leurs bienfaiteurs* (Bulletin de Correspondance Hellénique, Supplément 12), Paris.
Gawlinski, Laura (2012), *The Sacred Law of Andania. A New Text with Commentary* (Sozomena 11), Berlin/Boston.
Geppert, Karin (1996), "Gehöfte und Gräber im Gebiet des Avşar Tepesi", in: Frank Kolb (ed.), *Lykische Studien 3. Die Siedlungskammer von Kyaneai in Lykien. Bericht über Feldforschungen im Yanu-Bergland im Sommer 1992* (Asia Minor Studien 24), Bonn, 39–50.
Gherchanoc, Florence (2012), *L'*oikos *en fête. Célébrations familiales et sociabilité en Grèce ancienne* (Publications de la Sorbonne. Série Histoire ancienne et médiévale 111), Paris.
Graf, Fritz (1985), *Nordionische Kulte* (Bibliotheca Helvetica Romana 21), Vevey.
Graf, Fritz (1995), "Bemerkungen zur bürgerlichen Religiosität im Zeitalter des Hellenismus", in: Michael Wörrle and Paul Zanker (eds.), *Stadtbild und Bürgerbild im Hellenismus*. Kolloquium München, 24.–26. Juni 1993 (Vestigia 47), Munich, 103–114.
Grieb, Volker (2008), *Hellenistische Demokratie. Politische Organisation und Struktur in freien griechischen Poleis nach Alexander dem Großen* (Historia, Einzelschriften 199), Stuttgart.
Hailer, Ulf (2000), "Gehöfte, Land- und Viehwirtschaftliche Installationen, Gräber", in: Frank Kolb (ed.), *Lykische Studien 5. Die Siedlungskammer des Yanu-Berglandes. Berichte über die*

Ergebnisse der Feldforschungskampagne 1995 auf dem Territorium der Zentrallykischen Polis Kyaneai (Asia Minor Studien 41), Bonn, 59–78.

Hammerstaedt, Jürgen (2009), "Die Bedeutung inschriftlicher Zeugnisse für die Bauforschung", in: Martin Bachmann (ed.), *Bautechnik im antiken und vorantiken Kleinasien*. Internationale Konferenz Istanbul, 13.–16. Juni 2007 (Byzas 9), Istanbul, 243–250.

Harrison, Alick Robin Walsham (1968), *The Law of Athens*, vol. 1, Oxford.

Harter-Uibopuu, Kaja (2010), "Erwerb und Veräußerung von Grabstätten im griechisch-römischen Kleinasien am Beispiel der Grabinschriften aus Smyrna", in: Gerhard Thür (ed.), *Symposion 2009. Vorträge zur griechischen und hellenistischen Rechtsgeschichte*, Seggau, 25.–30. August 2009 (Akten der Gesellschaft für Griechische und Hellenistische Rechtsgeschichte 21), Vienna, 247–270.

Harter-Uibopuu, Kaja (2011), "Money for the Polis. Public Administration of Private Donations in Hellenistic Greece", in: Onno M. van Nijf and Richard Alston (eds.), *Political Culture in the Greek City after the Classical Age* (Groningen-Royal Holloway Studies on the Greek City after the Classical Age 2), Leuven/Paris/Walpole (MA), 119–139.

Hegyi, Dolores (1976), "Τεμένη ἱερὰ καὶ τεμένη δημόσια", in: *Oikumene* 1, 77–87.

Hellmann, Marie-Christine (1992), *Recherches sur le vocabulaire de l'architecture grecque, d'après les inscriptions de Délos* (Bibliothèque des Écoles Françaises d'Athènes et de Rome 278), Paris.

Hellmann, Marie-Christine (1999), *Choix d'inscriptions architecturales grecques traduites et commentées* (Travaux de la Maison de l'Orient 30), Lyon.

Hellmann, Marie-Christine (2006), *L'architecture grecque*, vol. 2: *Architecture religieuse et funéraire* (Manuels d'art et d'archéologie antiques), Paris.

Herrmann, Peter/Polatkan, Kemal Ziya (1969), "Das Testament des Epikrates und andere neue Inschriften aus dem Museum von Manisa", in: *Sitzungsberichte – Österreichische Akademie der Wissenschaften, Philosophisch-Historische Klasse* 265 (1), 1–63.

Herzog, Rudolf (1928), *Heilige Gesetze von Kos* (Abhandlungen der preussischen Akademie der Wissenschaften 1928, philosophisch-historische Klasse 6), Berlin.

Hiller von Gaertringen, Friedrich (1914), "Ἐπιγραφαὶ Ῥόδου, Θήρας, Νάξου, Ἀρκαδίας", in: *AEph*, 130–135.

Horster, Marietta (2004), *Landbesitz griechischer Heiligtümer in archaischer und klassischer Zeit* (Religionsgeschichtliche Versuche und Vorarbeiten 53), Berlin/New York.

Horster, Marietta (2010), "Religious Landscape and Sacred Ground: Relationships between Space and Cult in the Greek World", in: *RHR* 227 (4), 435–458.

Huebner, Sabine R. (2011), "Household Composition in the Ancient Mediterranean – What Do We Really Know?", in: Beryl Rawson (ed.), *Companion to Families in the Greek and Roman Worlds* (Blackwell Companions to the Ancient World), Malden (MA), 73–91.

Hupfloher, Annette (2012), "Kultgründungen durch Individuen im klassischen Griechenland", in: Jörg Rüpke and Wolfgang Spickermann (eds.), *Reflections on Religious Individuality. Greco-Roman and Judaeo-Christian Texts and Practices* (Religionsgeschichtliche Versuche und Vorarbeiten 62), Berlin/Boston.

Isager, Signe/Skydsgaard, Jens Erik (1992), *Ancient Greek Agriculture. An Introduction*, London/New York.

Isager, Signe (2004), "Halikarnassos and the Ptolemies I. Inscriptions on Public Buildings", in: Signe Isager and Poul Pedersen (eds.), *The Salmakis Inscription and Hellenistic Halikarnassos*, (Halicarnassian Studies 4), Odense, 133–144.

Jameson, Michael H. (1994), "*Theoxenia*", in: Robin Hägg (ed.), *Ancient Greek Cult Practice from the Epigraphical Evidence*. Proceedings of the Second International Seminar on Ancient Greek Cult Athens, 22.–24. November 1991, Stockholm, 35–57.

Jones, Christopher P. (1998), "Joint Sacrifices at Iasus and Side", in: *JHS* 118, 183–186.

Jost, Madeleine (1992), *Aspects de la vie religieuse en Grèce. Du début du Ve siècle à la fin du IIIe siècle avant J.-C.* (Regards sur l'Histoire 86), Paris.
Kader, Ingeborg (1995), "*Heroa* und Memorialbauten", in: Michael Wörrle and Paul Zanker (eds.), *Stadtbild und Bürgerbild im Hellenismus*. Kolloquium München, 24.–26. Juni 1993 (Vestigia 47), Munich, 199–229.
Kamps, Werner (1937), "Les origines de la fondation cultuelle dans la Grèce ancienne", in: *Archives d'Histoire du Droit Oriental* 1, 145–179.
Kindt, Julia (2012), *Rethinking Greek Religion*, Cambridge.
Klaffenbach, Günther (1930), "Rudolf Herzog: *Heilige Gesetze von Kos*", in: *Gnomon* 6, 212–217.
Klaffenbach, Günther (1955), "Zu zwei jüngst veröffentlichten griechischen Inschriften", in: Ladislav Varcl (ed.), *Studia antiqua Antonio Salač septuagenario oblata* (Československá Akademie Věd, Sbornik Filologický 3.1), Prague, 122–124.
Kloppenborg, John S./Ascough, Richard S. (2011), *Greco-Roman Associations: Texts, Translations, and Commentary*, vol. 1 (Beihefte zur Zeitschrift für die neutestamentliche Wissenschaft und die Kunde der älteren Kirche 181): *Attica, Central Greece, Macedonia, Thrace*, Berlin/New York.
Konuk, Koray (2004), "The Ptolemaic Coins in the Bodrum Underwater Archaeology Museum", in: Signe Isager and Poul Pedersen (eds.), *The Salmakis Inscription and Hellenistic Halikarnassos*, (Halicarnassian Studies 4), Odense, 165–175.
Kubińska, Jadwiga (1968), *Les monuments funéraires dans les inscriptions grecques de l'Asie Mineure* (Prace Zakładu Archeologii Śródziemnomorskiej Polskiej Akademii Nauk 5), Warsaw.
Lambrinoudakis, Vassilis et al. (2005), "Consecration, Foundation Rites", in: *Thesaurus Cultus et Rituum Antiquorum*, vol. 3, 303–346.
Landskron, Alice (2011), "Zur Komposition der Friese auf der Nord- und Ostseite des Heroons von Trysa", in: *Adalya* 14, 31–56.
Laum, Bernhard (1914), *Stiftungen in der griechischen und römischen Antike. Ein Beitrag zur Antiken Kulturgeschichte*, vols. 1–2, Leipzig/Berlin.
Lauter, Hans (1985), *Der Kultplatz auf dem Turkovuni* (Attische Forschungen 1), Berlin.
Le Roy, Christian (1986), "Un règlement religieux au Létôon de Xanthos", in: *RA n.s.* 2, 279–300.
Lewis, Norman N./Sartre-Fauriat, Annie/Sartre, Maurice (1996), "William John Bankes. Travaux en Syrie d'un voyageur oublié", in: *Syria* 73, 57–95.
Leypold, Christina (2008), *Bankettgebäude in griechischen Heiligtümern*, Wiesbaden.
Liddell, Henry George/Scott, Robert/Jones, Henry Stuart (1968), *A Greek English Lexicon with Supplement*, Oxford.
Lohmann, Hans (1999), "Survey in der *Chora* von Milet. Vorbericht über die Kampagnen der Jahre 1996 und 1997", in: *AA* 1999, 439–473.
Lupu, Eran (2005), *Greek Sacred Law. A Collection of New Documents* (Religions in the Graeco-Roman World 152), Leiden/Boston/Cologne.
Maillot, Stéphanie (2013), "Les associations à Cos", in: Pierre Fröhlich and Patrice Hamon (eds.), *Groupes et associations dans les cités grecques (IIIe siècle av. J.-C.– IIe siècle apr. J.-C.)*. Actes de la table ronde Paris, 19.–20. Juin 2009 (Hautes études du monde gréco-romain 49), Geneva, 199–226.
Mannzmann, Anneliese (1962), *Griechische Stiftungsurkunden. Studie zu Inhalt und Rechtsform* (Fontes et Commentationes 2), Münster.
Marchiandi, Daniela (2011), *I periboli funerari nell'Attica classica: lo specchio di una 'borghesia'* (Studi di Archeologia e Topografia di Atene e dell'Attica 3), Athens/Paestum.
Migeotte, Léopold (2006), "Téménè dèmosia" in: Pascale Brillet-Dubois and Edith Parmentier (eds.), Φιλολογία. *Mélanges offerts à Michel Casevitz* (Collection de la Maison de l'Orient et de la Méditerranée 35), Lyon, 187–195.
Migeotte, Léopold (2010), "La fondation d'Aristoménès et Psylla à Corcyre: dispositions administratives et financières", in: Biagio Virgilio (ed.), *Studi ellenistici* 24, Pisa, 63–69.

Migeotte, Léopold (2012), "À propos de la fondation d'Eudèmos à Milet: questions administratives et numismatiques", in: *Phoenix* 66 (1–2), 1–10.
Mikalson, Jon D. (2005), *Ancient Greek Religion* (Blackwell Ancient Religions), Oxford.
Modrzejewski Mélèze, Joseph (1963), "À propos des fondations en droit grec", in: *Revue historique de droit français et étranger* 4e série 41, 82–92.
Modrzejewski Mélèze, Joseph (2011), *Droit et justice dans le monde grec et hellénistique* (Journal of Juristic Papyrology, Supplement 10), Warsaw.
Naerebout, Frits G. (2009), "Territorialität und griechische Religion. Die aufgeteilte Landschaft", in: Eckart Olshausen and Vera Sauer (eds.), *Die Landschaft und die Religion*. Stuttgarter Kolloquium zur historischen Geographie des Altertums 9 (Geographica Historica 26), Stuttgart, 191–213.
Nilsson, Martin Persson (1955–1961), *Geschichte der griechischen Religion* (Handbuch der Altertumswissenschaft 5. 2. 1–2), vols. 1–2, Munich.
Oberleitner, Wolfgang (1994), *Das* Heroon *von Trysa: ein lykisches Fürstengrab des 4. Jahrhunderts v. Chr.* (Antike Welt, Sondernummer 1994/1), Mainz.
Olshausen, Eckart/Sauer, Vera (eds.) (2009), *Die Landschaft und die Religion*. Stuttgarter Kolloquium zur historischen Geographie des Altertums 9 (Geographica Historica 26), Stuttgart.
Papazarkadas, Nikolaos (2011), *Sacred Land and Public Land in Ancient Athens* (Oxford Classical Monographs), Oxford.
Parker, Robert (1983), Miasma: *Pollution and Purification in Early Greek Religion* (Clarendon Paperbacks), Oxford.
Parker, Robert (2005), *Polytheism and Society at Athens*, Oxford/New York.
Parker, Robert (2008), "πατρῷοι θεοί: The Cults of Sub-Groups and Identity in the Greek World", in: Anders Holm Rasmussen and Susanne William Rasmussen (eds.), *Religion and Society: Rituals, Resources and Identity in the Ancient Graeco-Roman World*. The Bomos–Conferences 2002–2005 (Analecta Romana Instituti Danici, Supplementum 40), Rome, 201–214.
Parker, Robert (2010), "A Funerary Foundation from Hellenistic Lycia", in: *Chiron* 40, 103–121.
Patera, Ioanna (2010), "Éspace et structures cultuels du sanctuaire grec: la construction du vocabulaire", in: *RHR* 227 (4), 535–551.
Paton, William Roger/Hicks, Edward Lee (1891), *The Inscriptions of Cos*, Oxford.
Paul, Stéphanie (2013), *Cultes et sanctuaires de l'île de Cos* (Kernos, Supplément 28), Liège.
Pedersen, Poul (2004), "Halikarnassos and the Ptolemies II. The Architecture of Hellenistic Halikarnassos", in: Signe Isager and Poul Pedersen (eds.), *The Salmakis Inscription and Hellenistic Halikarnassos* (Halicarnassian Studies 4), Odense, 145–164.
Pernin, Isabelle (2007), "La question des baux dans la Grèce des cités", in: Patrice Brun (ed.), *Economies et sociétés en Grèce classique et hellénistique*. Actes du Colloque Bordeaux, 30.–31. Mars 2007 (Pallas 74), Toulouse, 43–76.
Petrakos, Vasileios Ch. (1999), *Ο δήμος του Ραμνούντος: σύνοψη των ανασκαφών και των ερευνών 1813–1998*, vol. 2: *Οι επιγραφές* (Vivliothēkē tēs en Athēnais Archaiologikēs Hetaireias 182), Athens.
Picard, Charles (1921), *Exploration archéologique de Délos 6. L'établissement des Poseidoniastes de Bérytos* (Exploration archéologique de Délos 6), Paris.
Pippidi, Dionisie Mihail (1966), "Sur un décret des thiasites de Callatis", in: *StudClas* 8, 87–96.
Poland, Franz (1909), *Geschichte des griechischen Vereinswesens* (Preisschriften gekrönt und herausgegeben von der Fürstlich Jablonowskischen Gesellschaft zu Leipzig 38), Leipzig.
Polinskaya, Irene (2009), "Fifth-Century *Horoi* on Aigina: A Reevaluation", in: *Hesperia* 78, 231–267.
Pomeroy, Sarah B. (1997), "Family Values: The Uses of the Past", in: Per Bilde, Troels Engberg Pedersen, Lise Hannestad and Jan Zahle (eds.), *Conventional Values of the Hellenistic Greeks* (Studies in Hellenistic Civilization 8), Aarhus, 204–219.

Pomeroy, Sarah B. (1997), *Families in Classical and Hellenistic Greece. Representations and Realities*, Oxford.
Price, Simon (1999), *Religions of the Ancient Greeks* (Key Themes in Ancient History), Cambridge.
Pugliese Carratelli, Giovanni (1957), "Gli Asclepiadi e il sinecismo di Cos", in: *PP* 12, 333–342.
Pugliese Carratelli, Giovanni (1963–1964), "Il *damos* coo di Isthmos", in: *ASAA* 41/42, 147–202.
Pugliese Carratelli, Giovanni (1991), "La norma etica degli *Asklapiadai* di Cos", in: *PP* 46, 81–94.
Pugliese Carratelli, Giovanni (1994), ΑΠΟΓΟΝΟΙ ΑΣΚΛΗΠΙΟΥ ΚΑΙ ΗΡΑΚΛΕΟΥΣ, in: *Storia poesia e pensiero nel mondo antico: studi in onore di Marcello Gigante* (Saggi Bibliopolis 46), Naples, 543–547.
Purvis, Andrea (2003), *Singular Dedications. Founders and Innovators of Private Cults in Classical Greece* (Studies in Classics 1), New York/London.
Robert, Louis (1937), *Études Anatoliennes. Recherches sur les inscriptions grecques de l'Asie Mineure* (Études orientales 5), Paris.
Ross, Ludwig (1845), *Inscriptiones Graecae Ineditae*, vol. 3, Berlin.
Roux, Georges (1987), *Fouilles de Delphes 2. Topographie et architecture. La terrasse d'Attale I*, Paris.
Rudhardt, Jean (1992), *Notions fondamentales de la pensée religieuse et actes constitutifs du culte dans la Grèce classique* (Antiquité/Synthèses 3), Paris.
Sartre-Fauriat, Annie (2001), *Des tombeaux et des morts. Monuments funéraires, société et culture en Syrie du Sud du Ier s. av. J.-C. au VIIe s. apr. J.-C.*, vol. 2 (Bibliothèque archéologique et historique 158), Beirut.
Schazmann, Paul (1934), "Das *Charmyleion*", in: *JDAI* 49, 110–127.
Schmitt-Pantel, Pauline (1982), "Évergétisme et mémoire du mort. À propos des fondations de banquets publics dans les cités grecques à l'époque hellénistique et romaine", in: Gherardo Gnoli and Jean-Pierre Vernant (eds.), *La mort, les morts dans les sociétés anciennes*, Cambridge/Paris, 177–188.
Scholl, Andreas (1994), "ΠΟΛΥΤΑΛΑΝΤΑ ΜΝΗΜΕΙΑ. Zur literarischen und monumentalen Überlieferung aufwendiger Grabmäler im spätklassischen Athen", in: *JDAI* 109, 239–271.
Schuler, Christof (1998), *Ländliche Siedlungen und Gemeinden im hellenistischen und römischen Kleinasien* (Vestigia 50), Munich.
Segre, Mario (1993), *Iscrizioni di Cos*, vol.1 (Monografie della Scuola Archeologica di Atene e delle Missioni Italiane in Oriente 6), Rome.
Sherwin-White, Susan (1977), "Inscriptions from Cos", in: *ZPE* 24, 205–217.
Sherwin-White, Susan (1978), *Ancient Cos. An Historical Study from the Dorian Settlement to the Imperial Period* (Hypomnemata 51), Göttingen.
Sinn, Ulrich et al. (2005), "Kultorte", in: *Thesaurus Cultus et Rituum Antiquorum*, vol. 4, 1–127.
Sosin, Joshua D. (2001), "Accounting and Endowments", in: *Tyche* 16, 161–175.
Sourvinou-Inwood, Christiane (1988), "Further Aspects of Polis Religion", in: *AION(archeol)* 10, 259–274.
Sourvinou-Inwood, Christiane (1990), "What is Polis Religion?", in: Oswyn Murray and Simon Price (eds.), *The Greek City from Homer to Alexander*, Oxford, 295–322.
Stavrianopoulou, Eftychia (2006), *'Gruppenbild mit Dame'. Untersuchungen zur rechtlichen und sozialen Stellung der Frau auf den Kykladen im Hellenismus und in der römischen Kaiserzeit* (Heidelberger althistorische Beiträge und epigraphische Studien 42), Stuttgart.
Stowers, Stanley K. (1998), "A Cult from Philadelphia: *Oikos* Religion or Cultic Association?", in: Abraham J. Malherbe, Frederick W. Norris and James W. Thompson (eds.), *The Early Church in Its Context. Essays in Honor of Everett Ferguson* (Supplements to Novum Testamentum 90), Leiden/Boston/Cologne, 287–301.
Svenson-Evers, Hendrik (1997), "ΙΕΡΟΣ ΟΙΚΟΣ. Zum Ursprung des griechischen Tempels", in: Wolfram Hoepfner (ed.), *Kult und Kultbauten auf der Akropolis*. Internationales Symposion

Berlin, 7.–9. Juli 1995 (Schriften des Seminars für Klassische Archäologie der Freien Universität Berlin), Berlin, 132–151.

True, Marion et al. (2004), "Processions", in: *Thesaurus Cultus et Rituum Antiquorum*, vol. 1, 1–20.

Trümper, Monika (2002), "Das Sanktuarium des 'Établissement des Poseidoniastes de Bérytos' in Delos. Zur Baugeschichte eines griechischen Vereinsheiligtums", in: *BCH* 126 (1), 265–330.

Vallarino, Giulio (2009), "I dedicanti di Cos in età ellenistica: il caso dei magistrati eponimi tra polis e demi", in: John Bodel and Mika Kajava (eds.), *Dediche sacre nel mondo greco-romano. Diffusione, funzioni, tipologie / Religious Dedications in the Greco-Roman World. Distribution, Typology, Use*, Roma, 19.–20. Aprile 2006 (Acta Instituti Romani Finlandiae 35), Rome, 181–207.

van Bremen, Riet (1996), *The Limits of Participation: Women and Civil Life in the Greek East in the Hellenistic and Roman Periods* (Dutch Monographs on Ancient History and Archeology 15), Amsterdam.

van Bremen, Riet (2003), "Family Structures", in: Andrew Erskine (ed.), *A Companion to the Hellenistic World* (Blackwell Companions to the Ancient World), Malden (MA)/Oxford, 313–330.

van Straten, Folkert (2005), "Ancient Greek Animal Sacrifice: Gift, Ritual Slaughter, Communion, Food Supply, or What? Some Thoughts on Simple Explanations of a Complex Ritual, in: Stella Georgoudi, Renée Koch Piettre and Francis Schmidt (eds.), *La cuisine et l'autel. Les sacrifices en questions dans les sociétés de la Méditerranée ancienne* (Bibliothèque de l'École des Hautes Études, Sciences Religieuses 124), Turnhout, 15–29.

Vatin, Claude (1974), "Jardins et vergers grecs", in: *Mélanges helléniques offerts à Georges Daux*, Paris, 345–357.

Velissaropoulos-Karakostas, Julie (2008), "Les *nothoi* hellénistiques", in: Edward Harris and Gerhard Thür (eds.), *Symposion 2007. Vorträge zur griechischen und hellenistischen Rechtsgeschichte*, Durham, 2.–6. September 2007 (Akten der Gesellschaft für Griechische und Hellenistische Rechtsgeschichte, 20), Vienna, 253–274.

Velissaropoulos-Karakostas, Julie (2011), *Droit grec d'Alexandre à Auguste (323 av. J.-C. – 14 ap. J.-C.). Personnes – Biens – Justice* (Meletēmata 66), vols. 1–2, Athens/Paris.

Vérilhac, Anne-Marie/Vial, Claude (1998), *Le mariage grec du VIe siècle av. J.-C. à l'époque d'Auguste* (Bulletin de Correspondance Hellénique, Supplément 32), Athens.

Veyne, Paul (1976), *Le pain et le cirque. Sociologie historique d'un pluralism politique* (L'Univers Historique), Paris.

Weber, Berthold F. (2004), *Die römischen* Heroa *von Milet* (Milet 1.10), Berlin/New York.

Weiß, Peter (1998), "Festgesandschaften, Städtisches Prestige und Homonoiaprägungen", in: *Stadion* 24 (1), 59–69.

Wittenburg, Andreas (1990), *Il testamento di Epikteta* (Pubblicazioni del Dipartimento di Scienze dell'Antichità 4), Trieste.

Wittenburg, Andreas (1998), "Grandes familles et associations cultuelles à l'époque hellénistique", in: *Ktema* 23, 451–456.

Wörrle, Michael (1988), *Stadt und Fest im kaiserzeitlichen Kleinasien. Studien zu einer agonistischen Stiftung aus Oinoanda* (Vestigia 39), Munich.

Wycherley, Richard Ernest (1960), "*Neleion*", in: *ABSA* 55, 60–66.

Zelnick-Abramovitz, Rachel (2005), *Not Wholly Free. The Concept of Manumission and the Status of Manumitted Slaves in the Ancient Greek World* (Mnemosyne, Supplementum 266), Leiden/Boston.

Zuiderhoek, Arjan (2011), "Oligarchs and Benefactors. Elite Demography and Euergetism in the Greek East of the Roman Empire", in: Onno M. van Nijf and Alston Richard (eds.), *Political Culture in the Greek City after the Classical Age* (Groningen-Royal Holloway Studies on the Greek City after the Classical Age 2), Leuven/Paris/Walpole (MA), 185–195.

Flavia Manservigi and Melania Mezzetti

The Didyma Inscription: Between Legislation and Palaeography

1 Research Aims

The Research aims to investigate the shift from a majuscule to a minuscule writing system in the Roman world, which started in the 3rd century A.D. This change has been extensively analysed within historiography, and it has been at the centre of an intense debate.[1] Recent discoveries allow a reinvestigation of this complex problem, and this article will examine the survival of the majuscule writing, the so-called Ancient Roman cursive, in its chancery results.

The article will focus in particular on the relationship between the chancery writing system and a law issued in A.D. 368[2] by the emperors Valentinian and Valens, which was incorporated into the Theodosian Code as CTh. 9.19.3. Historiography recognises the crucial role of this law in the regulation and development of Late Empire chancery scripts: for the first time, imperial power intervened directly in a matter of writing, and imposed a clear prohibition on the use of a type of script.

The sources commonly used to chart the development of writing systems are papyri. However, epigraphic sources have also recently been included in this analysis, particularly thanks to the studies of Denis Feissel, who investigated an inscription that is thus far the only epigraphic evidence relating to the problem of the shift from the majuscule to the minuscule writing system.

This inscription was found in the small town of Didyma (Caria) in 1991, and it contains a rescript of Justinian I dated to A.D. 533. The relevance of this source was highlighted by Feissel's analysis, which relied on a historic and epigraphic approach.

This article emerged from the Heidelberg Collaborative Research Center 933 "Material Text Cultures. Materiality and Presence of Writing in Non-Typographic Societies". The CRC 933 is financed by the German Research Foundation (DFG). The present study started some years ago at the University of Bologna; it initially focused on papyri, but at a later stage extended to the analysis of epigraphic witnesses. A benchmark of the research is the work of Lorenza Iannacci, Maddalena Modesti and Annafelicia Zuffrano, published in 2012, where they analysed the matter in connection with papyrological sources (Iannacci/Modesti/Zuffrano 2012). The article is divided into two parts: the first, concerning the historical frame, was managed by Melania Mezzetti. The second part, concerning the analysis of the Didyma inscription from a palaeographic viewpoint, and the conclusions were written by Flavia Manservigi.

1 About this argument see in particular: Perrat 1955; Mallon/Marichal/Perrat 1939; Mallon 1948, 1952 and 1982; Marichal 1948; Cencetti 1950, 1956/57, 1961, 1993 and 1997; Tjäder 1955, 1977, 1979, 1982a and 1985; Innacci/Modesti/Zuffrano 2012.

2 About the issue date of the law see Iannacci/Modesti/Zuffrano 2012, 96, fn. 21.

DOI 10.1515/9783110417845-006,

Nevertheless, it seems useful to analyse the inscription again, in particular from a palaeographic perspective. Furthermore, the analysis will focus on lines 36 and 37, which are written in a special enlarged cursive script directly copied from an original papyrus model.

Therefore, the initial focus of this research will be the law CTh. 9.19.3. Subsequently, the graphical context of the period will be briefly discussed, and the relationship between this context and the law will be investigated. Then, the article will analyse previous research on papyrological sources; finally, the problem will be analysed in the light of the Didyma inscription.

1.1 The Law CTh. 9.19.3

During the Roman Empire, a few laws were issued which directly or incidentally concerned the script and shape of documents, and they were incorporated into the most important compilations of Roman Laws: the Theodosian Code, promulgated in A.D. 438, and the *Corpus iuris civilis*, issued between A.D. 528 and 534.[3]

These laws are very interesting as they regulate writing, predominantly of public documents, in many different ways. They may regard both the type of writing and the form of the documents or their preservation: all these regulations were functional in assuring the necessary authenticity and reliability of official documents.

CTh. 9.19.3 is the most ancient of these laws, and it was issued by the emperors Valentinian and Valens in Trier, where the court was held.
This is the text of the law:

> Impp. Valentinianus et Valens AA. ad Festum proconsulem Africae. Serenitas nostra prospexit inde caelestium litterarum coepisse imitationem, quod his apicibus tuae gravitatis officium consultationes relationesque complectitur, quibus scrinia nostrae perennitatis utuntur. Quam ob istius sanctionis auctoritate praecipimus, ut posthac magistra falsorum consuetudo tollatur et communibus litteris universa mandentur, quae vel de provincia fuerint scribenda vel a iudice, ut nemo stili huius exemplum aut privatim sumat aut publice. Dat. V id. iun. Treviri Lupicino et Iovino conss.

> Emperors Valentinian and Valens Augustuses to Festus, Proconsul of Africa. Our Serenity has observed that the practice of imitating Our celestial imperial letters has arisen from the fact that the office of Your Gravity, in composing references of cases to the Emperor and reports to Him, uses the same kind of script as that which the bureaus of Our Eternity use. Wherefore, by the authority of this sanction, We command that hereafter this custom, a teacher of forgery, shall be abolished and that everything which must be written either from a province or by a judge shall be

3 The laws concerning the writing and shape of documents are: CTh. 9.19.3, CTh. 9.35.1, C. 23.3, C. 23.6, Nov. 47.

entrusted to commonly used letters, so that no person shall have the right to appropriate a copy of this style, either privately or publicly. Given on the fifth day before the ides of June at Trier in the year of the consulship of Lupicinus an Jovinus.[4]

In A.D. 368 the imperial chancellery ordered the proconsul of Africa, *Iulius Hymetius Festus*, to stop imitating the *litterae caelestes* in composing reports, since the emperors had reserved that type of script for themselves. The stated purpose of this law was to remove the risk of falsification. The emperors commanded that this particular script had to be used only by the imperial chancery; the provincial chanceries, as well as every administrative, bureaucratic and judiciary office, had to use another type of script, the so-called *litterae communes*.

1.2 *Litterae caelestes* and *litterae communes*

In order to understand what distinguished these two types of writing, the wider context of documentary writings of the period and their development should be taken into account.

The terms *litterae caelestes* and *litterae communes* are used to refer to two types of script, deriving from different graphic models, both commonly used in the Roman Empire, but in different periods.

The first script (*litterae caelestes*) is the result of a writing system that, for convenience, we can consider to be of the majuscule type,[5] the so-called Ancient Roman cursive, which was normally used for documents from the 1st century A.D. until the end of the 3rd century. The second (*litterae communes*) is the result of a new minuscule script, the so-called Later Roman cursive, which progressively circulated from the 2nd century A.D. onwards.

After a period of coexistence in the sources (2nd and 3rd centuries), the Ancient Roman cursive was gradually used only in some chanceries. From the end of the 3rd century on, the Later Roman cursive had completely replaced the Ancient script, not only for private documents, but also for those of the public bureaus, whereas the majuscule script was restricted to a specific use: the documentary production of the Imperial chancery.

4 *The Theodosian Code*, p. 241. For the bibliography on this law see Iannacci/Modesti/Zuffrano 2012, 96.

5 Jean Mallon, as will be discussed below, was the first to question the current opinion, developed in the 19th and 20th centuries, which hypothesized a gradual development of the Later Roman cursive from the ancient script (see in particular Schiaparelli 1921, 118: "La maiuscola corsiva finisce gradatamente in minuscola corsiva"). He rejected the previous theories and identified a uniform common script (*écriture commune classique*) that did not directly give rise to the Later Roman cursive (*nouvelle écriture commune*), see Mallon 1952 passim.

For the purposes of the current study, the extensive debate over the reasons and origins of this change, which involved many scholars in the last century, will not be reviewed. Nevertheless, it should suffice to remember that nowadays, especially following Jean Mallon's contributions to the field, scholars are unanimous in recognising that the shift from a majuscule to a minuscule system was a graphic *caesura*. As it was pointed out, this change took place in the field of the "usual script", according to the definition of Giorgio Cencetti, or, in other words, in the one of the "common script";[6] furthermore, it can be affirmed that this was a very complex phenomenon, determined by different and concurrent causes and aspects[7] that progressively effected the change.

What is important to highlight is that in the 4th century, when CTh. 9.19.3 was issued, the Later Roman cursive had already become the usual script for some time. It was indeed the referential script commonly used to write every kind of public and private document.

This is obviously only a short review of the complex graphic context in which the imperial chancery issued CTh. 9.19.3. Scholars of Roman jurisprudence and legal historians have not dealt with this law; studies of the 4th century have focused in particular on the relationship between paganism and Christianity, rather than on the history of writing. Those scholars who specialised in the legislation of Valentian and Valens have also said nothing about CTh. 9.19.3. On the other hand, palaeographers have often dealt with this law, always in relation to the problem of chancery writings, because they connected the development and the chancery outcomes of the writing to what CTh. 9.19.3 commanded.

Over the 19th and 20th centuries, great attention was given to new evidences that could offer a better interpretation of the development of the Late Empire chancery writings. After these discoveries and the studies of Brandi, Saint Martin, Champol-

---

6 "In ciascuna epoca e in ciascun luogo gli atteggiamenti delle scritture spontanee dei singoli individui [...] possono essere più o meno diversi: hanno, peraltro, tutte qualche cosa in comune, se non altro il modello ideale, lo schema, lo stampo [...] Questa comunità, questa costanza delle scritture individuali, che in certo modo le comprende tutte epperciò non può essere costretta e configurata in regole precise e inderogabili, ma pure ha caratteri suoi propri e uniformi" (Cencetti 1997, 53); see also Cencetti 1967/69 (= rpt. 1993, 277). According to Cencetti, the script used in the 2nd and 3rd centuries was the "official" branch of the cursive; there was also a branch for common use, though it is not documented by any source. Cencetti states that the Later Roman cursive developed from the latter branch: on this issue see in particular Cencetti 1948; Cencetti 1965, 18–24; Cencetti 1997, 60–66; Nicolaj 1973, 18, fn. 46; for the development of the library script see Cencetti 1965, 25.

7 As Cencetti said, it was "[...] un fatto complesso, tecnico, estetico e culturale insieme, e sua sede è l'intera scrittura 'usuale', non l'uno o l'altro dei rami (documentario e librario) nei quali è per lo più divisa da coloro che non riconoscono la sostanziale unità e non la distinguono dalle scritture canonizzate [...] e il processo evolutivo consiste in una serie di mediazioni fra quei due parametri, ora adattandosi al tracciato calligrafico forme nate dalla spontaneità corsiva, ora tracciandosi rapidamente e corsivamente segni di formazione calligrafica." (Cencetti 1997, 65).

lion-Figeac, Massmann, De Wailly and Mommsen, a fundamental work on this matter was *L'écriture de la chancellerie impériale romaine*.[8] Published in 1948 by Jean Mallon, it was the first complete study of the writings of the Roman imperial chancery.

Mallon was the first scholar to associate the *litterae caelestes* with the Ancient Roman cursive; as a consequence, all subsequent historiography connected the *litterae communes* with the writing system of the Later Roman cursive.

Furthermore, Mallon recognised the script used in two papyri (one conserved in Leiden and the other in Paris)[9] as an evidence of the *litterae caelestes* mentioned in the law, a type of script that he considered special not only for its external features (for example, its exaggerated and forced calligraphic shape), but above all for its anachronistic classicism.[10]

Those papyri, found in Egypt, contain two imperial rescripts of the 5th century. They are very important sources, since they are the only surviving documents of the Roman imperial bureau, whereas all the other documents are copies. Today they are edited in the collection of the *Chartae Latinae Antiquiores*.

With the law of A.D. 368, the imperial chancery reserved for itself the exclusive use of this script, characterised by a majuscule model no longer in common use. According to Mallon, the *litterae caelestes* were the last surviving branch of a majuscule writing that had already disappeared from common use in the 4th century.

This scholar states that, once the law took effect, the *litterae caelestes* were used only by the imperial chancery, and abandoned by all other offices, which adopted the *litterae communes*, characterizing them with chancery artifices. This use was maintained until the Early Middle Ages.

The conclusions of Jean Mallon have generally been accepted by the historiographical community, which, with a few exceptions, did not return to the issues of the law CTh. 9.19.3 and the *litterae caelestes*.

---

8 Mallon 1948, in particular 22–23 (= rpt. 1982, 176) see also Mallon 1952, 116–117.

9 ChLA XVII 657: Paris, Bibliothèque nationale de France, lat. 16915 + Paris, Louvre, P. 2404 + Leiden, Rijksmuseum van Oudheden, P. 421. For other possible witnesses see Iannacci/Modesti/Zuffrano 2012, 97.

10 Mallon 1948, 15 (= rpt. 1982, 172): "En quoi consiste donc exactement, la 'spécialité', si mal dégagée jusqu'ici, de l'écriture de Leyde-Paris? Du point de vue graphique, sa particularité extrêmement superficielle: elle réside seulement en ceci que les lettres sont démesurément allongées et associées en autant de ligatures qu'il est possible, qu'elles sont calligraphiées, de la sorte, sans aucun souci de lisibilité, mais au contraire dans un esprit qui laisse percer des intentions décoratives et esthétiques, tous caractères qui conviennent fort bien à une écriture fossilisée dans une chancellerie. Ceci dit, par tout sa structure fondamentale, par tous ses ductus, cette écriture est, purement et simplement, de l'ancienne cursive romaine, encore plus évoluée, et dans le même sens, que celle des derniers exemples que nous en possédons antérieurement, et qui sont du IIIème siècle. Ce qui fait essentiellement la 'spécialité' de l'écriture de Leyde-Paris, c'est qu'elle est du Vème siècle. Son classicisme anachronistique et suranné, d'une part, et le fait qu'elle est employée pour des originaux de rescrits impériaux, d'autre part, sont deux données qui se soutiennent mutuellement et s'expliquent l'une par l'autre".

For this reason, every modern palaeography book states that this law sanctioned the definitive appropriation of the Ancient Roman cursive by the imperial chancery; consequently, this script, which had already fallen into disuse in common practice, was also abandoned by the other chanceries and public offices. This, according to the generally accepted view, entailed the end of its working life, and heralded its gradual death.

At the same time, the smaller Roman chanceries, which were not allowed to use the majuscule script, were forced to adopt the Later Roman cursive, the minuscule model that, in this way, was transmitted to the Middle Ages. From that perspective, it is due to CTh. 9.19.3 that the chancelleries of the Early Middle Ages adopted scripts of the minuscule type; for example, the Merovingian one.

In fact, Giorgio Cencetti and, more recently, Otto Kresten, demonstrated that the writing systems of the early medieval chanceries were a direct result of the chancery writings of the Roman provinces.[11]

After Jean Mallon, other scholars investigated, more or less directly, the relationship between CTh. 9.19.3 and the chancery writing system of the Later Roman Empire, focusing their attention on new papyrus sources that had been discovered in the meantime, or on already-known sources that were connected with this question. Some of these sources cast new light on this complicated graphical issue. Some very interesting work can be done, in particular, with regard to the studies of Robert Marichal, Jan-Olof Tjäder and more recently some scholars at the University of Bologna.

Marichal's research on reports of proceedings from Egypt, and Tjäder's on the "*misteriosa scrittura grande*"[12] of Ravenna, though having different purposes, highlighted new aspects of the study of the Late Empire chancery scripts and the relationships between these and CTh. 9.19.3. These scholars found evidence of majuscule scripts, which could be linked to the shape of the *litterae caelestes*, outside the imperial chancery.

In particular, Marichal analysed some proceedings where the initial script, marking the consular dating, presents elongated characters similar to the *litterae caelestes*:[13] it can be recalled, for example, in P.Sakaon 34, dated to A.D. 321,[14] and P.Oxy. XVI 1879[15] dated to A.D. 434, to which we will return later.

11 See also Cherubini/Pratesi 2010, 131ff.

12 That is "mysterious and big script".

13 Marichal 1952.

14 P.Sakaon 34 = P.Thead. 13 = ChLA XLI 1204. Marichal affirmed that the characters of the script were surely elongated, even though he could not see the reproduction that was available only a few years later. See Marichal 1952, 348–349.

15 P.Oxy. XVI 1879 = ChLA XLVII 1409.

An opening script similar to the one studied by Marichal on papyri from Egypt can be found in a group of Ravenna papyri studied by Tjäder, and dated between the 6th and 7th centuries, containing *gesta municipalia*.[16]

In those documents, a particular script characterises only the beginning of the first column of the papyrus, while the rest of the text is written in Later Roman cursive and its final part (with the final statements of the magistrate) is written in a tightened and elongated script, belonging, in turn, to the shapes of the Later Roman cursive.

The initial script analysed by Tjäder is characterised by an enlarged form and a complex and overwritten shape that created several problems for the scholars who tried to decipher it.[17]

The particular and complicated shape of the initial script of the Ravenna papyri led Tjäder to call it "*misteriosa scrittura grande*".[18]

The diplomatic analysis of this script, also based on a comparison with the almost coeval initial script of the reports of proceedings from Egypt,[19] allowed Tjäder to understand that it contained merely the date and the place where the act was issued.[20]

Therefore, the opening scripts of the documents from Egypt and Ravenna, although not used for the entirety of the document, can be connected with the majuscule model used in the *litterae caelestes*.[21]

What this paper wants to emphasise, as some scholars of the University of Bologna recently have,[22] is that evidence of this special opening script can be found in sources dated both before A.D. 368, when the law was promulgated, and after this date.

It is not unusual to find examples of this majuscule opening script before 368: even though the majuscule script was no longer in use, it was still adopted by some chanceries, without contravening any legislative ban.

---

16 P. Tjäder 29 (A.D. 504 = ChLA XLV 1332), 8 (A.D. 564 = ChLA XVII 652), 14–15 (A.D. 572 = ChLA XXIX 889) and 21 (A.D. 625 = ChLA XXII 720), see Tjäder 1982a.

17 See in particular Marini 1805, fn. 2; see also Tjäder 1952, 186ff.

18 Tjäder 1952.

19 "La corsiva iniziale dei papiri dell'Egitto è [...] senza alcun dubbio uno svolgimento alquanto artificiale della corsiva antica, ed è altresì eguale, in sostanza, alla scrittura dei rescritti imperiali, definita, da ultimo dal Mallon, appunto 'L'écriture de la cancellerie impériale romaine'" (Tjäder 1952, 205). These sources were analysed by Marichal in the same period, but for other purposes: see Marichal 1952, passim and Marichal 1950b. For a list of the Egyptian papyri analysed by Tjäder 1952, 201–203 and Feissel 2004, 331, fn. 219 (= rpt. 2010, 295, fn. 219), and see also Kresten 1965, Kresten 1966 and the bibliography quoted in ChLA XLI 1188.

20 I.e. indication of the exact day of the month and the place where the act was issued, which were repeated in Later Roman cursive in the following line (for this matter see below).

21 Marichal had come to the same conclusion in the same period through the analysis of the reports of proceedings from Egypt, Marichal 1952, 348–349.

22 Iannacci/Modesti/Zuffrano 2012, 104–105.

Nonetheless, Marichal and Tjäder[23] demonstrated that the use of an overstated writing continued after that date. It is visible in the aforementioned reports of proceedings, but also in the opening lines of the Ravenna papyri. This is a very important consideration, which recalls the question whether CTh. 9.19.3 had been effectively applied. This is a matter that has not been sufficiently considered yet.

Only Marichal mentioned this issue, stating that the diffusion of the special opening scripts in the proceedings confirms that the use of *litterae caelestes* was legal, since it was limited to consular dating, and it represented a sign of respect to high government offices.

For this reason, it is necessary to examine this matter again, and the research will be based not only on papyri, but above all on an epigraphic source, the Didyma inscription.

In this paper, this inscription will be analysed starting from the graphic data, and great attention will be given to the analysis of how this source is connected to the general theme of the relationships between CTh. 9.19.3 and the Late Empire chancery scripts.

Furthermore, the analysis will concentrate in particular on lines 36 and 37, containing the date and the place of issue of the *gesta praefectoria*, which were clearly copied from an original model written on papyrus.[24]

2 The Didyma Inscription

The discovery of the Didyma inscription, which is currently kept in the Museum of Miletus (Balat), occurred during archaeological surveys in the southwest area of the 'sacred path'. The stone had been re-used as building material.

The inscription is divided into three texts: the upper half of the stone (lines 1–35) bears an imperial rescript of Justinian I, qualified as a *pragmatica sanctio*. The lower half is occupied by two complementary texts: an act of the prefect of the Orient (lines 36–56, *gesta praefectoria*) and an act of the governor of Caria (lines 57–64).

The rescript is written in Greek (except for the date, which is in Latin), while the *gesta praefectoria* and the act of the governor are written in Greek and Latin (Latin is used especially for the titulature). The alternation of the two alphabets does not

23 Marichal 1952; Tjäder 1952, 5–21.

24 Even if the object of the research is an inscription, SEG 54, 1178 (= AÉ 2004, 1410), the particular script of lines 36–37 need a palaeographic approach. It must be stated, nevertheless, that often the terminology and classifications used by palaeographers and epigraphists are not uniform and homogeneous. Concerning the development of the Latin script in the epigraphic field see Calabi Limentani 1974, 147 and Casamassima/Staraz 1977, 9.

correspond to that of the languages: some Greek words are transcribed into the Latin alphabet.[25]

In 2004, as already mentioned, Denis Feissel analysed this inscription from an epigraphical, palaeographical and historical point of view.[26] He highlighted the importance of a document reporting the complete text of a Constitution of Justinian that is, furthermore, anchored in its original context.[27]

Feissel states that the general features of the inscription (particularly the equilibrium of the *mise en page* and the researched contrast between the alphabets and the regularity of each stroke) makes this inscription "une sorte de chef d'œuvre de ce qu'on peut appeler la renaissance épigraphique justinienne."[28]
On the Didyma inscription are written three different texts, which can be differentiated from each other by their original function and their date of issue.

In this area of Asia Minor it is not rare to find acts edited on the basis of this treble scheme: another example is an inscription found at Miletus, a contemporary of the Didyma inscription, which bears an act of Justinian I, a decree of a high officer, and the act of issuance by the governor of Caria.[29]

In the inscription of Didyma, the three acts are in chronological order, which also corresponds to the hierarchical succession of the offices that had promulgated the acts themselves.

The emperor's act was issued at Constantinople on 1st April 533; the *gesta praefectoria* were promulgated in the same place the following day (2nd April 533); the governor's act was issued in the spring of the same year. In this case, the place of issue is not indicated, but the act was probably promulgated in the provincial metropolis of Aphrodisias of Caria.

Thanks to the Didyma inscription, Feissel retraced the complex procedure which led to the issuing of the documents that, written originally on papyrus, were later copied onto stone. In other words, the Didyma inscription represents the transposition onto stone of an original papyrus text.

---

25 In the 6th century the practice of writing official documents both in Greek and Latin (also transliterating Greek characters into Latin), which developed in reports of proceedings, imperial constitutions and officers' seals, can be considered a mark of identity of the government apparatus; this practice was also followed by the notaries of Byzantine Egypt. The continuing use of the Latin alphabet in Byzantine writing clearly attests the great regard in which this language was held in Late Antiquity; see Cavallo 1970; Cavallo/Magistrale 1985; Cavallo 2008; Radiciotti 1998; Radiciotti 2008; Radiciotti 2009.

26 Feissel 2004.

27 See Feissel 2004, 287 (= rpt. 2010, 252).

28 Feissel 2004, 290 (= rpt. 2010, 255).

29 Feissel 2004, 235, fn. 7 (= rpt. 2010, 288, fn. 7). For other examples found in this area see Feissel 2004, 288 (= rpt. 2010, 253).

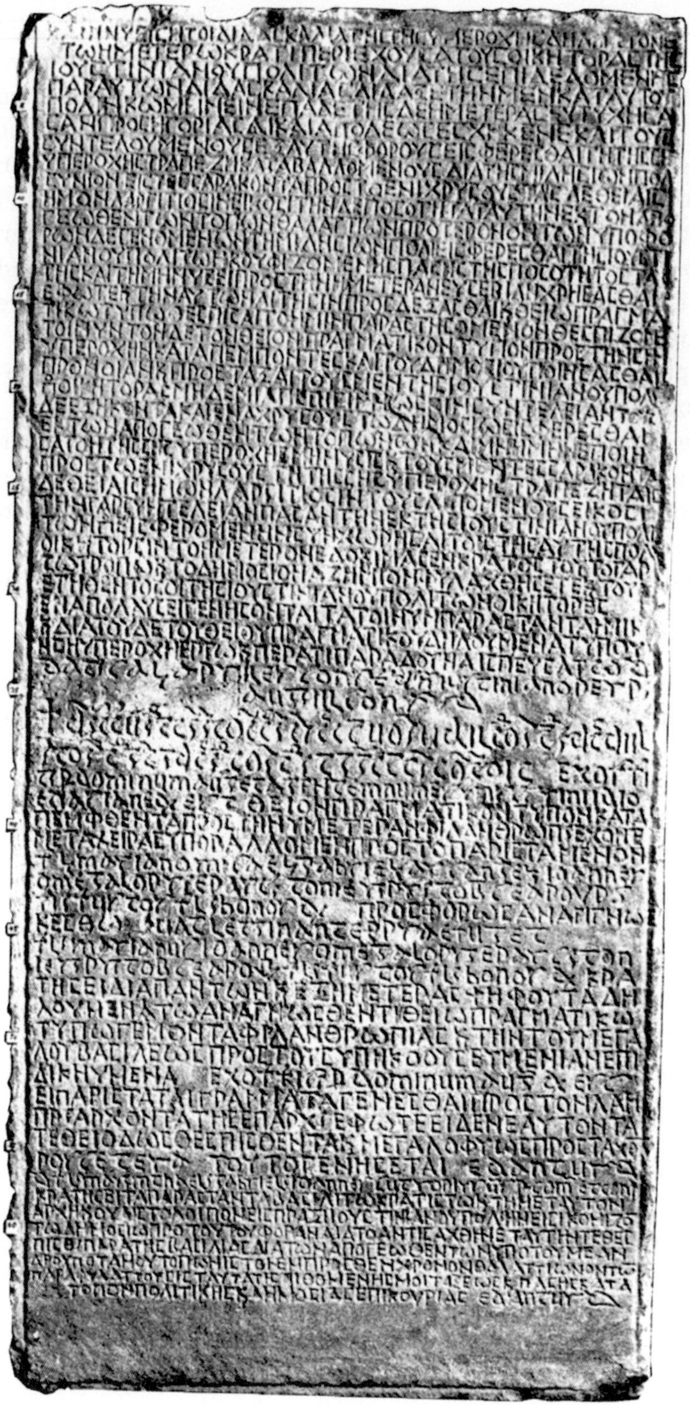

Fig. 1: Didyma inscription (taken from Feissel 2004) © Deutsches Archäologisches Institut.

The procedure which led to these different acts was as follows: after a request made by the citizens of Justinianopolis to the prefect of the Orient (phase 1), the prefect's report was taken to Constantinople by an embassy (phase 2); a brief paraphrase of this phase is reported in the emperor's act.

This was followed by the emperor's answer, the *pragmatica sanctio* (phase 3), which was preserved in the archives of the prefecture of the Orient. The original of this document was not copied onto the inscription, which contains only the copy of the act edited by the prefect's office.

The fourth phase consisted in the verbal process of the prefecture of the Orient (2nd April 533; phase 4); on the stone is reproduced the verbal process (*gesta praefectoria*) that the city embassy had obtained in the form of an authentic abstract, provided by the prefect's subscription (*edantur*).

The further phase was the ordinance of the prefect to the governor of Caria (phase 5), followed by the issue of the act of the governor himself (phase 6), which was provided (like the *gesta praefectoria*) by the official subscription (*edantur*).

This complex procedure was written on the inscription in the form given to the document in the prefect's curia (where the original version of the petition and of the emperor's act were also preserved[30]).

This document was owned by the city embassy that had requested it at the offices of the *gesta praefectoria*; the governor had only the copy of the imperial act and of the prefect's letter.[31]

This means that the inscription was copied from the document owned by the city embassy, and was not produced on the initiative of the local governor: the verbal process reproduced on the inscription had been requested by the embassy in Constantinople, and then carried out at Didyma. It represented, therefore, the statement of the achievement of the citizenry's request. This is the main reason for the great care put into the reproduction of the original shape of the papyrus on the stone, as will be discussed.

Even if these acts concern three different phases of the same procedure, they can nevertheless be considered, from a diplomatic point of view, as two documents: from line 1 to 56 there is the *edictio actorum*, which consists in the order to give a copy of the *pragmatica sanctio* to the citizenry who had presented the petition; this part concludes with the verb *edantur*.

The second document (lines 57–64) is the act of the governor of Caria, and it concludes with a second *edantur*.

---

30 The prefect's report was probably preserved in the imperial archives, while the original version of the prefect's letter had to be preserved in the provincial archives of Caria.

31 Feissel 2004, 306–307 (= rpt. 2010, 273–274).

The following section will briefly recall some technical data of the inscription, already pointed out by Feissel.[32]

The inscription has been carved by a single hand.[33] The stonecutter had to carve the letters on a stone whose height was twice its width, so, in comparison with the papyrus, he had to superimpose texts that, originally, were written from the left to the right, filling more numerous columns: the lines of text on the papyrus were probably longer than those on the stone.

According to Feissel, this is especially true for the titulature of the prefects' college (lines 42–44), which, on the papyrus, were probably recorded on a single long line. The most striking fact, however, is that the same happens for the date and place of issue of the prefecture's verbal process, which are developed in 61 letters extended over almost two lines (36–37): on the original papyrus, they probably did not occupy more than two lines and they had to finish with a *vacat*.[34]

As Feissel properly emphasised, the main characteristic of the inscription is a great formal care, which appears in particular on the expedients used to make the text more easily legible and in the *mise en page*, where the various parts of the text are clearly separated.[35]

The internal structure of the text, although it does not show marked separation signs like the whole *mise en page*,[36] is also characterised by expedients that make clearer the partitioning of the various documents. For example, the prefect's act is clearly articulated by the alternation of Greek and Latin alphabets, and the different parts of the text are marked by a more-or-less large *vacat* (lines 37, 39, 44, 45, 52, 56).[37]

---

32 The inscription is 173.0 cm tall, 79.0 cm wide and 49.0 cm deep; the written face is surrounded by a relief archetto of 2.0 cm width. The writing surface is developed on 64 lines, separated by an unwritten space 11.0 cm tall. On the stone some thin horizontal lines are visible, separated by spaces of about 2.5 cm (while on lines 57–64 the space is 2.0 cm). The letters are about 2.0 cm tall on lines 1–35 and 38–65, while on lines 36–37 they are about 3.7 cm; on lines 57–64 the letter height is 1.5 cm. The interlinear space is 0.3 and 0.4 cm.

33 The original papyrus copy was written by different hands, Feissel 2004, 293, fn. 36 (= rpt. 2010, 257, fn. 36).

34 In light of these elements it can be inferred that the imperial rescript (1–35) was probably written over two columns, while the prefect's act (which occupies 21 lines, 36–56) could easily have been written on seven or eight lines.

35 The emperor's act is preceded by a cross (this cross was probably a draft, as can be seen from comparison between this stroke and the great Latin cross on line 36, and also from the fact that the first letters of lines 1 and 2 are slightly backward in respect to the beginning words of the other lines). The third part of the text starts with a leaf of ivy (line 57); the same sign also marks the end of the imperial act (lines 33 and 35) and the two following parts of the text (lines 56 and 64).

36 For example, in the imperial act there are no spaces indicating the beginning or end of the ruling (16–31); according to Feissel, the original text probably had a more articulated *mise en page*, Feissel 2004, 291, fn. 25 (= rpt. 2010, 256, fn. 25).

37 Furthermore, the prefect's titulature always starts at the beginning of the line (lines 42 and 46); at the end of each intervention of the prefect, the stonecutter left a space of two or three letters (lines 46, 52, 56).

Furthermore, the stonecutter put great care into writing the various parts of the text according to the hierarchy of the people involved: the words of the president of the session are written in bigger letters than the ones of his interlocutor. Also the act of the governor (lines 57–64) was written in letters which were smaller than those used for the prefect, in order to emphasise the hierarchy of the administrative rulers.

The shortening system is composed of ligatures and abbreviations.[38] The stonecutter used abbreviation signs to differentiate the Latin text from the Greek one, and, according to Feissel, this expedient was probably present in the original manuscript.[39]

In the various parts of the text there are some variations in the number of letters, which differentiate the single lines, but also the different acts and the two languages.[40] In particular, the Latin alphabet on lines 36 and 37 is enlarged, and is developed over 36 characters.

With regard to orthography, the Greek part of the inscription shows a remarkable linguistic level.[41] Many graphic mistakes can be found, however, in the governor's act: Feissel states that the responsible for these oversights (at least as concerns the Greek words) was not the stonecutter, but the writer of the original manuscript. It is not surprising that linguistic accuracy can be found in the emperor's act and in the *gesta praefectoria*, which were edited in Constantinople, while many mistakes distinguish the governor's act, whose original had been written by a minor official of the Province.

The Greek alphabet is written in epigraphic capital letters, in a mixed squared and rounded style.[42]

---

38 For the abbreviation system, over a total of 64 lines, 40 end with a complete word, 6 with a contracted one and 18 with an abbreviated one. See Feissel 2004, 291 (= rpt. 2010, 256). Sometimes the stonecutter extended the writing until the right moulding of the stone (lines 1–4, 44, 55), or used some ligatures alternatively at the end of the words (lines 16, 17, 31, 40) or in different parts of them (lines 53, 60, 62, 63) and abbreviations at the end of the word (lines 8, 16, 19, 27, 55, 59, 62) or in other positions (lines 20, 27, 40, 47, 54bis, 59, 61). The abbreviations of the Greek part are all by suspension; the others are all common use; the words in the Latin alphabet are all abbreviated (as are the Greek words transcribed into Latin, lines 42–44, 46–47).

39 Feissel 2004, 292–293 (= rpt. 2010, 257).

40 The emperor's act is composed of 45 letters on line 1, 37 on line 3 (where the term Ιουστινιανουπολιτῶν is enlarged); on lines 2 and 6 there is enlarged writing, which grows compact again on lines 9 and 10. The prefect's act is composed of 39 letters on lines 40 and 41, 34 on lines 48 and 49. The governor's act, written in smaller characters, reaches a maximum of 53 letters only at line 59.

41 For an analysis of the orthography of the inscription, see Feissel 2004, 330ff. (= rpt. 2010, 266ff.). The few orthographical mistakes of this part of the text can be ascribed to oversights of a phonetic type.

42 For an analysis of the Greek alphabet of the inscription see Feissel 2004, 293–294 (= rpt. 2010, 258).

The Latin text[43] (apart from lines 36–37, which Feissel considers "un alphabet à part"[44], as will be discussed) is written in semi-uncial style (which is also used for the Greek terms transcribed into Latin[45]), and shows stylistic uniformity;[46] sometimes it takes on features that are similar to those of the literary minuscule used in the Orient.

The contrast between the two alphabets is highlighted by the alternation of the angular forms of the Greek words and the rounded forms of the Latin text. Nevertheless, this does not lead to a lack of equilibrium in the general composition: apart from lines 36–37, the size of the Latin words is the same as the size of the Greek ones. The governor's act (which, as already mentioned, is written in smaller letters), is also composed of Greek and Latin letters of the same size.

Above all, the inscription of Didyma represents a document of the greatest importance for the special alphabet shown on lines 36–37, which, although written on a stonen epigraph, cannot be analysed using epigraphic standards of research, since this is a faithful copy of a papyrus writing. As has been observed, Feissel considered it as a distinct alphabet from the Greek and the Latin ones, because of the enlarged and very complicated shape of the strokes, which probably made it almost illegible for most of people who lived in the 5th century.[47]

As has already been pointed out, after the 368 A.D. issuance of CTh. 9.19.3, the Ancient Roman cursive also survived outside imperial chancery uses, for example, in the Ravenna papyri and in the reports of proceedings studied by Tjäder and Marichal.

Lines 36 and 37 of the Didyma inscription illustrate another feature of the survival of this script. Feissel was able to decipher those lines, which correspond to the consular dating and place of issue in reports of proceedings. These elements recall the data that emerged from Marichal's studies of proceedings and Tjäder's investigation of the papyri of Ravenna.[48]

---

43 For an analysis of the Latin text see Feissel 2004, 294–295 (= rpt. 2010, 259).

44 Feissel 2004, 292–293 (= rpt. 2010, 257).

45 For the problem of the transcription of Greek names into Latin see Feissel 2004, 295ff. (= rpt. 2010, 259ff.). According to Feissel, the names transcribed into Latin were based entirely on the Greek model.

46 Feissel states that in the original manuscript the Latin text was not so uniform as the one on the stone; in particular, the prefect's act probably showed a variety of styles similar to the one of the contemporary reports of proceedings, which graphically underscored the importance of the president of the session (Feissel 2004, 294 [= rpt. 2010, 259]). On the use of different kinds of scripts to distinguish the hierarchy of a session's participant, see also Tjäder 1982b, 124–128.

47 Feissel 2004, 330 (= rpt. 2010, 294).

48 According to Feissel 2004, 331 (= rpt. 2010, 295f.): "Preuve que la majuscule, surchargée de fioritures, des documents de 564 et de 625 représentait en fait l'ultime développement de la cursive latine ancienne, Tjäder reconnut dans les gesta ravennates les éléments constitutifs d'un en-tête de procès verbal, à savoir la date et le lieu de la séance—qui dans ce cas ne pouvait être que Ravenne. Une hypothèse analogue permet de déchiffrer de même la 'grande écriture' de l'inscription de Didymes".

Some documents will be considered as terms for comparison in the analysis of this script:[49]

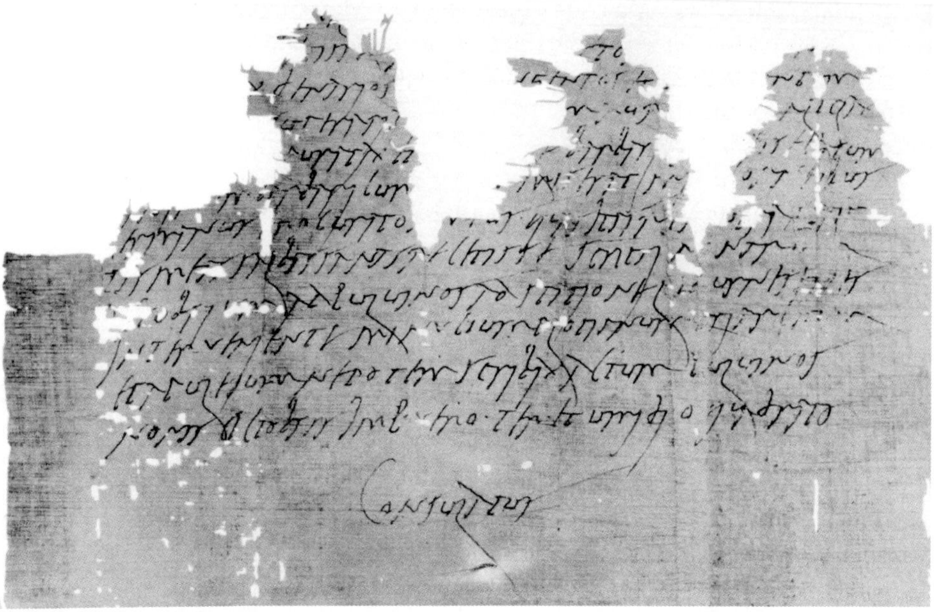

Fig. 2: London, British Library P. 730 = ChLA III 204; 7 October, A.D. 167 (Table 1, Col. II).[50] © The British Library Board, P.730.

49 The documents are considered in chronological order, to provide a general framework of the development of Latin script from the 2nd to the 6th century.

50 Receipt for goods in Ancient Roman cursive, provenance unknown. Picture taken from ChLA III 204.

Fig. 3: Oxford, Ashmolean Museum P.Oxy. XX 2269r = ChLA IV 262; 30 April–14 July, A.D. 269 (Table 1, Col. III).[51] (Courtesy of the Egypt Exploration Society and Imaging Papyri Project, Oxford).

51 Register of sales by auction, found at Oxyrhynchus; the first line is written in an "elongated" Greek hand, while the rest of the document is written in Ancient Roman cursive.

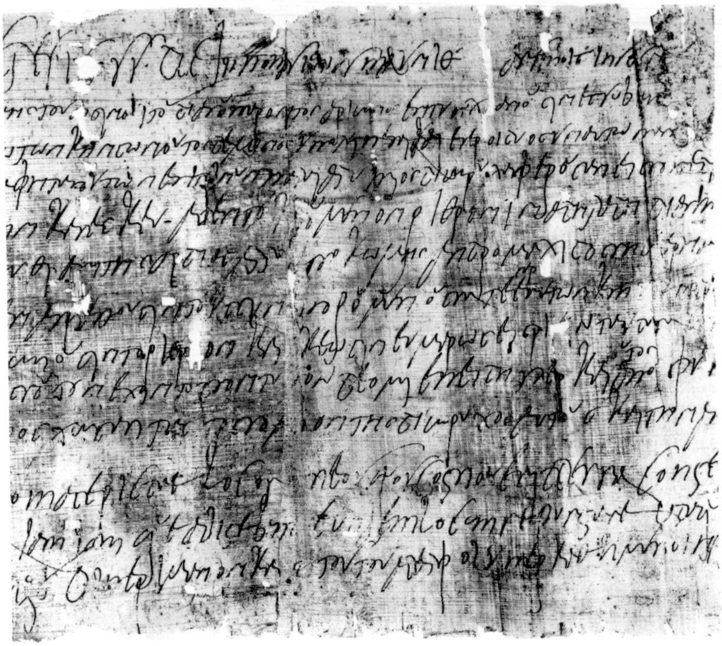

Fig. 4: Manchester, John Rylands Library P.Ryl. IV 653 = P.Sakaon 33 = ChLA IV 254 Ptolemais Euergetis; 3 June, A.D. 318–320 (Table 1, Col. IV).[52]

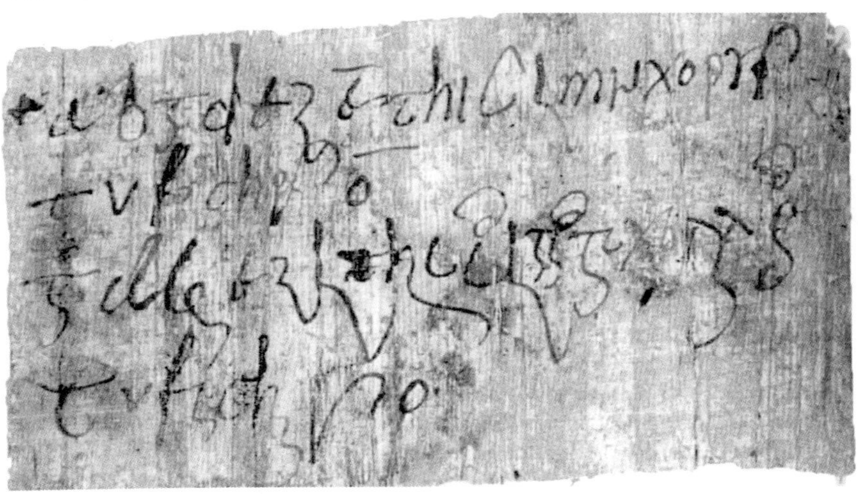

Fig. 5: University of Leuven, P.Lugd. Bat. XXXIII 11 = P. Worp 11, first half of the 5th century A.D. (alphabet B: Table 1, Col. V; alphabet A: Table 2, Col. IV), now lost.[53]

---

52 A report of proceedings held at Arsinoë before the praeses of Herculia, written in Greek and Latin. Picture from ChLA IV 254.

53 This papyrus was studied by Feissel (Feissel 2008 = P.Lugd. Bat. XXXIII 11 = P. Worp 11), who pro-

Fig. 6: Oxford, Ashmolean Museum P.Oxy. XVI 1879 = ChLA XLVII 1409; 8 Sept.–31 Dec., A.D. 434[54] (Table 1, Col. VI).[55] (Courtesy of the Egypt Exploration Society and Imaging Papyri Project, Oxford).

posed a new edition starting from that of Willy Clarysse and Bruno Rochette (Clarysse/Rochette 2005). The document, which had been preserved at the University of Leuven, was lost in a fire in 1940, and was analysed on microfilm reproductions. This papyrus bears two alphabets, called A and B by Feissel and written by a single hand, which represented the transliteration of the Greek alphabet in Latin characters; the first alphabet "représente la cursive récente, devenue l'écriture commune à partir du IIIe s."; the second is "une forme évoluée de l'ancienne cursive latine, dont l'archaïsme contraste délibérément avec l'écriture commune" (Feissel 2008, 53 [= rpt. 2010, 541]). Feissel explains this double alphabet as a kind of instrument for those who could no longer read the forms of the Ancient Roman cursive: "Un document comme le nôtre pouvait leur fournir une aide au déchiffrement. En particulier, la maîtrise de l'alphabet B étant sûrement plus rare que celle de l'alphabet A, ce double modèle pouvait rendre service à quiconque savait lire l'écriture commune, mais peinait à décrypter une date de procès-verbal d'ancien style" (Feissel 2008, 61–62 [= rpt. 2010, 549]). Picture from Feissel 2008, 54 (= rpt. 2010, 542) and available online at http://lhpc.arts.kuleuven.be/img/pleuven/descr/alphabet.html (last accessed: 2015/03/04).

54 P.Oxy. XVI 1878 and P.Oxy. XVI 1879 are two examples of the initial script analysed by Tjäder.

55 Report of proceeding from Egypt written in Greek and Latin, with the initial script in Ancient Roman cursive. It was analysed by Tjäder in his research into the "misteriosa scrittura grande", see Tjäder 1952 passim.

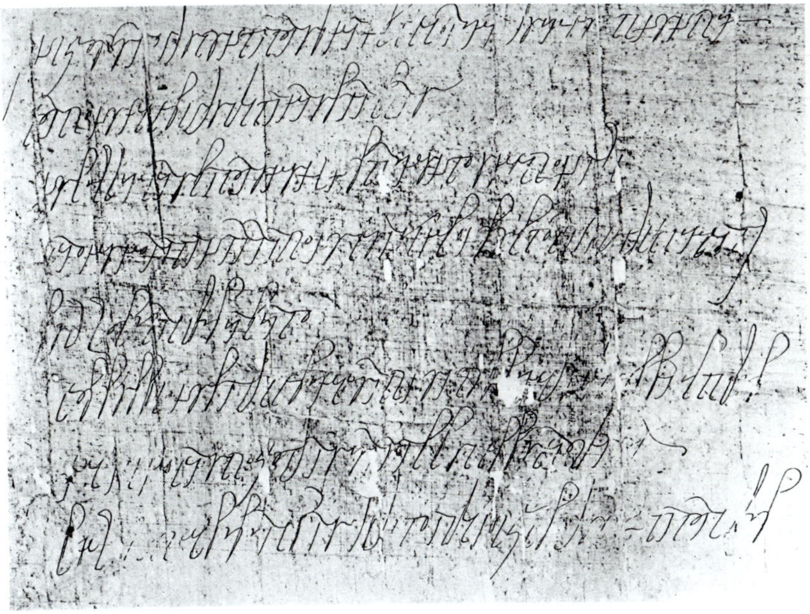

Fig. 7: Paris, Louvre, P. 2404 = ChLA XVII 657, A.D. 436–450 (Table 1, Col. VII).[56]

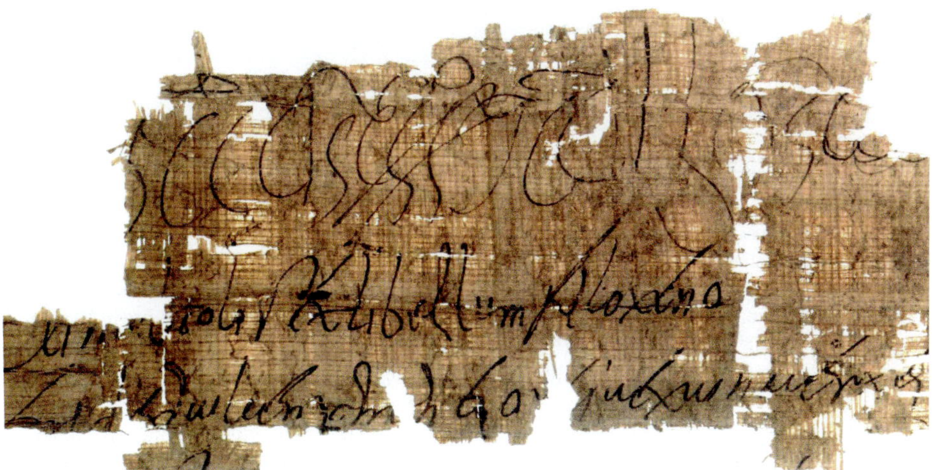

Fig. 8: Oxford, Ashmolean Museum P.Oxy. XVI 1878 = ChLA XLVII 1408, Heracleopolis; 1 Sept., A.D. 461[57] (Table 1, Col. VIII).[58] (Courtesy of the Egypt Exploration Society and Imaging Papyri Project, Oxford).

56 *Litterae caelestes*; this is the only preserved example of this script besides the papyri in Leiden and Paris, see fn. 10. Picture from Steffens 1910, pl. 16.

57 These papyri can be considered examples of Ancient Roman cursive, while those that follow are cases of Later Roman cursive.

58 Report of proceeding from Egypt written in Greek and Latin, with the initial script written in An-

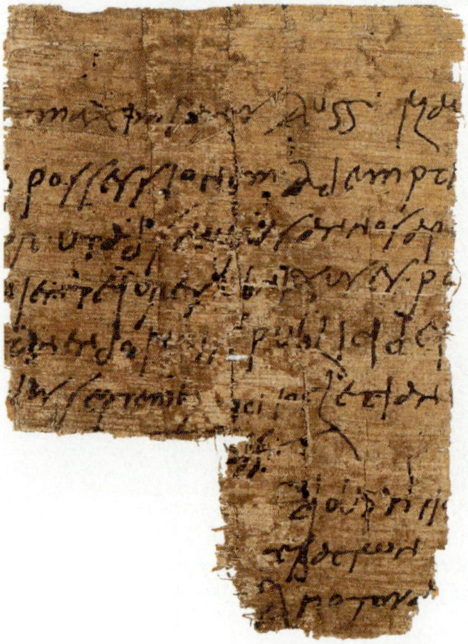

Fig. 9: Florence, Biblioteca Laurenziana PSI I 111 = ChLA XXV 780, A.D. 287–304 (Table 2, Col. II).[59] (Reproduced with permission of MiBACT. Further reproduction by any means is prohibited).

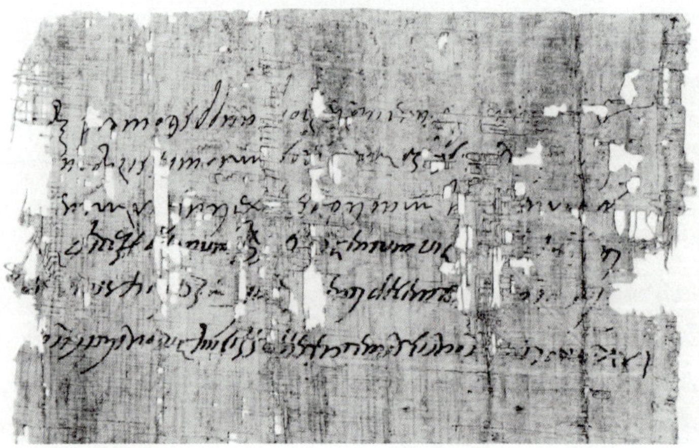

Fig. 10: London, British Library P. Lond. 731 = ChLA III 205, A.D. 293 (Table 2, Col. III).[60] © The British Library Board, P.731.

cient Roman cursive. It was analysed by Tjäder in his research on the "misteriosa scrittura grande", see Tjäder 1952, passim.

59 Private copy of an imperial rescript; although this alphabet can be identified with an example of the Later Roman cursive, it is still characterised by the presence of a few elements belonging to the system of the Ancient Roman cursive, Casamassima/Staraz 1977, 71.

60 Military receipt written in Later Roman cursive; exact provenance unknown. Picture taken from ChLA III 205.

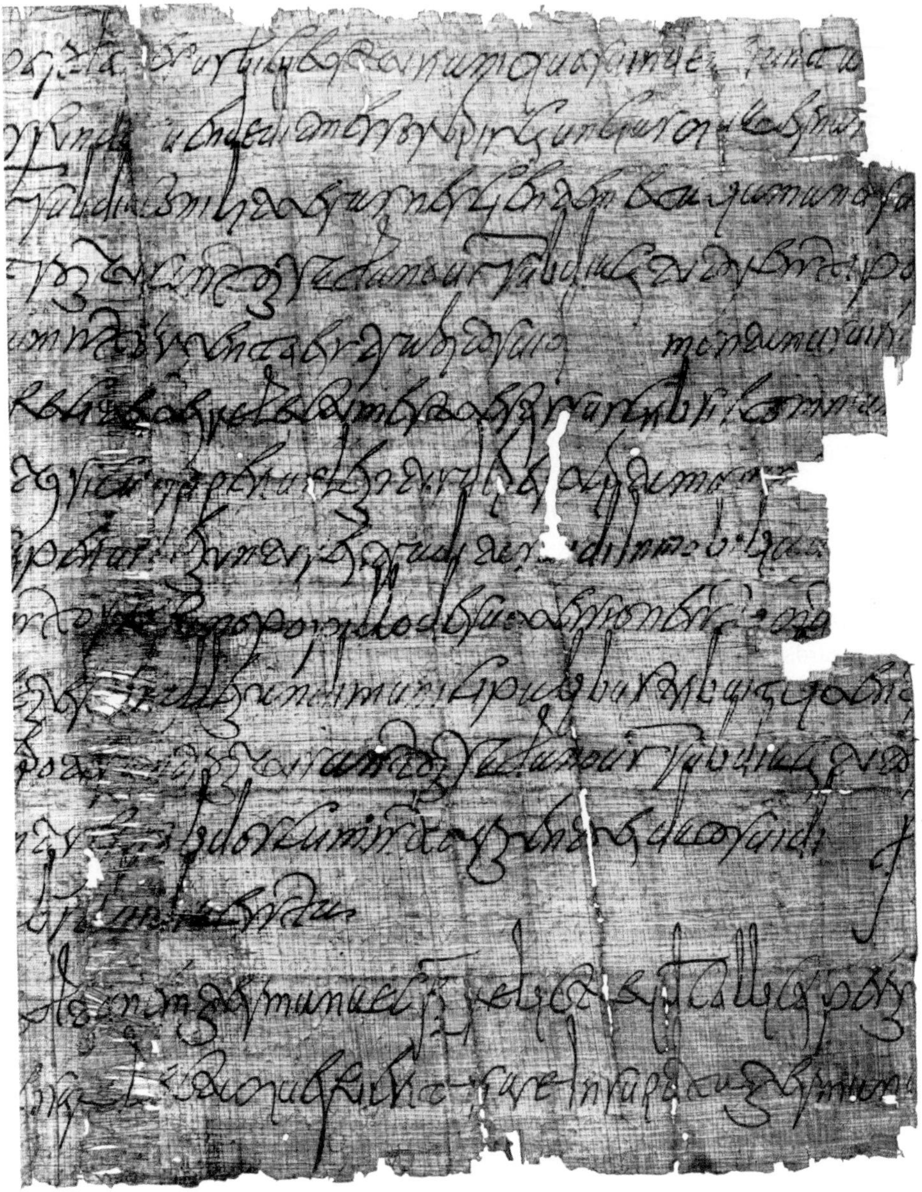

Fig. 11: Paris, Bibliothèque nationale de France, lat. 4568 A (col. III:3) = P.Ital. I 8 = ChLA XVII 652, 17 July, A.D. 564 (Table 2, Col. V).[61] (Courtesy of the Bibliothèque nationale de France).

61 Registration protocol of a *charta plenariae securitatis* of the *gesta municipalia* of Ravenna; this document bears an initial script in the shape of the "misteriosa scrittura grande", while the rest of the document is written in Later Roman cursive. Picture taken from ChLA XVII 653.

We propose the following edition of lines 36–37:

Fig. 12: Lines 36–37 in the inscription of Didyma (taken from Feissel 2008, 56) (Table 1, Col. I; Table 2, Col. II).

Although Feissel has already carried out a close examination of this special script, it seems useful to return to it, in order to highlight some further data.
This is, first of all, the edition of the text:

36 + D(OMINO) N(OSTRO) IUSTI<NI>ANO PERPETUO AUG(USTO) IIII CONS(ULE) DIE III

37 NON(AS) APRILES CONSTANTINOPOLI

A: formed by a vertical stroke similar to an *S*, and from a second slanted stroke, sometimes separated from the top of the letter (IUSTI<NI>ANO), or joined to it and elongated towards the left (see AUGU(USTO); here the slanted sign is wavy and extended over the previous letter, while in CONSTANTINOPOLI it is short and truly horizontal). In the first case, it is very similar to ChLA III 204, P.Oxy. XX 2269[62] and P.Oxy. XVI 1878, both related to the Ancient Roman cursive model.[63] Tjäder also states that "l'esecuzione della *a* e la *r* in due tratti, l'uno in sostanza verticale e l'altro più o meno orizzontale, sono due delle più tipiche lettere della corsiva antica"[64]; the clear proximity of this letter to the *a*s of the *litterae caelestes* of ChLA XVII 657[65] can be effectively observed.

62 Tjäder 1985, 91, states that the script of this papyrus, where the *a*s take on different forms, can be considered "a stage of transition".
63 Feissel 2008, 56 (= rpt. 2010, 544).
64 Tjäder 1952, 197; on the shape of *a* in the Ancient Roman cursive, see also Tjäder 1979, 57–59 (P.Mich III 164 = ChLA V 281): "più tardi delle altre lettere dell'alfabeto latino la *a* arrivò alla forma che avrebbe avuto nella corsiva nuova romana. Fino al II sec d.C. incluso, si trovano di fatto soltanto forme 'antiche', o quella, che non sembra andare oltre il I sec. d.C., che ha la traversa attaccata alla fine del secondo tratto, o quella nota a due tratti che fu normale nella scrittura ufficiale, o, infine, altre forme meno corsive, che in generale avevano conservato, in un modo o nell'altro, la traversa [...] La nuova *a* [...] appare per la prima volta nel P.Mich. III, 164, datato all'anno 242 o non molto dopo, poi nel P. Dura 81 del 250 circa e nel P. Dura 95, dell'anno 250 o 251[...] Paleograficamente, questa nuova forma potrebbe benissimo essere la trasformazione della antica *a* in due tratti, ma è oramai chiaro che mancavano le premesse per tale trasformazione. Insieme alle prime *a* del nuovo tipo che si sono rinvenute infatti non si trova la *a* in due tratti ma invece quella 'onciale arcaica'"; see also Tjäder 1979, 52, fn. 54.
65 See in particular the fragment P. Louvre 2404 that is a part of the same roll.

C[66]: formed by a unique curved stroke; it is not clear whether, in the word CONS(ULE), the sign similar to a Ω written above *c* is part of this letter (the equivalent of the "boucle" on P.Lugd. Bat. XXXIII 11 = P.Worp. 11, which, according to Feissel, "fait [...] partie intégrante du caractère"[67]; this is a typical sign of the Ancient Roman cursive,[68] also clearly visible in the *litterae caelestes* of ChLA XVII 657) or an extension of the horizontal stroke of *n*. Although the other *c*s of the inscription lack this particular sign, this stroke is probably part of the letter *c*, since in all of the other occurrences the horizontal stroke of *n* ends with a bow only on the left part of the letter (see the second *n* of NON(AS) and the second *n* of CONSTANTINOPOLI). Feissel highlights the risk of confusion between *c* and *i* at this stage of evolution of the writing.[69]

D: formed by two strokes, a curved one and a vertical one slightly slanted at the top. The downward curve is joined to the bar in the lower part. There is a wavy abbreviation stroke above the letter. Tjäder states that the form of ancient *d* was fully recovered in the Later Roman cursive:

> There can be no doubt that ancient *d* was an official letter in the period of the Later Roman Cursive. There are a few instances from Egypt; there are several from Italy, including about 100 in the [...] record of *gesta* [...] It is clear that this ancient *d* [...] could have had a future even after the 6th century.[70]

E: formed by two curved, overlapping strokes, not totally joined in the word APRILES, and totally disjointed in the word DIE (see below). Although the difference between this shape and the *litterae caelestes* of ChLA XVII 657 (P. Louvre 2404) is evident, the letter is not very far from the Ancient Roman cursive shapes of P.Oxy. XX 2269, P.Sakaon 33, P.Oxy. XVI 1878 and P.Oxy. XVI 1879. Nevertheless, there is also a strong resemblance to PSI I 111, which is related to the Later Roman cursive. *E* for this reason can be considered a "mixed shape" between the Ancient and the Later Roman cursive.

G: this letter has a smaller module than the others, and it is formed by a short, rounded stroke completed at the top with a very small sign similar to a *comma*. This shape resembles that of the Later Roman cursive (see in particular PSI I 111), as stated

---

66 Tjäder considers *c* as one of those signs that are not very meaningful for the comprehension of the development of writing because, in the cursive script, they maintain more or less the same shape, Tjäder 1952, 198–199; *c, h, i, l, o* and *t* are also considered less meaningful in the comparison of the two systems in Casamassima/Staraz 1977, 26, fn. 27.

67 Feissel 2008, 58 (= rpt. 2010, 546).

68 See above fn. 66.

69 Feissel 2004, 333 (= rpt. 2010, 297).

70 Tjäder 1982a, 16.

also by Feissel,[71] and represents one of the most meaningful attestations of the fact that this script is not a closed system, but is open to the influences of the *litterae communes*.

I: formed by a unique stroke, similar to a complete *c*. Feissel states that "la forme incurvée du *I* caractéristique de ce type de cursive contraste avec la haste droit de l'alphabet A",[72] which, as we have seen, was an example of a Later Roman cursive alphabet. The shape of this letter is very similar to that of P.Oxy. XVI 1879, even if it is slightly less curved in the papyrus. As in the case of *c*, it is not clear whether the small sign above the first *i* of IUSTI<NI>ANO[73] is the equivalent of the "boucle" of P.Lugd. Bat. XXXIII 11 = P.Worp. 11[74] or the prosecution of the horizontal stroke of *n*.

L: formed by a vertical, slightly wavy stroke. Not meaningful.

N: formed by two vertical strokes (similar to *S*) and by a third horizontal, wavy stroke extending horizontally on the bars and dominating with its greater length. On three occurrences (CONS(ULE), second *n* of NON(AS), first *n* of CONSTANTINOPOLI) the horizontal stroke ends, on the left, with a sign similar to a Ω, or a bow. The horizontal stroke of the *n* of N(OSTRO) has a particularly undulating form: it may be the result of the stonecutter's attempt to reproduce the abbreviation sign that was surely present on the original text. In the second *n* of CONSTANTINOPOLI, the horizontal stroke is elongated above *o*. The superior stroke of the letter is often separated from the vertical bars (see in particular N(OSTRO) and the second *n* of NON(AS)); Feissel states that this development of the letter "se traduit par une croissance progressive de la partie droite de la lettre, tendant à la symétrie avec sa partie gauche".[75] For this reason, Feissel considers this letter as "le plus obstacle à la lisibilité", because, when the horizontal stroke is separated from the vertical one, there is a risk of reading this sign as if it were two letters.[76] The stonecutter put great care into reproducing the *n* of the

---

71 Feissel 2008, 57 (= rpt. 2010, 545); the derivation of this type of *g*, which can also be found in the Ravenna papyrus from the Later Roman cursive system, has been noticed also by Tjäder 1952, 214, fn. 50.
72 Feissel 2008, 58 (= rpt. 2010, 546).
73 Feissel states that the omission of NI is due to the fact that, in this particular alphabet, the group of letters NI and STI had a very similar shape: "Pareille confusion semble s'être produite dès la lecture de l'original, quand le lapicide a copié IUSTIANO en omettant les lettres NI. De fait, le groupe STI est si semblable au groupe NI que l'on a affaire ce cas à une quasi-haplographie" (Feissel 2004, 333, fn. 231 [= rpt. 2010, 297, fn. 231]).
74 Feissel observes that in the *litterae caelestes* "cette boucle est commune à c, f et s" (Feissel 2008, 58 [= rpt. 2010, 546]).
75 Feissel 2008, 58 (= rpt. 2010, 546).
76 Feissel 2004, 333 (= rpt. 2010, 297): "Un N ainsi désarticulé peut se confondre avec le group RT ou ST"; see also Tjäder 1979, fn. 33, and 56.

papyrus on the inscription: the marked wavy shape of the horizontal stroke, ending with a bow, seems to be a faithful copy of the *n* of P.Oxy. XVI 1878 and 1879 (see figs. 8 and 6). According to Tjäder, the shape of this letter originates from the Ancient Roman cursive.[77]

O: oval in shape (see in particular the first *o* of CONSTANTINOPOLI), slightly slanted towards the right, but there are some differences in proportion of width and height among the several *o*s of the inscription. According to Tjäder, the ancient way of writing this letter, "which implied disarticulated writing, survived in the period of the Later cursive, alongside the current way of writing, which was the Later Roman cursive".[78] This letter can be considered "as common to the two stages of handwriting".[79] Feissel connects this letter with the alphabet A of Feissel 2008 (= P.Lugd. Bat. XXXIII 11 = P. Worp 11).[80]

P: very similar to *a*; formed by a first vertical stroke similar to an *S*, and a second slanted stroke, sometimes separated from the top of the letter (first *p* of PERPETUO and CONSTANTINOPOLI; in this case the letter is very similar to the *a* of IUSTI<NI>ANO), or joined to the top and elongated towards the left (APRILES). The second *p* of PERPETUO takes on a stylized shape, with the slanted stroke truly straight, and lengthened upwards. It can be stated without doubt that the shape of *p* originates from the Ancient Roman cursive: a similar form can be observed in ChLA III 204, P.Lugd. Bat. XXXIII 11 = P.Worp 11 (alphabet B) and P.Oxy. XVI 1878. Feissel also states that this letter is typical of the Ancient Roman cursive alphabet.[81]

R: See the interpretation of PERPETUO below. The form of the letter certainly comes from the Ancient Roman cursive (see the observation of Tjäder about *r* and *a* and the strong resemblance between this letter and that of the *litterae caelestes* of ChLA XVII 657).

S: formed by a wavy, vertical stroke and a superior bow. The bow is an attempt to reproduce the "boucle" characterising this letter in the papyrus (P.Lugd. Bat. XXXIII 11 = P.Worp 11 and P.Oxy XVI 1879, first *s* of *conss.*), and it allows *s* to join the Ancient Roman cursive alphabet,[82] since this is a typical element of the *litterae caelestes*

77 Tjäder 1952, 195–196.

78 Tjäder 1982a, 12; see also Tjäder 1982a, 13: "In certain cases [...] the choice of the disarticulated way of writing could have been determined by the large size, but in other cases, there is no technical basis for the choice and it must [...] have been a question of conservative adherence to an older form".

79 Tjäder 1985, 189.

80 Feissel 2008, 59 (= rpt. 2010, 546).

81 Feissel 2008, 59 (= rpt. 2010, 547).

82 According to Feissel, the *S* of P.Lugd. Bat. XXXIII 11 (= P.Worp 11), which is very similar to that of

(ChLA XVII 657). According to Feissel, this sign "fait partie intégrante de la lettre".[83] In the abbreviation CONS(ULE) the *s* is barred.

T: in CONSTANTINOPOLI, the letter is formed by two strokes, a curved one (similar to a complete *c*) and a short, horizontal one. For the *t* of PERPETUO see below.

U: formed by two joined strokes, one curved and one vertical, slightly elongated in the lower line. This letter is not indicative of much; here it is more curved than in the Ancient Roman cursive shapes of the alphabet B in P.Lugd. Bat. XXXIII 11 = P.Worp 11 (Feissel states that the form of *u* in this papyrus can be reconnected to the alphabet A), rather similar to the Later Roman cursive shapes of PSI I 111.[84]

It seems appropriate to make some observations about Feissel's interpretation of the script. In the word PERPETUO, there are three strokes after PE, two of a vertical shape, similar to an *S*, and a short horizontal one, written above the first vertical stroke. Here the shape of *r* is different from the *r* of APRILES, where it is composed by only two strokes, a vertical one similar to an *S* and a superior, horizontal one (a shape that is similar to those of ChLA III 204 and P.Lugd. Bat. XXXIII 11 = P.Worp 11, alphabet B). It seems possible that in the case of PERPETUO the stonecutter saw on the papyrus a form of *r* similar to that of P.Oxy. XVI 1879, where the letter takes on a shape similar to an *n*, because its superior stroke is extended downwards on the right. On the stone, the stonecutter reproduced the latter sign with three strokes, a first, vertical one, corresponding to the vertical bar of *r*, and two separated ones, the first horizontal and the second vertical (see Tab. 4).

A similar interpretation also seems to be possible for the *t* in the same word: after the second *e* a curved sign can be observed, similar to a *c*; it can be supposed that this is truly a *c* (in this case, the word should be read as PERPECTUO, but it would be an evident mistake), but the stroke is less curved than the other *c*s, and more squared.

It could be supposed that this is the continuation of the superior stroke of *t*, interpreted by the stonecutter as a stroke aside. Maybe he saw, on the papyrus, a shape similar to the second *t* of P.Ital. I 8 where the horizontal stroke is slanted downwards (see Tab. 5; Feissel defines this stroke as a "boucle", which "se détache du corps de la lettre"[85]), and he reproduced it on the stone with two separated signs, as in the case of the *r*.

Didyma, is "en réalité indépendant de l'alphabet A" (Feissel 2008, 59 [= rpt. 2010, 547]).

83 Feissel 2008, 59 (= rpt. 2010, 547).

84 On the form of this letter in the Ancient and Later Roman cursives see Tjäder 1979, 48, fn. 43; Tjäder 1982a, 13–14; Iannacci/Modesti/Zuffrano 2012, 99.

85 Feissel 2008, 59 (= rpt. 2010, 547).

It also seems possible to propose an interpretation of the word *die* that differs from Feissel's: the scholar states that this word is conveyed only by a *d* (D<IE>), but this would not explain the presence of the three signs after *d* itself.

Fig. 13: The word DIE in the Didyma inscription (taken from Feissel 2008, 56 [= rpt. 2010, 544]).

It can be argued that the sign similar to an *e* after the *d* is actually an *i* (in the form of a complete *c*, very curved), followed by an *e* formed by two disjointed curved strokes. In other words, it is possible that the superior stroke of *e* is set before its normal place.

Feissel himself, analysing the shape of the *e* of F (alphabet B), points out the strong disarticulation of this letter in the Didyma epigraph: "l'évolution de cet alphabet tend à désarticuler le haut et le bas de la lettre, nettement séparés dans les acta de 533".[86]

One might also suppose that the stonecutter confused the *i* with the base of *e*, and so he relocated the superior part of the latter to the *i*.

The global shape of the alphabet raises some interesting remarks, which recall some observations made by Tjäder on the Ravenna papyri containing the "*misteriosa scrittura grande*".

The analysis of the special alphabet of the Didyma inscription highlights the presence of a remarkable number of letters that can be connected with the Ancient Roman cursive alphabet, similar to the opening script of Ravenna's *gesta praefectoria*: it can be stated without a doubt for *a*, *n*, *p* and *r*, as we have already noted in the analysis of the single letters.

There are also some letters that, at a first sight, seem to be insufficiently distinct for assigment to one alphabet or another (like *i* and *s*), but actually show elements of connection with the Ancient Roman cursive: in particular, this can be stated for *s*, characterised by a superior sign similar to the "boucle" typical of the Ancient Roman cursive and clearly visible also in the *litterae caelestes*.

The Didyma inscription thus represents an epigraphic statement of the preservation of the Ancient Roman cursive alphabet beyond its use in the imperial writing.

Nevertheless, the alphabet in lines 36 and 37 cannot be completely ascribed to the influence of Ancient Roman cursive: some letters appear connected to the writing system of the Later Roman cursive. This is especially true for *d*, *e*, *g* and *o*.

86 Feissel 2008, 57 (= rpt. 2010, 545).

The alphabet found in the Didyma inscription, therefore, is not a pure writing form, but the result of a complex graphical situation, and it testifies to the complex writing panorama developed after the 4th century.

It is interesting to recall the observations made by Iannacci, Modesti and Zuffrano[87] on the "misteriosa scrittura grande" analysed by Tjäder: the shape of the initial script of the Ravenna papyri and those from Egypt papyri[88] was not the pure shape of the Ancient Roman cursive, and some letters could also be connected to the Later Roman cursive shapes.[89]

In those scripts, there is not "una situazione di contrapposizione netta e radicale tra due sistemi grafici opposti (*corsiva antica* e *corsiva nuova*)", but they are the mirror of a "condizione più sfumata di quanto non possa sembrare in prima istanza".[90]

The system of a special initial script in the papyri from Ravenna, like that of lines 36 and 37 in the Didyma inscription, is not, therefore, a closed one, unlike the *litterae caelestes*,[91] but is in part open to the influences of the *litterae communes*, and it shows the complexity of the graphic panorama of this period.[92]

The script of lines 36–37 of the Didyma inscription seems to represent an intermediate step between those in the papyri from Ravenna and Egypt.

A closer comparison of these writing systems shows that the differences are not substantial: Feissel states that the script of the inscription shows a "caractère relativement conservateur".[93]

Feissel does not consider the lack of ligatures characterising the inscription to be the biggest dissimilarity between this script and the one of the papyri. The biggest difference is due not only to the writing material, but also to a phenomenon already highlighted by Tjäder:

---

87 Iannacci/Modesti/Zuffrano 2012.

88 For the Later Roman cursive elements in the papyri from Egypt see Marichal 1952, 349, fn. 1, Kresten 1966, 8ff.; Iannacci/Modesti/Zuffrano 2012, 100.

89 Innacci/Modesti/Zuffrano 2012, 99–100. Also Tjäder reached this conclusion in 1982a, 8, where he states that "[...] the Ancient Roman cursive [...] remained in a pure form, presumably until about A.D. 600, as a monopoly script in the Imperial chancery, and, in a sometimes slightly mixed form, it was in use in the first half of the 7th century [...] for the opening line [...] in records of *regesta* and *acta*".

90 Iannacci/Modesti/Zuffrano 2012, 100.

91 Innacci/Modesti/Zuffrano 2012, 110: "Intorno alla metà del V secolo le *litterae caelestes* testimoniate nei due rescritti imperiali si presentano in forme ancora sostanzialmente incontaminate e vicine a quelle dei papiri del III secolo".

92 On this matter, it is also worth recalling the words of Casamassima and Staraz: "nel III secolo il sistema della scrittura romana nel trasformarsi [...] è divenuto straordinariamente più ricco e complesso nella morfologia dei segni e nella struttura, appare articolato in una varia e vivace dialettica di tipi grafici" (Casamassima/Staraz 1977, 11).

93 Feissel 2004, 332 (= rpt. 2010, 296).

> L'évolution de la cursive d'en-tête du Ve au VIe s., respectivement en Égypte et en Italie, affecte non seulement la forme individuelle des lettres, mais tend à passer d'une écriture liée a une quasi-capitale aux caractères distincts. La 'grande écriture' de l'inscription de 533, proche encore du ductus des deux siècles précédents, paraît amorcer déjà la tendance à en séparer les caractères.[94]

Furthermore, this characteristic can be considered one of the strongest elements attesting to the derivation of the epigraphic script from the ancient writing system: in the *litterae caelestes*, a system of ligatures did not develop.[95] Although the letters did overlap,[96] the only joining system was represented by narrowing the distance between small groups of letters, or segments of single letters, in correspondence to the final stroke of the letters themselves. This did not lead to a modification of the features of the characters.[97]

It is most likely that the stonecutter saw, on the papyrus, strokes that were not ligated (even though they probably overlapped, as in the "misteriosa scrittura grande"), and he reproduced them on the stone as separate and unconnected letters.

Feissel states that the main difference between the letters on the Didyma inscription and those in papyri is the great space that sometimes separates different parts of the same letter on the inscription.[98]

It is certainly true in the cases of *r* and *t* (but only in the word PERPETUO, as discussed), but it can not be stated in the case of *n*, which Feissel conversely considers "le cas plus net de ce défaut".[99] The left bar of *n* is clearly separated from the horizontal bar in P.Oxy. XVI 1878 (*cons.*) and P.Oxy. XVI 1879 (*Areobindo*), and it seems possible that the stonecutter intentionally reproduced this feature of the letter on the stone, in order to approach the original shape that he saw on the manuscript.

---

94 Feissel 2004, 332f. (= rpt. 2010, 297); the lack of ligatures also characterises the "misteriosa scrittura grande" of Ravenna, where "vi è una visibile distinzione generale tra la maggior parte delle lettere, tanto che si ha in qualche punto l'impressione di vedere piuttosto una iscrizione" (Tjäder 1952, 206); see also Iannacci/Modesti/Zuffrano 2012, 101f., which states that in the Later Roman cursive "la struttura essenziale della singola lettera rimane invariata senza quei cedimenti tipici invece della *corsiva nuova*, dove le lettere, pur di adeguarsi alle legature, modificano anche di molto la propria struttura fino ad assumere tre o quattro forme diverse."

95 Unlike the Later Roman cursive system, where the letters modify their structure according to the ligatures and take on different shapes.

96 This is an element that can give the impression of a woven script: Marini effectively defined those letters as "intralciate fra se" (Marini 1805, 267, fn. 2).

97 Mallon 1948, 9ff. (= rpt. 1982, 169ff.); Iannacci/Modesti/Zuffrano 2012, 101, fn. 43.

98 "La déformation la plus sensible entre le manuscrit et la pierre ne consiste pas dans la séparation des caractères entre eux, mais dans l'espace exagéré qui tend à disloquer les éléments d'une même lettre" (Feissel 2004, 333 [= rpt. 2010, 297]).

99 Feissel 2004, 333 (= rpt. 2010, 297).

This observation raises the question introduced by Feissel, who discussed whether the reproduction of the manuscript on the stone can be considered a faithful one.

It seems possible to state that the stonecutter not only reproduced the form of the single letters, but also some particular features typical of their shape on a papyrus support.

It is particularly evident in the case of *n*, which is provided with a sign similar to a Ω or a bow at the end of the horizontal stroke: it seems to be an attempt to reproduce faithfully the bows concluding the left part of the *n* in, e.g., P.Oxy. XVI 1879 and P.Oxy. XVI 1878 (see figs. 6 and 8).

But for what reason did the stonecutter put such great care into faithfully reproducing the script that he saw on the papyrus?

The craftsman should not have had any difficulty in transposing onto stone the Latin and Greek alphabets, written in the classic epigraphic capital shapes; there must have been greater difficulties in copying the opening lines of the *gesta praefectoria*, written in a complicated cursive and totally unrelated to all epigraphic forms.

The reproduction of those signs on a stone went against all epigraphic uses: the stonecutter had to adapt to a hard material a kind of script that was complicated even on papyrus, since it was full of curved and sinuous strokes that were technically difficult to reproduce.

Certainly the precision of the single strokes is a sign of the stonecutter's great craftsmanship.

Furthermore, the faithful reproduction of the papyrus shapes also seems to assume an important meaning for the function of this inscription: as has already been pointed out, the citizens of Didyma had obtained the original text of the *gesta praefectoria*, provided with the subscription of the prefect, and it was reproduced on the stone "sans copie intermédiaire".[100]

The effort made by the stonecutter in reproducing the original letter of the prefect on the stone can be attributed to the will to demonstrate the authenticity of this document, which represented the original attestation of the achievement of the citizenry's request to the emperor.

This could explain the great care which the stonecutter put into the reproduction of these two lines, and which represents a certification of guarantee for the whole document,[101] as Feissel has also observed: "Le soin scrupuleux mis par lapicide à calquer un en-tête en cursive archaïsante traduit l'intention de mettre en évidence

100 Feissel 2008, 332 (= rpt. 2010, 296).

101 This was also the function of the special scripts of the papyri from Ravenna and Egypt. On this see in particular Tjäder 1952, 212, Iannacci/Modesti/Zuffrano 2012, 108–110.

l'authenticité de l'acte, reconnaissable à cette calligraphie particulière et difficilement imitable".[102]

It is also possible to quote other examples similar to the Didyma inscription, where a special script was carved in stone in order to affirm the authenticity of the written document.

This is the case, for example, of two stone fragments found in Kairouan (Tunisia), which were studied by Charles Diehl in 1894.

In this case, the special script was limited to two words (*sancimus* and [*con*]*firmamus*), indicating the attestation of guarantee given by the emperor to the act.

Even if those words are not written in characters of a majuscule type, but in a minuscule cursive, in this case as at Didyma the stonecutter reproduced strokes that were totally unnatural for stone writing, and put great care into reproducing a papyrus script on a hard surface, with the clear purpose of giving a mark of authenticity to the document:

> au lieu de les transcrire dans le même caractère que le reste de l'inscription, il a voulu faire éclater aux yeux la différence, jadis exprimée d'une manière en quelque sorte abstraite par la mention: *et alia manu*. Il s'est appliqué à reproduire sur la pierre, *en un véritable fac-similé*, les notes impériales inscrites dans l'original; et ainsi l'inscription de Kairouan nous conserve, dans le *Sancimus* et le *Confirmamus*, écrits en cursive, une copie, plus ou moins imparfaite sans doute, mais incontestable, de la souscription émanant du basileus byzantin.[103]

Fig. 14: Detail: [Con]firmamus (taken from Diehl 1894, pl. III).

102 Feissel 2008, 332 (= rpt. 2010, 296).

103 Diehl 1894, 392f. A similar artifice can also be found in a fragment discovered at Salamis, and also studied by Denis Feissel, where the word *recognovi* presents some letters in cursive forms: Feissel 2001b, 192ff. (= rpt. 2010, 88ff.).

Fig. 15: Detail of Sancimus (taken from Diehl 1894, pl. II).

3 Conclusions

What emerged from this palaeographical analysis is that the opening lines of the *gesta praefectoria* of the Didyma inscription can be considered a missing link between the opening script of the reports of proceedings from Egypt and those of the Ravenna papyri, and this is not only for chronological reasons, as Feissel has also pointed out:

> Le document constitue [...] un nouveau jalon dans les dernières phases d'évolution de la cursive latine ancienne. Entre les procès-verbaux égyptiens précisément datés, qui ne descendent pas au-delà de 461 ou peut-être 483, et ceux de Ravenne en 564 et 625, l'inscription de Didymes apporte à l'histoire de cet alphabet une sorte de chainon manquant.[104]

The epigraphic script is close to the script in the Ravenna papyri because of the mixed shapes of its alphabet, which shows a remarkable number of letters originating from the Ancient Roman cursive system but, at the same time, shows some strokes clearly derived from the Later Roman cursive alphabet.

Nevertheless, a careful examination of the letters of the Didyma inscription reveals a closer proximity to the alphabet of the papyri from Egypt, in particular P.Oxy. XVI 1878 and P.Oxy. XVI 1879.

Therefore, if we try to reconstruct the shape of the script of the original papyrus that represented the model of the inscription, we can suppose, with a good degree of certainty, that the letters had probably assumed an intermediate shape between those

104 Feissel 2004, 332 (= rpt. 2010, 296).

of the papyri from Egypt and Ravenna, which is a script based on a majuscule type, but with a marked degree of elaboration.[105]

Although this script was of a majuscule type, it was also open to the influence of a minuscule model alphabet, and this is an element that clearly shows a situation of graphical complexity. Furthermore, this is proof of the fact that the minuscule model had penetrated common writing practices at such a high level that it also influenced conservative scripts, like those of the administrative and bureaucratic offices.

A further observation that emerged from this research is that the case of the Didyma inscription, where a special and complicated script is carved in stone in order to highlight the authenticity of the document, is not an isolated one. The inscription of Kairouan is a further example of a script formed by curvy and wavy strokes that are hard to reproduce in stone. This demonstrates that inscriptions were used not only to communicate a message, to give public evidence to a text, to represent the power of kings or emperors, or to represent examples of a *scrittura esposta* (exhibited writing):[106] they were also legal documents indeed, provided with a specific legitimacy.

Furthermore, in the case of the Didyma inscription we can observe not only the stonecutter's effort to reproduce difficult strokes on a hard material, but also the will to write in a script that must have been almost illegible for the majority of the people who lived at that time.

In this way, the main function of the inscription, that is, to show a public message to the people, failed, and the writing was the bearer of another function, which was to grant authenticity to a document.

In this sense, the Didyma inscription represents an exemplary case, since, from a palaeographical point of view, it is a witness to what Cencetti defined as the general unit of the graphic phenomenon in the Roman Era,[107] a phenomenon that today is often forgotten by historiography. Beyond the different material supports, we find a unique graphic phenomenon adapted to different variations and contexts and the Didyma inscription is a clear statement of this fundamental unit.

---

105 The exaggerated grade of elaboration of this script was probably one of the results of CTh. 9.19.3, which probably hastened a process of exaggeration of the graphic artifice that had already started with the scripts of the majuscule type and was due to the fact that they were falling into disuse. This involution was proportional to the diminishing legibility of this script (see Iannacci/Modesti/Zuffrano 2012, 105–106).

106 Petrucci 1985.

107 "[...] a parte la prevalenza dell'un materiale scrittorio sull'altro nelle varie epoche e nei diversi luoghi, per l'uso quotidiano la scrittura a inchiostro e quella scalfita dovevano essere adoperate contemporaneamente, o quasi, dalle medesime mani [...] Pur tenendo conto delle differenze reciproche, dobbiamo dunque non dimenticare che i due tipi di scrittura non solo sono interdipendenti e s'influenzano reciprocamente, ma costituiscono in realtà semplicemente due varietà di atteggiamento di una medesima scrittura usuale" (Cencetti 1950, 7 [= rpt. 1993, 55]).

From the standpoint of the function of writing, the Didyma inscription shows how going beyond disciplinary barriers and drawing on a more interdisciplinary approach leads to a deeper and better understanding of this phenomenon, because only a comparison of all the existing sources (palaeographic, papyrological, epigraphic and also diplomatic), from an interdisciplinary perspective, can explain the presence of such a peculiar script on the inscription of Didyma.

Returning to the theme of CTh. 9.19.3 and to the problem of the effective application of this law, we agree with Iannacci, Modesti and Zuffrano that the presence of the opening script in the cases of the papyri from Egypt and Ravenna (and to these we can now add the Didyma inscription) did not represent violations of the law of Valentian and Valens.[108]

Nevertheless, it appears very difficult to reach a definitive conclusion on this matter, since it would imply a deep exegesis of the law and of the graphic context where it was issued.

What can be pointed out in this debate is that the Didyma inscription demonstrates further that a script similar to the *litterae caelestes* survived not only in places of the Empire where the conservative tradition was very strong, like Ravenna or Egypt, but also in other areas, in the West and in the Orient as well.

This shows that the use of this script was not fortuitous or isolated, but was generalised, and this is an important consideration that will have to be taken into account in future studies of CTh. 9.19.3.

The Didyma inscription is thus a source of greatest importance, which illuminates a complex phenomenon not only from a graphic point of view, but also in the context of the relationship between legislation and writing.

108 Iannacci/Modesti/Zuffrano 2012, 104.

Tab. 1: Comparison between lines 36–37 of the Didyma inscription and some alphabets of the Ancient Roman cursive.

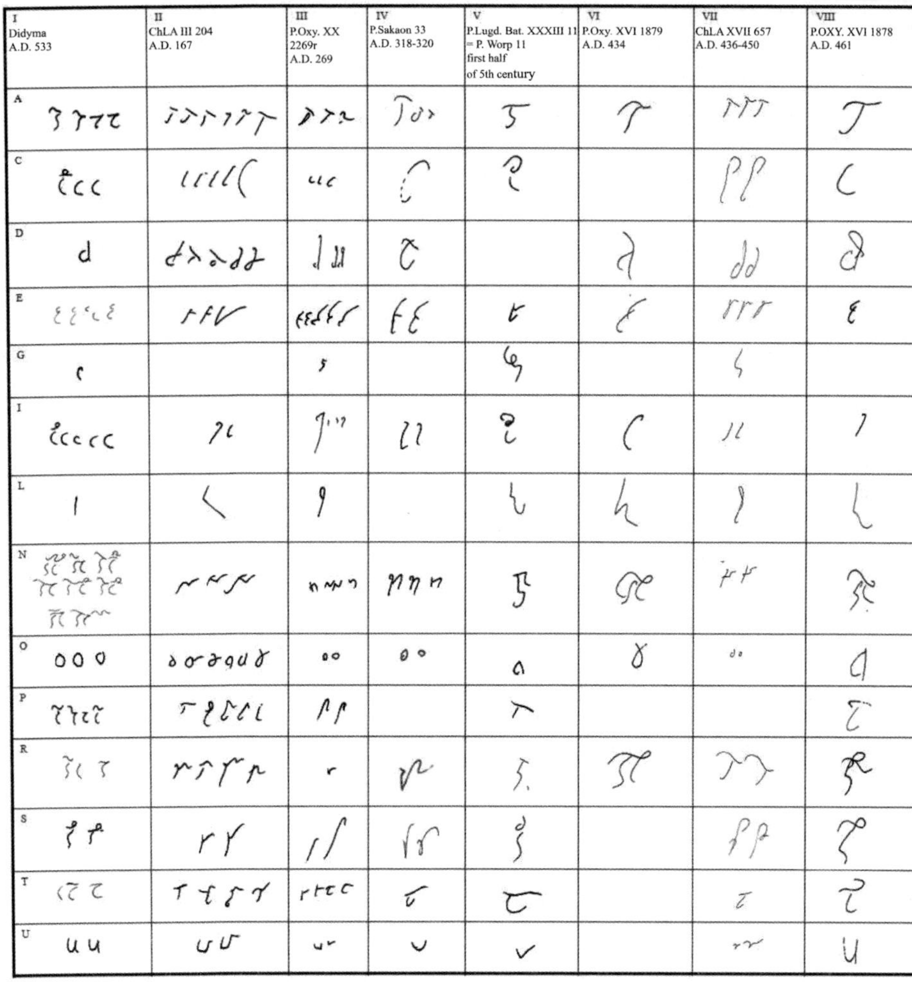

Tab. 2: Comparison between lines 36–37 of the Didyma inscription and some alphabets of the Later Roman cursive.

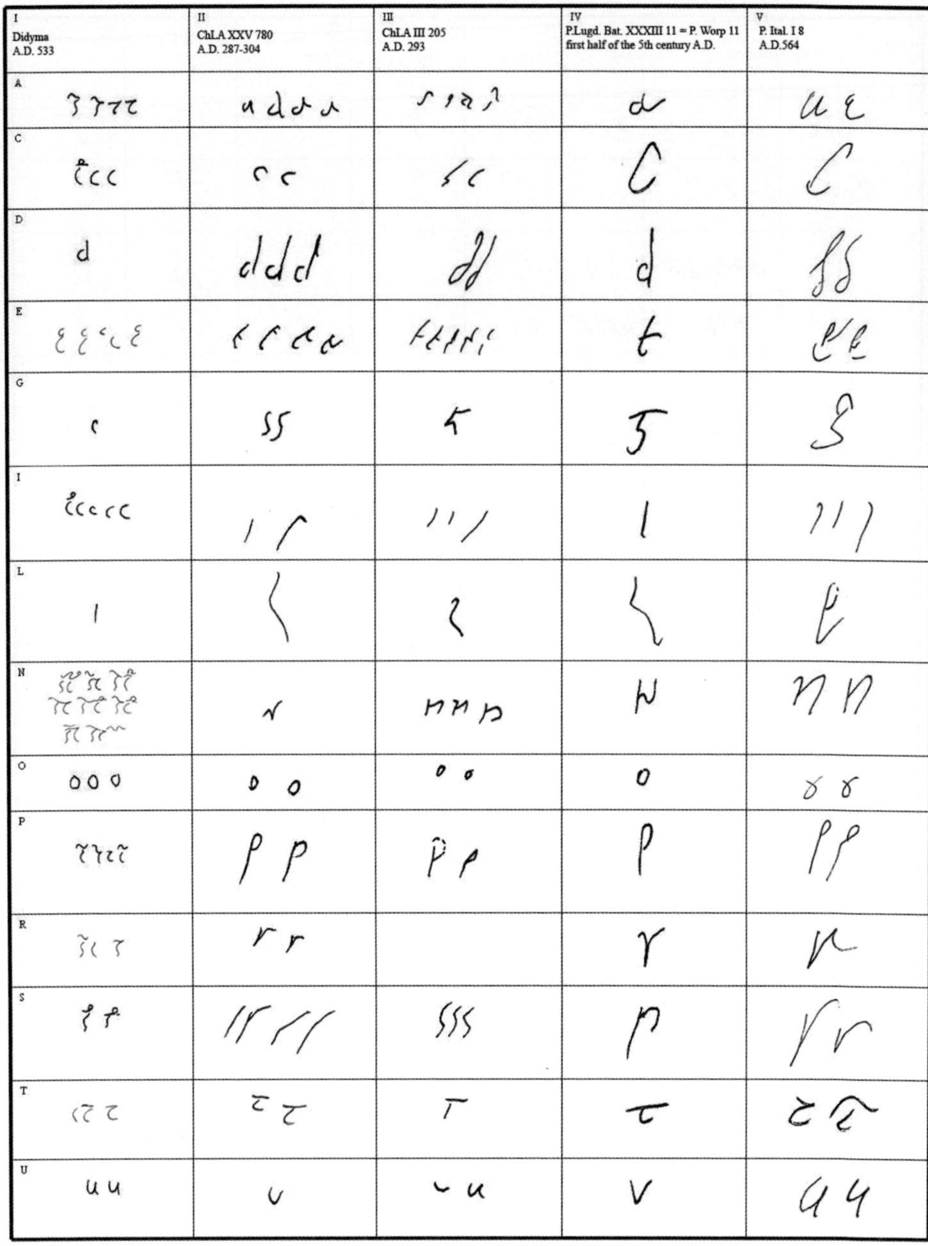

Tab. 3: Comparison between the superior part of N in lines 36–37 of the Didyma inscription and in P.Oxy. XVI 1878.

Didyma A.D. 533		P.Oxy. XVI 1879 A.D. 434	P.Oxy. XVI 1878 A.D. 461
	=		

Tab. 4: Comparison between the shape of R in lines 36–37 of the Didyma inscription and P.Oxy. XVI 1879.

Didyma A.D. 533		P.Oxy. XVI 1879 A.D. 434
	=	
	=	

Tab. 5: Comparison between the shape of T in lines 36–37 of the Didyma inscription and P.Ital. I 8 = Paris, Bibliothèque nationale de France, lat. 4568 A = ChLA XVII 652.

Didyma A.D. 533		P.Ital. I 8 A.D. 564
	=	
	=	

Abbreviations

AÉ	*L'Année épigraphique*
ChLA	Bruckner, Albert/Marichal, Robert (eds.) (1954 –...), *Chartae Latinae Antiquiores. Facsimile-edition of Latin Charters*, 1st series, Zurich.
C.	*Codex Iustinianus*: Krueger Paulus (ed.) (1967), *Corpus iuris civilis*, vol. 2: *Codex Iustinianus*, Berlin.
CTh.	*Codex Theodosianus*: Mommsen Theodor (ed.) (1971), *Theodosiani libri XVI cum constitutionibus*, Berlin.
Nov.	*Novellae*: Schoell Rudolfus/Kroll Guilelmus (1972), *Corpus iuris civilis*, vol. 3: *Novellae*, Berlin.
SEG	*Supplementum epigraphicum Graecum*, Leiden 1923–...
The Theodosian Code	Pharr, Clyde (ed.) (1952) *The Theodosian Code and Novels and Sirmondian Constitutions. A Translation with Commentary, Glossary and Bibliography* (The Corpus of Roman Law), Princeton.

Papyrological texts are quoted in accordance with the Checklist of Papyri, http://library.duke.edu/rubenstein/scriptorium/papyrus/texts/clist.html

Bibliography

Calabi Limentani, Ida (1974^3), *Epigrafia latina, con un'appendice bibliografica di Attilio Degrassi*, Milan.

Casamassima, Emanuele/Staraz, Elena (1977), "Varianti e cambio grafico nella scrittura dei papiri latini", in: *Scrittura e civiltà* 1, 9–110.

Cavallo, Guglielmo (1970), *La koinè scrittoria greco-romana nella prassi documentale di età bizantina*, in: *Jahrbuch der Österreichischen Byzantinistik* 19, 1–31.

Cavallo, Guglielmo/Magistrale, Francesco (1985), "Libri e scritture del diritto nell'età di Giustiniano", in: Gian Gualberto Archi and Mario Amelotti (eds.), *Il mondo del diritto nell'epoca giustinianea. Caratteri e problematiche*. Atti del Convegno tenuto a Ravenna nel 1983 (Biblioteca di 'Felix Ravenna' 2), Ravenna, 43–58.

Cavallo, Guglielmo (2008), *La scrittura greca e latina dei papiri. Una introduzione* (Studia erudita 8), Pisa.

Cencetti, Giorgio (1948), "Vecchi e nuovi orientamenti nello studio della paleografia", in: *La Bibliografia* 50, 4–23.

Cencetti, Giorgio (1950), "Note paleografiche sulla scrittura dei papiri latini dal I al III secolo d.C.", Bologna (= Cencetti, Giorgio (1993), "Note paleografiche sulla scrittura dei papiri latini dal I al III secolo d.C.", in: Giovanna Nicolaj (ed.), *Scritti di paleografia*, Zurich, 47–108).

Cencetti, Giorgio (1956–1957), "Ricerche sulla scrittura latina nell'età arcaica", in: *Bullettino dell'Archivio Paleografico Italiano*, Nuova Serie 2–3 (1), 173–207 (= Cencetti, Giorgio (1993), "Ricerche sulla scrittura latina nell'età arcaica", in: Giovanna Nicolaj (ed.), *Scritti di paleografia*, Zurich, 135–169).

Cencetti, Giorgio (1961), "Dall'unità al particolarismo grafico. Le scritture cancelleresche romane e quelle dell'alto medioevo", in: *Il passaggio dall'Antichità al Medioevo in Occidente*, Spoleto 6–12 Aprile 1961 (Settimane di Studio del Centro Italiano di Studi sull'Alto Medioevo 9), 237–273 (= Cencetti, Giorgio (1993), "Dall'unità al particolarismo grafico. Le scritture cancelleresche romane e quelle dell'alto medioevo", in: Giovanna Nicolaj (ed.), *Scritti di paleografia*, Zurich, 225–272).

Cencetti, Giorgio (1965), *Compendio di paleografia latina per le scuole universitarie e archivistiche*, Naples.

Cencetti, Giorgio (1967–1969), "La paleografia del bibliotecario. Lezione tenuta il 15 aprile 1967 e il 5 novembre 1969 ai corsi di aggiornamento per Bibliotecari di I classe (Università degli Studi di Roma, Scuola Speciale per Archivisti e Bibliotecari)" (= Cencetti, Giorgio (1993), "La paleografia del bibliotecario. Lezione tenuta il 15 aprile 1967 e il 5 novembre 1969 ai corsi di aggiornamento per Bibliotecari di I classe (Università degli Studi di Roma, Scuola Speciale per Archivisti e Bibliotecari)", in: Giovanna Nicolaj (ed.), *Scritti di paleografia*, Zurich, 273–278)

Cencetti, Giorgio (1993), *Scritti di paleografia, a cura di Giovanna Nicolaj*, Zurich.

Cencetti, Giorgio (1997²), *Lineamenti di storia della scrittura latina*, Bologna.

Cherubini, Paolo/Pratesi, Alessandro (2010), *Paleografia latina. L'avventura grafica del mondo occidentale* (Littera Antiqua 16), Vatican City.

Clarysse, Willy/Rochette, Bruno (2005), "Un alphabet grec en caractères latins", in: *Archiv für Papyrusforschung und verwandte Gebiete* 51 (1), 67–75.

Diehl, Charles (1894), "Une charte lapidaire du VIe siècle", in: *Comptes rendus des séances de l'Académie des inscriptions et Belles-Lettres* 38 (5), 383–393.

Feissel, Denis (2001b), "Un rescrit impérial et une consécration d'après une inscription du gymnase de Salamine", in: *Cahiers du Centre d'Études Chypriotes* 31, 189–207 (= Feissel, Denis (2010), "Un rescript imperial et une consecration d'après une inscription du gymnase de Salamie", in: Denis Feissel, *Documents, droit, diplomatique de l'Empire romain tardif* (Bilans de recherche 7) Paris, 85–100).

Feissel, Denis (2004), "Un rescrit de Justinien découvert à Didymes (1er avril 533)", in: *Chiron* 34, 285–365 (= Feissel, Denis (2010), "Un rescrit de Justinien découvert à Didymes (1er avril 533)", in: Denis Feissel, *Documents, droit, diplomatique de l'Empire romain tardif* (Bilans de recherche 7) Paris 2010, 251–324).

Feissel, Denis (2008), "Deux modèles de cursive latine dans l'ordre alphabétique grec", in: Hoogendijk, Francisca A. J. (ed.), *Sixty-five Papyrological Texts presented to Klaas A. Worp on the Occasion of His 65th Birthday* (Papyrologica Lugduno-Batava 33), Leiden, 53–64 (= Feissel, Denis (2010), "Deux modèles de cursive latine dans l'ordre alphabétique grec", in: Denis Feissel, *Documents, droit, diplomatique de l'Empire romain tardif* (Bilans de recherche 7) Paris 2010, 541–552).

Feissel, Denis (2010), *Documents, droit, diplomatique de l'Empire romain tardif* (Bilans de recherche 7), Paris.

Gordon, Joice S./Gordon, Arthur E. (1957), *Contributions to the Palaeography of Latin Inscriptions* (University of California Publications in Classical Archaeology 3.3), Los Angeles.

Iannacci, Lorenza/Modesti, Maddalena/Zuffrano, Annafelicia (2012), "La misteriosa scrittura grande dei papiri ravennati, tra prassi documentaria e pubblica legislazione", in: *Legal Roots. The International Journal of Roman Law, Legal History and Comparative Law* 1, 89–120.

Kresten, Otto (1965), "Zur Frage der 'Litterae Caelestes'", in: *Jahrbuch der Österreichischen Byzantinischen Gesellschaft* 14, 13–20.

Kresten, Otto (1966), "Diplomatische Auszeichnungsschriften in Spätantike und Frühmittelalter", in: *Mitteilungen des Instituts für Österreichische Geschichtsforschung* 74, 1–50.

Mallon, Jean/Marichal, Robert/Perrat, Charles (1939), *L'écriture latine de la capitale romaine la minuscule*, Paris.

Mallon, Jean (1948), "L'écriture de la chancellerie impériale romaine", in: *Acta Salamanticensia, filologia y letras* 4 (2), 5–43. (= Mallon, Jean (1982), "L'écriture de la chancellerie impériale romaine", in: Jean Mallon, *De l'écriture. Recueil d'études publiées de 1937 à 1981*, Paris, 167–189).

Mallon, Jean (1952), *Paléographie romaine* (Scripturae 3), Madrid.
Mallon, Jean (1982), *De l'écriture. Recueil d'études publiées de 1937 à 1981*, Paris.
Marichal, Robert (1948), "De la capitale romaine à la minuscule", in: Marius Audin, *Somme typographique*, Paris, 63–111.
Marichal, Robert (1950), "Paléographie précaroline et Papyrologie. II: l'écriture latine du Ier au VIIe siècle: les sources", in: *Scriptorium* 4 (1), 116–142.
Marichal, Robert (1952), "L'écriture latine de la chancellerie impériale", in: *Aegyptus* 32, 336–350.
Marini, Gaetano (1805), *I papiri diplomatici raccolti e illustrati*, Rome.
Nicolaj, Giovanna (1973), "Osservazioni sul canone della capitale libraria romana fra I e III secolo", in: *Miscellanea in memoria di Giorgio Cencetti*, Turin, 3–27.
Perrat, Charles (1955), "Paléographie romaine", in: *Relazioni del X Congresso internazionale di scienze storiche*, vol. 3, Florence, 345–384.
Petrucci, Armando (1985), "Potere, spazi urbani, scritture esposte: proposte ed esempi", in: *Culture et idéologie dans la genèse de l'état moderne*. Actes de la table ronde organisée par le Centre national de la recherche scientifique et l'École française de Rome (Rome, 15–17 ottobre 1984), (Collection de l'École Française de Rome 82), Rome, 85–97.
Radiciotti, Paolo (1998), "Manoscritti digrafici grecolatini e latinogreci nell'antichità", in: *Papyrologica Lupiensia* 6, 107–146.
Radiciotti, Paolo (2008), "Il problema del digrafismo nei rapporti fra scrittura latina e greca nel medioevo", in: Otto Kresten and FranzLackner (eds.), *Régionalisme et internationalisme. Problèmes de paléographie et de codicologie du Moyen Âge.* Actes du XVe colloque du comité international de paléographie latine (Vienne, 13–17 septembre 2005) (Österreichische Akademie der Wissenschaften, Philosophisch-historische Klasse, Denkschriften 364), Vienna, 19–33.
Radiciotti, Paolo (2009), "I papiri e le scritture greca e Latina", in: *Atene e Roma, seconda nuova serie* 3 (1/2), 97–119.
Sandys, John Edwin (1969), *Latin Epigraphy. An Introduction to the Study of Latin Inscriptions*, Groningen.
Schiaparelli, Luigi (1921), *La scrittura latina nell'età romana. Note paleografiche* (Auxilia ad res italicas medii aevi exquirendas in usum scholarum instructa et collecta 1), Como.
Steffens, Franz (1910), *Paléographie latine*, Paris.
Tjäder Jean-Olof (1952), "La misteriosa 'scrittura grande' di alcuni papiri ravennati e il suo posto nella storia della corsiva latina nella diplomatica romana e bizantina dall'Egitto a Ravenna", in: *Studi Romagnoli* 3, 173–221.
Tjäder, Jan-Olof (1955), *Die nichtliterarischen lateinischen Papyri Italiens aus der Zeit 445–700*, vol. 1: *Papyri 1–28* (Skrifter utgivna av Svenska institutet i Rom, Serie in 4° 19.1), Lund.
Tjäder, Jean-Olof (1977), "Latin Palaeography, 1975–77", in: *Eranos, Acta philologica Suecana* 75, 131–161.
Tjäder, Jean-Olof (1979), "Considerazioni e proposte sulla scrittura latina nell'età romana", in: *Palaeographica, Diplomatica et Archivistica. Studi in onore di Giulio Battelli*, Rome, 31–62.
Tjäder, Jean-Olof (1982a), "Some Ancient Letter-forms in the Later Roman Cursive and Early Medieval Script of the Notarii", in: *Scrittura e Civiltà* 6, 5–21.
Tjäder Jean-Olof (1982b), *Die nichtliterarischen lateinischen Papyri Italiens aus der Zeit 445–700*, vol. 2: *Papyri 29–59* (Skrifter utgivna av Svenska institutet i Rom, Serie in 4° 19.1), Stockholm.
Tjäder Jean-Olof (1985), "Later Roman (Common) Script. A Tentative Definition in Anticipation of a Forthcoming Monograph", in: Jacques Lemaire and Emile van Balberghe (eds.), *Calames et cahiers. Mélanges de codicologie et de paléographie offerts à Léon Gilissen*, Brussels.

Anastasia Grib

The Symbolic Repertoire of the Qur'anic Board in Islamic Africa

Abstract

The Qur'anic board is a central object in the material culture of the Qur'an in the local Islamic centres of Africa. The board features ornaments and calligraphy, a symbolic language which one needs to learn and interpret. This article provides a transcription of the symbolic code of the Qur'anic board based on the study of 124 samples from the Brooklyn Museum, the Gallery of Sam Fogg, the Musée du quai Branly, and other collections. The boards are dated to the end of the 13th and 14th centuries Hijri/19th and 20th centuries CE and represent several regional groups: Mali, Morocco, Mauritania, Nigeria, Sudan, Ethiopia, Somalia and Senegal. A functional typology of the boards is discussed, i.e., festive, training, magical, and healing boards; to this a classification is added that proceeds from the consideration of material, form, ornamentation, and script. Four different types of ornamentation are identified: architectural, geometric (cosmic symbols), carpet, and figurative. Some ornamental motifs can be described as an *ex libris* connecting the board to an important local centre or shrine; among such motifs are those with pre-Islamic origin. Two boards are localized to the Ceremony Mountains in Northern Nigeria; they feature anthropomorphic and zoomorphic images and their iconography traces back to Neolithic rock graffiti. Playing a crucial role in initiation, the Qur'anic board is a symbol of masculinity, as exemplified by the wedding ceremony *mba* and the Qur'anic school graduation fest *saukar fari* in Nigeria.

1 Qur'anic Board as an Emblem of the Material Culture of the Qur'an

The Material Culture of the Qur'an (MCQ) refers to artefacts and their associated practices focused on the Holy Qur'an and its representations. I distinguish between three domains of the MCQ. The domain of primary objects includes Qur'anic boards (QBs) and manuscripts. The domain of secondary objects includes architectural surfaces, wooden objects, pottery, calabashes, textiles, and rock graffiti featuring a pictorial

This article emerged from the Heidelberg Collaborative Research Center 933 "Material Text Cultures. Materiality and Presence of Writing in Non-Typographic Societies". The CRC 933 is financed by the German Research Foundation (DFG).

DOI 10.1515/9783110417845-007,

image of the QB. The domain of ritual refers to the reified existence of the sacred text: primary objects participate in this domain directly, whereas secondary objects point to a ritual function. The third domain involves the use of the Qur'anic text and of its symbolic representations (in the form of 'false calligraphy'), as well as the participation of the QB in the following ritual activities and objects: magic, sorcery, talismans, and various protections against the evil eye.

In this paper I attempt a decoding of the symbolic language of the MCQ in the local Islamic centres of West and North Africa, where the most striking example of the MCQ is the Qur'anic board *allo* (pl., *alluna*, from Arabic *al-lawḥ*). The board is used extensively: it is the major emblem of the *malam's* profession and is used by students in Qur'anic schools in learning Arabic grammar, the holy text of the Qur'an, and the art of calligraphy.

The background of the Qur'an board is the transhistorical concept of the Hidden Tablet (*al-Lawḥ al-Maḥfūẓ*), the Mother of the Book (*'Umm al-Kitab*), or proto-Qur'an. Islamic tradition holds that Allah had inscribed upon this tablet the totality of His decrees concerning creation.[1]

Until recently, the board was the main writing material for teaching the Qur'an, Arabic literacy, and calligraphy in Qur'anic schools (*khalwa*, *muhadra*, *makarantar allo*, etc.) and *madrasa* (*madrasah*, pl. *madāris*) throughout Sub-Saharan Africa. The majority of such local traditions in Pakistan, Malaysia, Iran, and various regions of Central Asia were lost in the course of the 15th century Hijri/20th century CE after the introduction of modern schooling systems, but in Africa use of the board has persisted in Islamic learning and can still be observed in, e.g., Mali (Timbuktu), Sudan (Umdurman), Nigeria (Kano, Sokoto, Marghi, Zaria, Bauchi), Mauritania (Tichit, Oualata, Ouadan, Chinguetti), Algeria, Morocco (Rabat, Fes), Guinea, Somalia, Ethiopia (Harar), the Comoro Islands, and Senegal. Below I classify the boards in accordance with their regional variations.

1.1 Local Traditions

Islamic learning and writing centres exist all over North and West Africa, so I did not limit my investigation to one particular locality. The geographical scope has been

1 On the Hidden Tablet Allah has written all the destinies of His creatures: "For every term there is a book prescribed; Allah erases out whatever He pleases and writes (whatever He pleases); and with Him is the Mother of the Book" (13:39); "Nay, it is the glorious Qur'an, in the Guarded/Hidden Tablet" (85:22). In the Qur'an, the plural form of the word "tablet", *alwah*, refers to the Tablets of Moses.

deliberately broadened to allow a comparative examination of the objects under discussion.

Being a synthetic cultural product, a QB displays characteristics of regional Islam found in a variety of local practices—in contrast to the so-called 'normative Islam' of the Central Islamic Lands. In Nigeria, Mauritania, Mali and Sudan, there are multi-religious, multicultural communities that have preserved the tradition of copying the Qur'an. Here one finds permanent and itinerant schools known variously as *makarantar allo* (in Nigeria), *muhadra* (in Mauritania) and *khalwa* (in Sudan). The schools are established and run by local or itinerant scholars through whom the tradition of sacred text copying is largely preserved and disseminated. Students write the assigned Qur'anic verses on a flat surface of a wooden board; at the end of the exercise they wash the text off upon receiving their *malam*'s (teacher's) permission. The water is often collected and used for ritual purposes and is therefore treated with care and piety. At the same time, the Qur'anic board is an integral element in indigenous initiation (male initiation and wedding) ceremonies—in parallel to the board's use in Qur'anic schooling, where it functions to initiate a child into the Muslim faith.

One of the most intriguing traditions is found among the Hausa. The common occupation of a *Hausa malam* is that of a craftsman, artist, and scribe simultaneously, which influences their unique school of writing. The calligraphic tradition of Hausa in Northern Nigeria represents a distinctive version of the *Sūdānī Kūfī*.

In Mauritania, there are two scripts for copying the Qur'an, one traditional and one modern. The *Sūdānī* script is widespread, but the most developed and perfected script today is the *Andalusī* (of the Maghribian branch, see below).

In Sudan, Qur'anic schools are commonly called *khalwa* and the boards are known as *loah*. Such schools are usually ruled by *sufi sheikhs*. For example, *khalwa* in Omdawanban is a

> religious institution that today welcomes more than 1250 students from Sudan, Chad, Cameroon, and Nigeria. It takes from three to six years to acquire a perfect knowledge of the Qur'an and to master written Arabic [...]. The children learn to write Arabic for the first time on the wooden tablet using a tube of sugar cane filled with an ink made of charcoal or soot.[2]

Wooden boards are used for writing lessons and copies of the Qur'an are used for reading.[3] In Sudan, as in some other regions, one also finds *khalwa* for girls.[4]

2 Cifuentes 2008, 50–59, esp. 58.

3 Cifuentes 2008, 56.

4 "In addition to the learning and practicing of religion, the women are trained in health care, nutrition, cooking, sewing, and handicrafts in order to help a living" (Cifuentes 2008, 56).

1.2 The Script and the Ornament

The Qur'anic boards and the manuscripts share similar types of calligraphy. The majority of the boards discussed in this article exhibit several variations of the *Sūdānī* hand (the most prominent is the *Hausa ductus*); most of the rest feature the *Andalusī* hand, and a few are in the *'Ifrīqī* hand.

With regard to the genealogy of the script, the West-African tradition is an offshoot of the *Kūfī* school.[5] The very term *Kūfī* thus pertains to the shape of letters and the diacritical marks that are representative of the *Maghribī* system of writing. These diacritics differ from those of the Middle-Eastern system by: (1) letter ڡ (*fā'*), which is marked by a single dot in the underscript; (2) letter ڧ (*qāf*), which has only one dot in the superscript; and (3) letter ص / ض (*ṣād/ḍād*), which doesn't have a final tooth. While remaining the principle Qur'anic script, the *Maghribian* branch of sacred *Kūfī* has been modified into a *cursive* hand, and beyond that, into a false calligraphy (observed as an architectural decoration). The '*Demonumentalised Kūfī*' is the result of this evolution; it is still practiced today as the script of the Qur'anic board.

There is a whole set of distinctions between the *Maghribī* styles (in the concrete geographic application of the term, the hands of North Africa and Spain) and the West African school. Such differences can be observed both in the graphic features of the script and in the iconography of the ornament. West African styles are characterized by sharp angularity and geometrical shape, whereas the *Maghribī* ones have ligatures and plasticity. This difference is very much apparent in the decoration of the Qur'anic boards. In terms of ornamentation the distinguishing characteristic of the *Maghribian* Qur'anic tradition is the predominance of the Islamic and Arabic features. On the other hand, in the remote parts of Africa where indigenous cultures prevail, the board and the manuscript have switched roles: compared with the manuscript, the board is a more potent object, both semantically and ritually.

'Ifrīqī (fig. 1) is an early and more simple version of the *Maghribī* style; it has been used on QBs in the past, but very randomly. The script is characterized by angularity and flatness, and the proportions are also random. Different modifications of *'Ifrīqī* have assimilated with the Spanish *Andalusī*.

Andalusī (fig. 2) is an elegant cursive type of *Maghribī*. It is commonly used on Qur'anic boards in North Africa, especially in Morocco, Mauritania, and Algeria. The script has the following characteristics:

1. final letters ص / ض (*ṣād/ḍād*) and س (*sīn*) sometimes have an extended shape
2. lines are not sharp but rather waved and curled

5 Cf. Bivar 1959, 324–349; Hassan 1992; Blair 2008, 59–75; and Mohameden Ould Ahmad Salem, calligrapher from Mauritania, with whom I corresponded by e-mail.

Fig. 1: Fragment of the Mukhtalitah of Sahnun from Qairawan, 11th century CE (MSS303 fol. 8b–9a) © Nasser D. Khalili Collection.

Fig. 2: Fragment of a Qur'anic manuscript from Morocco, 18th century CE (Qur. 149, part I, fol. 157b–158a) © Nasser D. Khalili Collection.

Fig. 3: Fragment of a Qur'anic manuscript, end of the 19th century CE (Qur.109 fol. 2b) © Nasser D. Khalili Collection.

3. final forms of the letters ى، ن ،ف ،ق (*yā', nūn, fā', qāf*) frequently have no diacritics.

The text is visually divided into groups of verses: every five *āyāt* are stamped with the pattern known as *khamisa* (or *khamsa*); every ten *āyāt* are marked with the *'ashira* (*'ashra*). *Sūdānī* style (fig. 3) is widespread in the area of Western Sudan; one of the most spectacular variations is the *Hausa ductus* which is also the style of the Hausa *malams* (s. *malam*, pl. *malamai*) in Northern Nigeria. The Qur'anic text is written in dark brown or black ink (usually made from charcoal), illuminations are polychromatic. *Malams* use natural colours such as red, yellow, blue, and green. Diacritics are executed in a different colour, whereas vocalization is done by means of short horizontal strokes. The majority of the manuscripts and a number of boards contain colourful round medallions (fig. 4) as well as *sajjāda* (a prayer rug ornament, which stands symbolically for *sajda*, prostration); these are painted on the margins with the help of compasses.

At the beginning of each new section of a manuscript there is an *'unwān* (or 'carpet page') called *zayyana* (figs. 5–6). A typical *zayyana* consists of a rectangular textile ornamental panel featuring colourful geometrical patterns. *Zayyana* is also a characteristic feature of the festive QB. In the Sudanic tradition the same type of ornament is called *charafa*.[6] *Sūra* headings are written in a different colour, usually red, and are marked off by a frame.

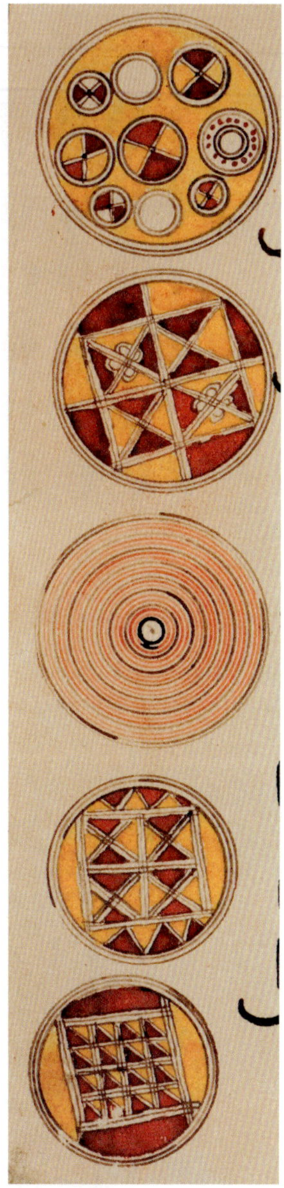

Fig. 4: Fragment of a Qur'anic manuscript from Western Sudan, end of the 19th century CE (Qur.109 fol. 3a) © Nasser D. Khalili Collection.

Among marginal signs one finds *ḥizb*—division into sixty sections, as well as further subdivisions: *nisf* (1/2 of *ḥizb*) (fig. 6), *rub'* (1/4), and *thumn* (1/8), following the numerical meaning of the letters *nūn*, *bā'*, and *thā'* (fig. 8). Each verse is marked off by a pattern of three dots forming a trefoil (red or yellow with a red outline) with a short 'antenna' on the top. Every five verses are divided

6 Thus, in Sudan, in the *khalwa* Qur'anic schools, "to complete their education, students must write a passage from the Qur'an in their best script and draw a holy illustration on the wooden tablet. This *charafa* is presented to the master during the last few days of schooling" (Cifuentes 2008, 59).

Fig. 5: Fragment of a Qur'anic manuscript from Western Sudan, end of the 19th century CE (Qur.109 fol.10) © Nasser D. Kahlili Collection.

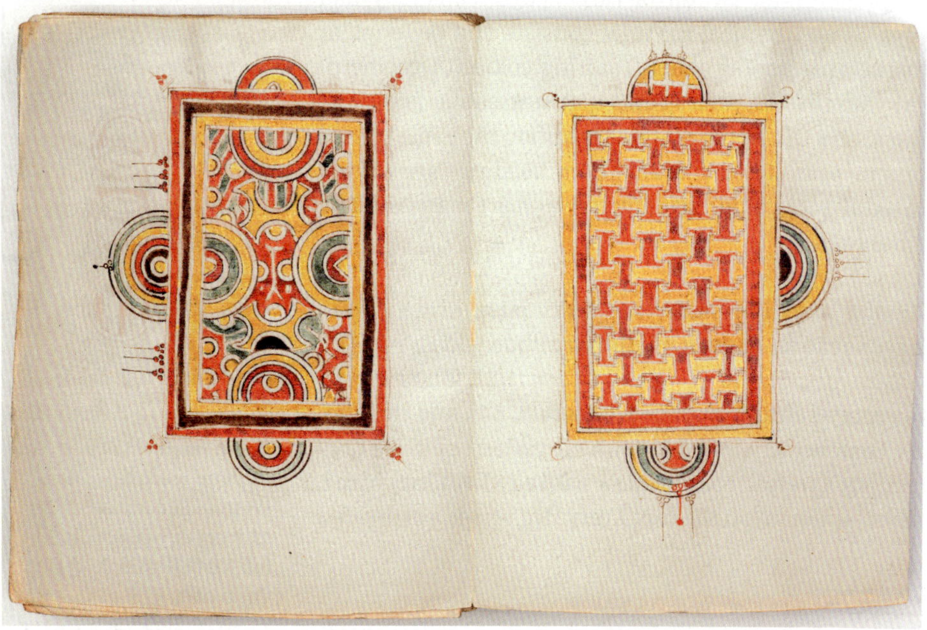

Fig. 6: Fragment of a Qur'anic manuscript from Western Sudan, end of the 19th century CE (Qur.110 fol.1b-2a) © Nasser D. Kahlili Collection.

by letter *hā'* modified into a tear-shaped loop. Every ten verses are divided by a double round medallion with three or four dots or strokes inscribed within the text, or sometimes with a small wheel-shaped medallion on the margin.

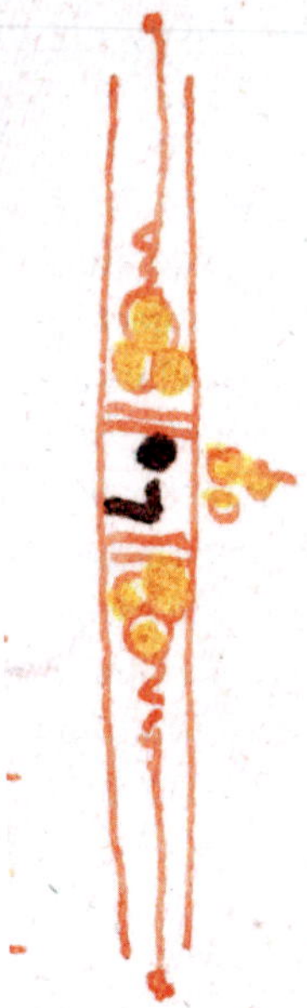

Fig. 7: Fragment of a Qur'anic manuscript (Qur.0110 fol. 65b) © Nasser D. Khalili Collection.

There are variations within the *Hausa* tradition: some scripts are more cursive, whereas other scripts are more monumental and can be described as heavy and colourful. The common script of Kano is almost square; the text is a lot denser than in the *Maghribī* Qur'ans of North Africa; letters are compact and much higher and heavier; lines are fat-faced and steady. One may regard this style as a version of *Hausa ductus*.[7]

There are notable differences between boards and manuscripts as regards illumination. Qur'anic boards have no marginal markers. Along with the strong tradition of *zayyana* in Nigeria and of *'unwān* in other areas, there is a greater variety of ornamentation on the Maghribian and West-African boards. For example, festive boards from Morocco feature a whole series of ornaments: colourful solar ornaments with geometric patterns, architectural ornaments portraying a mosque, various kinds of floral *'unwāns*. Such decorations usually cover the entire surface of the board on one side; on the other side the calligrapher inscribes a text from the Qur'an, or some other ornamental composition. Since Qur'anic boards differ from one African region to another, the type of this decoration may also provide clues for attribution.

2 Analysing the Symbolic Repertoire of the QB

2.1 The Dialect Approach

In my study, ornament and calligraphy have been analysed by applying the 'Dialect Approach' to Islamic art. Just like the Arabic language, one may talk about the classical, or 'standard', language of Islamic art and about its regional 'dialects' related

7 *Ductus* is "a visually recognizable aspect of a concrete graphic tradition [...]. In the domain of the Arabic script, there are clearly several graphic modifications that do not always coincide with the 'script' in the traditional classification." (Dobronravin 1999, 8;the translation of the quote from Russian is mine).

primarily to ethnic background. I choose to call the calligraphic traditions of West and North Africa the West- and North-African interpretations of *Kūfī* (including in the term *Kūfī* both the script and the ornament). I also replace the two commonly used academic terms *arabesque* and *geometric ornament* with the term 'motif-chain'.[8] The classification goes by genre: architectural motif-chain, textile motif-chain, calligraphic motif-chain, etc. Motif-chain has a composite nature: it is organized around a particular pictogram such as tree, column, tent, etc.; it also features a subject line, e.g., zoomorphic, cosmogonic, vegetative, etc. An additional element of classification proceeds from the consideration of individual artistic elements such as straight or curved line, zigzag, triangle, circle, dot, grid, leaf, fruit, etc. The character of the ornaments is determined by the object's purpose, usage, and ownership. Ornaments may speak of the board's provenance in a certain place; it may relate to the owner's personal need such as healing; or it may be part of a training board used in a Qur'anic school. The depictive canon of the board can be symbolically transcribed assuming one reads the object from the perspective of the time and the place of its (1) creation, (2) editing, (3) storing, and (4) interpretation/use.

2.2 Typology of the QB

The functional typology of the Qur'anic board distinguishes between festive, training, magical, and healing boards. The function determines the format of the board and the presence or absence of ornamentation.

Festive boards usually contain the first *Sūra al-Fātiha* and a short excerpt from the second *Sūra al-Baqara*. This symbolises the completion of the program of study of the Qur'an, from the first *Sūra* to the final one. The student starts with *Fātiha*; after that he/she moves to the very last *Sūra*, gradually working his/her way back toward the beginning.

Boards are vertical or square-shaped, with or without a handle.
There are several types of ornament:

- carpet type (geometrical, textile)—*zayyana* (*Hausa type*);
- architectural type—architectural motifs or the image of a mosque: the so called *reduced Temple image*;
- cosmogonic type (geometrical);
- floral type;
- figurative type, characterized by zoomorphic and anthropomorphic images.

8 I have borrowed the two terms, 'motif-chain' and 'chronotope' (below), from Mikhail Bakhtin's discussion of the cultural roots of the Western European novel, see Bakhtin 1984; Emerson/Holquist 1981.

Certain ornamental motifs can be described as an *ex libris* connecting the board to an important local centre or shrine; among such motifs are those with pre-Islamic (indigenous) origin.

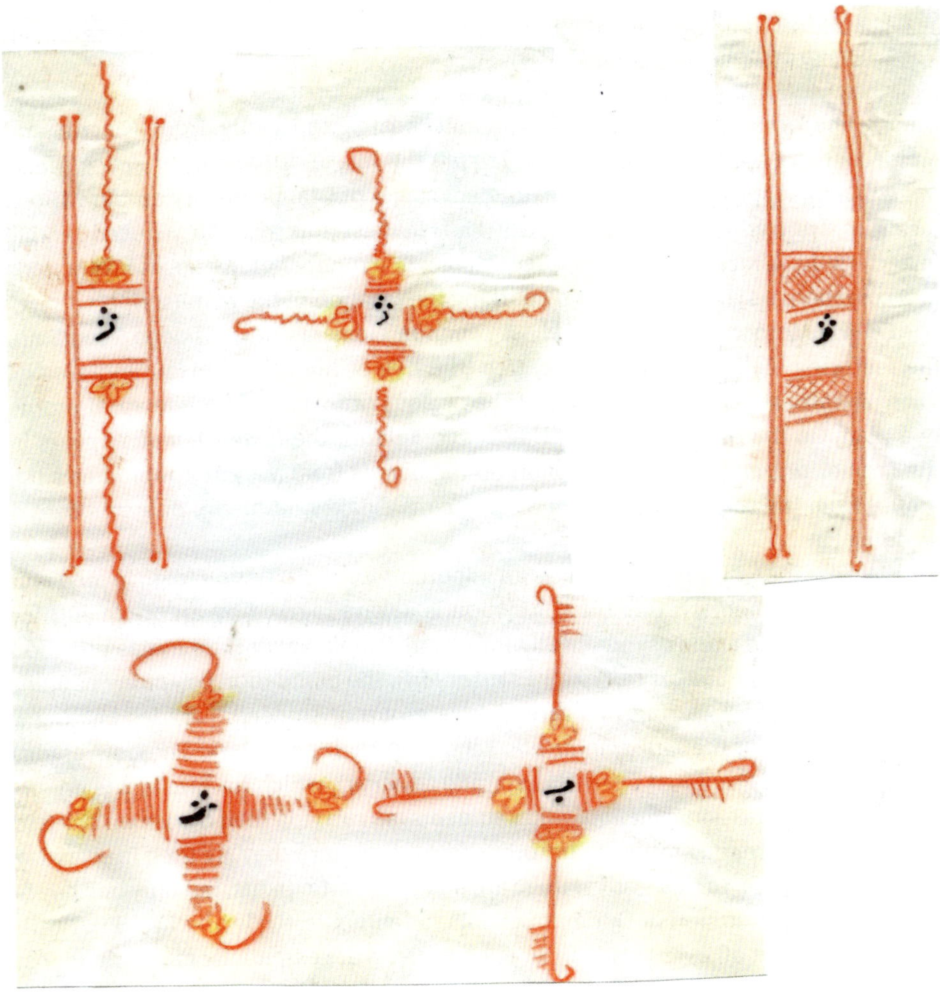

Fig. 8: Hand writing copy of the Marks from a Nigerian Qur'an C-1689 (Institute of Oriental Manuscripts, Russian Academy of Sciences, St. Petersburg).

The most striking Qur'anic board tradition is that of the Hausa in Northern Nigeria. There are several types of *Hausa allo*, including:

- a draft *allo* for practice (pl. *alluna*, study boards): the board allows corrections of both the text and the ornament

- *allo zayyana*: this board is of particular importance as it becomes the graduation certificate from the Qur'anic school (during the *saukar fari* festival), and is also used as a symbol of male initiation.

2.2.1 Imago Templi

Among the ornamentations of the Qur'anic boards especially from Nigeria and Morocco, there are versions of *Imago Templi*, the image of the sacred space/place. *Imago Templi* appears in the MCQ in different stages of Islamic history and in various localities. The tradition goes back to early *Kufic* manuscripts with the 'reduced Temple image', or the *genizah* (place of storage) image; in a number of instances it functions as a colophon, a symbolic place-tag marking the location of the manuscript.[9]

On the Moroccan boards, the most common elements of the *Imago Templi* are the following: column, lamp, niche, tree (the tree of life in Paradise), and arcade. There are also schematic depictions of the mosque drawn as an architectural profile or a plan. On the Nigerian boards (Hausa and Fulani), the *Imago Templi* appears in the form of *zayyana* or a mosque plan; on the boards from Marghi it reproduces the graffiti from the sacred caves (such as scorpions, snakes, hunters), or symbolizes a wedding tent.

Because the Qur'anic board plays an important role in initiation (e.g., in *mba* and *saukar fari* ceremonies), this ritual function is always marked by an ornament; for example, *zayyana* stands symbolically for *Imago Templi*. Salah Hassan has provided a description of the *saukar Qur'ani* (or *saukar fari*), the graduation ceremony popular among the Hausa.[10] During this ceremony, a group of students parades bearing a happy graduate of the *makarantar allo* school on their shoulders. The new graduate is awarded a decorated festive board (*allo zayyana*), which his family has pre-ordered from a professional *malam*. Other students are carrying their own boards: these *alluna* have no ornamentation or any other decoration since they are used in everyday study and practice. In this festival there are plenty of participants, but only one boy is celebrated. The tradition of Qur'anic school graduation festivals originated in North Africa and occurs all over West Africa. Such festivals are essentially male initiations, as is the case in Gambia (the *nau mini* ceremony) or Sierra Leone.[11] These ceremonies are of comparable meaning and symbolism even as they may differ in details.

9 Grib 2009b, 110–125.
10 Salah Hassan 1992, 95.
11 Hassan 1992, 99.

The *genizah* image is featured on the 'Brooklyn Board' (see below). The decorations on this board allow its place of origin to be identified: the sacred caves in Marghi.[12] In these caves, the wedding/engagement ceremony *mba* takes place, during which various objects of male prestige are paraded.

2.3 Surface, Body, and Iconic Symbolism

The symbolic repertoire of the QB accumulates the artistic language of African Islam; in particular, it consists of 1) images and symbols common for the decoration of the boards; 2) the board itself and its features when the board is used as a symbol; and 3) the functioning of the board.

I identify three symbolic aspects of the Qur'anic board:
- surface symbolism: the symbols on the surface of the board, such as the various ornamental motifs and script;
- body symbolism: the overall shape of the board, which often has anthropomorphic features;
- iconic symbolism: the use of the Qur'anic board as a pictorial icon in a different material context; particularly, all the instances when the pictorial image of the board appears on other surfaces (such as stone, textile, calabashes, etc.).

The first aspect of the symbolic repertoire of the QB emerges from a consideration of the board's surface (*surface symbolism*). There is a semantic unity between Qur'anic manuscripts and boards in this particular aspect. Both use similar sets of ornaments, though the board exhibits a much wider spectrum. For example, the *zayyana* ornament, which is common to the manuscripts and the boards, is a modification of the famous *'unwān*, or 'carpet page'.

Surface symbolism is built from the following artistic elements:
1. basic motif-chains: architectural, textile, calligraphic;
2. alternative motif-chains: floral/vegetative, animal, cosmological, ritual and the Temple;
3. motifs: arch, column, niche, *palmette*, etc.;
4. units: direct line, waving line, zigzag, circle, square, triangle, point, vase, spiral flower, net, border, etc.

The second aspect of the symbolic repertoire of the board relates to its *body symbolism*. There are various 'bodily' features and functions of the Qur'anic board. The board is often shaped anthropomorphically: it consists of a head, a neck, two shoulders, a body, and two feet (fig. 10). This is everything but the hands, since the board

12 Grib 2009a, 23.

and the copyist (*malam*) have but a single pair of hands between them. Moreover, one is not allowed to flip the board over, making it even more 'human'. Also in the Hausa tradition, one finds a special type of 'legless' board, the *gurgun allo*.[13]

Fig. 9: Qur'anic board from Western Sudan N° 9568 © Sam Fogg.

13 See Hassan 1992, 157. I am grateful to Dr. Abubakar Sule (University of East Anglia), a Hausa native and a graduate of a Nigerian *makarantar allo*, for the following personal communication: "'*Gurgu*' refers to disable disabled?? as in without leg. A lot of boards with sharp convex arches and elongated legs end up that way".

Fig. 10: Mosque in Zaria, decorated by Musa Yola (taken from Moughtin 1985, 130).

Fig. 11: a–c. Wedding textiles (taken from http://www.adireafricantextiles.com).

In addition to the body, the 'skin' of the board is very important, which is to say its ornamented surface. Various symbols used in the ornamentation (e.g., 'the Hand of Fatima', the 'leper hand') are bodily allegories full of magical meaning. Being an initiation object, the board communicates important changes in the life of the body. One can also include in this aspect the anthropomorphic and zoomorphic figures that may be inscribed on the board—given the fact that such figures carry distinct gender characteristics.

In sum, *body symbolism* is built from:

1. body-shaped parts of the board;
2. *gurgun allo*;
3. *allo zayyana/charafa*: a special board for the graduation ceremony (*saukar fari/ walīma khatmil-Qur'an*) at the end of the student's time at the Qur'anic school. This board functions as a graduation certificate.

The third aspect of the symbolic repertoire of the Qur'anic board is *iconic symbolism*. The pictorial image of the Qur'anic board can be seen on an architectural surface, e.g., above the portal of a mosque (fig. 10).[14] In wedding textiles (fig. 11a–c), the image of the QB stands symbolically for 'man'. On the other hand, among the Fulani, the calabashes featuring images of festive objects and, in some cases, engraved zodiac tablets (including four Qur'anic boards), relate the idea of 'femaleness' (fig. 12).[15]

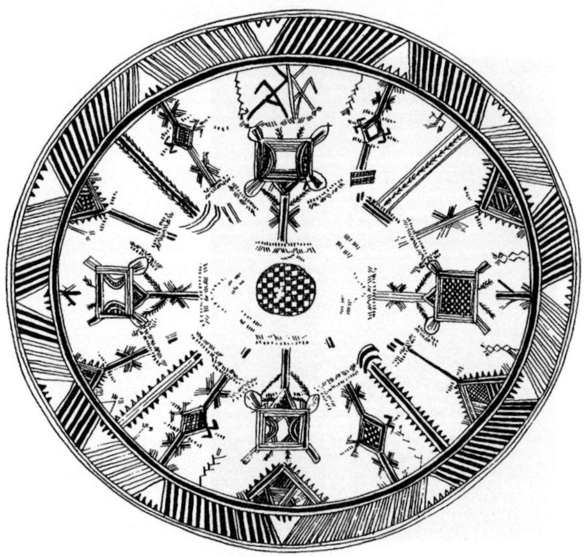

Fig. 12: Engraved zodiac tablet (taken from Blackmun Visona 2001).

14 As described by Kirk-Green 1963, 18: "Hausa walls may show the Koranic emblems of the slate (*allo*) [...] and the satchel for carrying the Holy Koran (*gafaka*)."
15 Blackmun Visona 2001, 102.

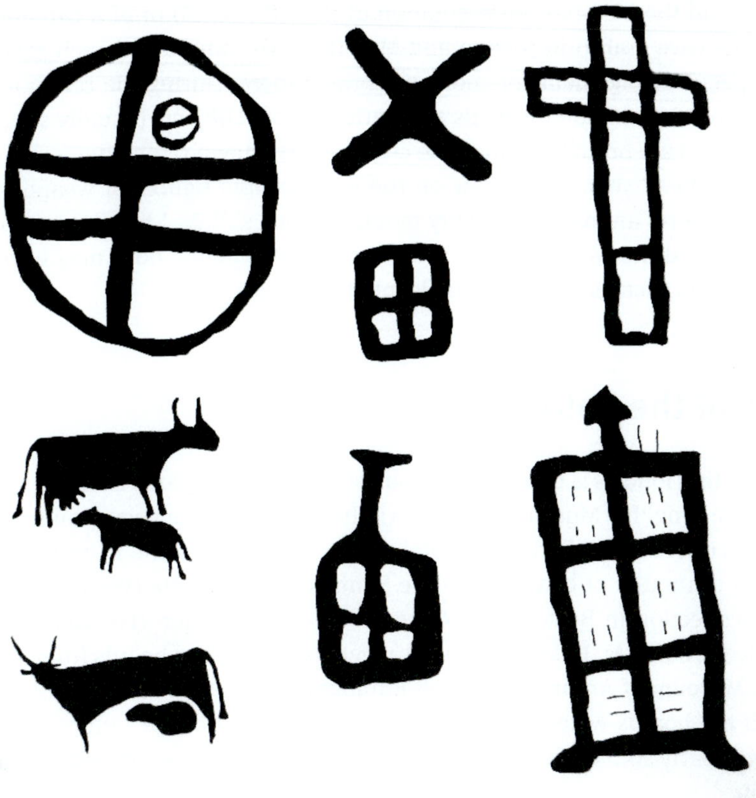

Fig. 13: Pictorial representation of the Qur'anic board (taken from Le Quellec 2004).

Fig. 14: Cave close to Songo village, Mali © Kirill Prokhorov.

Finally, one can find the pictorial representation of the QB (fig. 13) or of a Qur'anic satchel among the cave graffiti in Nigeria and Mali (fig. 14).[16] Studies of such caves take note of a special engagement ceremony celebrated there. During the ceremony various images are painted upon the walls and various sacred objects (usually signifiers of male prestige) are brought inside. The ceremony is known by the name *mba*. The boys (who are 14–17 years old) 'draw on the walls simple figures of weapons, shields, and pectiniform animals, especially mounted horses'.[17] As I have proposed elsewhere,[18] along with the weapons and the musical instruments, boys may bring other valuable objects such as Qur'anic boards into the caves.

3 Review of the Collections

The classification that follows is based on my analysis of Qur'anic boards from the following collections: Brooklyn Museum (New York); the Gallery of Sam Fogg (London), and the Musée du quai Branly (Paris). The total number of boards is 124. In addition, I have examined a number of boards from the Smithsonian Institute (Washington, D.C.), a private collection in Boston, the Nasser D. Khalili collection (London) and the photo-archive of Salah M. Hassan (Cornell University), as well as boards from the Peter the Great Museum of Anthropology and Ethnography (*The Kunstkamera* of St. Petersburg). All the boards were manufactured in the first half of the 14th century Hijri/late 19th and early 20th century CE, and represent various regions of West Africa and the Maghrib.

Each board has been analysed based on the following characteristics: script, decoration, form, colour, and ink. The resulting classification is presented by region, collection, and type of ornament used. In my typology, the boards can be grouped as follows: decorated and non-decorated (those containing only a text or individual letters); training and festive boards (the latter are used as certificates of graduation from the Qur'anic school).

Qur'anic manuscripts and ritual objects with 'magic' ornaments have provided additional information and were used as supporting material.

The largest collection of boards belongs to the Musée du quai Branly; it includes 114 samples. Another important collection was exhibited in 2007 at the Sam Fogg Gallery in London.[19] Consisting of only nine boards, the collection featured clearly distinguishable regional types, which has helped me to arrive at my own classifica-

16 See Grib 2009a, 22–34.

17 Le Quellec 2004, 74.

18 See Grib 2009, 22–34.

19 *Islam in Africa, An Exhibition on View: 9–26 October 2007* (London: Sam Fogg).

tion of the Qur'anic boards. For the present article, I have made a selection of the most representative samples of the regional types.

While Sam Fogg has exhibited four regional types of Qur'anic boards, the Mauritanian and Malian samples from the collection of the Musée du quai Branly look different; one board from Mali (at the Musée du quai Branly) has the same shape as the boards from Southern Morocco and Mauritania in Sam Fogg's collection, but it is turned upside down.

The most common ornaments painted on the boards I have studied belong to cosmogonic and textile motif-chains: this is the *palmette* as well as other solar symbols such as Nigerian 'Northern knot'. The same is true for Qur'anic manuscripts: African Qur'ans also feature a *palmette*, but here the design carries a somewhat different meaning, being known as the 'leper hand', or 'the Hand of Fatima', the sign of protection from evil.

Two objects present a special case as they do not fit into my main regional classification of Qur'anic boards. The first board is from the Brooklyn Museum in New York; the second is from the Musée du quai Branly in Paris. Both boards feature zoomorphic and anthropologic images of similar provenance. I discuss these two boards in a special section below.

3.1 Boards from the Collection of Sam Fogg (fig. 9; figs. 15–19)

The Sam Fogg Gallery has exhibited nine Qur'anic boards, the images and the description of which have been published in a catalogue.[20] These boards are examples of four regional types:

- Western Sudan (or Nigeria): N° 95682, N° 12303—two boards from the thirteenth century Hijri/nineteenth century CE
- Southern Morocco and Mauritania: N° 11987001, N° 9305—two boards from the first half of the fourteenth century Hijri/early twentieth century CE
- Somalia: N° 9306, N° 12304, N° 1770—three boards from the first half of the fourteenth century Hijri/early twentieth century CE
- Ethiopia: N°12301, N° 9569—two boards from the first half of the fourteenth century Hijri/early twentieth century CE

Type 1 (fig. 9): The board has a wider surface, which makes it more convenient to write the Qur'anic text. The base features two legs; in a number of instances it is concave. The handle has a horn- or a crescent-like top; in a number of instances the top rep-

20 Ibid.

licates the curve of the base. The top is fashioned from a different material such as leather, wood, or ivory; it attaches to the handle or directly to the board. The shape of the board is square-like and symmetrical. There are no holes in the handle, but it is nicely made, with a high degree of accuracy. The first type is representative of the boards from Western Sudan, and is also known as Nigerian.

In Nigeria and Sudan, *festive boards* are very popular. These boards are rather elegant and may be used during initiation ceremonies. Among the distinguishing characteristics of the festive board is the polished surface of the wood, use of dark black/brownish ink, text written in Western *Sūdānī* script, and the *zayyana* ornament.

Type 2 (figs. 15–16): The board is more narrow and elongated in shape; occasionally, the shape is symmetrical. The base is smoothly curved. The handle is long, thin, and jagged-edged, and has a hole for hanging. This type is representative of Southern Morocco and Mauritania.

Fig. 15: Qur'anic board from Mauritania N° 9305 © Sam Fogg.

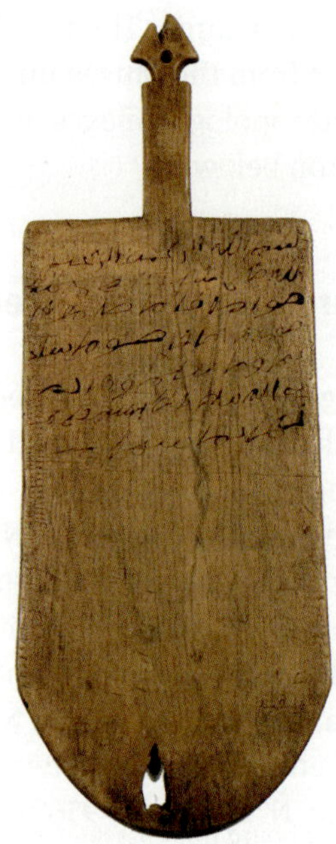

Fig. 16: Qur'anic board from Southern Morroco N° 11987001 © Sam Fogg.

Types 3 and 4: These two types include study boards from Ethiopia and Somalia. Both sets feature a handle with a hole.

Type 3 (figs. 17–18): Study boards from Somalia. The shape of the board is narrow and greatly elongated along the vertical axis. There can be single- and double-legged boards, as well as single- and double-headed boards. The script is relatively large, clear, and primitive on all of the boards examined; it betrays the non-professional hand of a pupil. It is apparent that the ink has been washed off repeatedly: one can see through to the bottom layer, plus there is a black tincture on the surface of the board which usually results from ink. The board is not straight but tilted. This is probably because wood is rather expensive and they use any wooden tablet available no matter what its quality. The single leg allows the rotation of the board during the class exercise. The student pushes the leg firmly into the ground: it is much easier to work it this way than to hold it horizontally on one's knees; also, the ink dries much faster and the hands do not get dirty as it is possible to reverse the board at any time. On the other hand, double-legged boards require repositioning, but they are better suited for practicing calligraphy. The boards from Somalia are the simplest ones among all the local groups examined; their sole purpose is to serve as training material in the Qur'anic school.

Fig. 17: Qur'anic board from Somalia N° 12304 © Sam Fogg.

Fig. 18: Qur'anic board from Somalia N° 1770 © Sam Fogg.

Type 4 (fig. 19): Study boards from Ethiopia. The shape of the board is more 'traditional': a wide rectangle with a single handle on top. These boards can have one or two legs. The proportions are similar to the boards from Nigeria and Western Sudan, 1:1.8.

Fig. 19: Qur'anic board from Ethiopia N° 9569 © Sam Fogg.

All four types (1–4) of the Qur'anic board have a vertical format.

Proportions (width:height):
- Somalia: 1:5, 1:8, 1:4.5
- Ethiopia: 1:1.8
- Western Sudan (or Nigerian type): 1:1.8
- Sothern Morocco and Mauritania: 1:5, 1:3

3.2 Further Observations

The proportions of the Ethiopian boards are identical: the size varies by 5 cm at most, while maintaining the same width-to-height ratio. The height of Somalian boards varies from 80 to 130 cm, and these boards are greatly elongated.

The number and size of the letters differ from one study board to another. This means that there is no universal manufacturing canon, in contrast with the festive board *allo zanyyana*. On some boards the handwriting is large, like in copy-books, whereas on other boards it is small. Excepting the boards from Nigeria (the so-called Western Sudanic or Nigerian type—Type 1), almost all the samples are the non-ornamented boards used for study. On the Western Sudanic board, the *zayyana* ornament and the text have been washed away.

Ethiopian boards all have the proportions 1:1.8; the height is almost twice the width. One possible conclusion is that there may be a formal rule behind this. In Northern Nigeria, the requirements for format, size, ornamentation, and script are fairly strict, whereas in Ethiopia and Somalia the rather primitive style indicates the hand of an unprofessional crafter.

The two boards from Nigeria/Western Sudan fall exactly into the type described by Hassan in his study; one can argue therefore that these two boards (N° 2 and N° 4) are typical *alluna*. The only difference is in the type of script used: it is another version of the *Maghribī* rather than *Hausa* hand; this script is not as geometrical and angular and has no sharp diagonals in the loops.

3.3 Boards from the Musée du quai Branly (figs. 20–24)

The collection consists of 114 Qur'anic boards from West Africa and the Maghrib.

For this article, five examples of the local types have been selected:

(1)

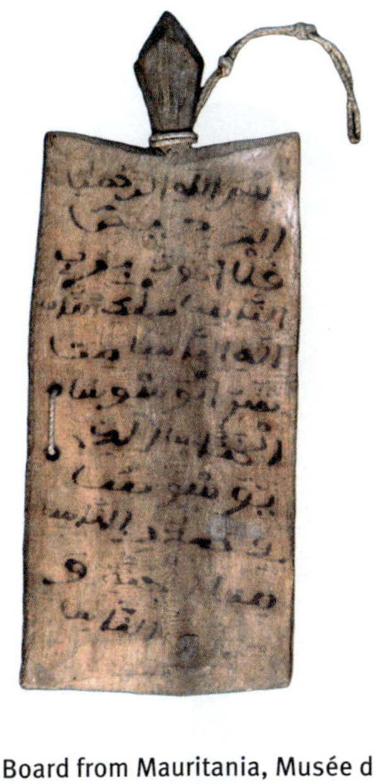

Fig. 20: Board from Mauritania, Musée du quai Branly Inv. N° 74.1962.0.158 © Musée du quai Branly.

(2)

Fig. 21: Board from Morroco, Musée du quai Branly Inv. N° 74.1962.0.148 © Musée du quai Branly.

(3)

Fig. 22: Board from Morroco, Musée du quai Branly Inv. N° 74.1962.0.127 © Musée du quai Branly.

(4)

Fig. 23: a–b. Board from Timbuktu, Musée du quai Branly Inv. N° 74.1962.0.1049 © Musée du quai Branly.

(5)

Fig. 24: a–b. Board from Hausa/Nigeria, Musée du quai Branly Inv. N° 73.1963.0.268 © Musée du quai Branly.

The most interesting board is N° 73.1963.0.268 from Hausa, Nigeria (fig. 24a–b). It is catalogued as a 'Planchette pyrogravée' rather than a 'Qur'anic board,' most likely because its distinct ritual character goes beyond Islamic usage.

The majority of the boards in this museum collection are localized to North Africa/Maghrib. Most were manufactured in Morocco and Algeria, but there are also a number of specimens from Mali, Nigeria, Sudan, Guinea, Somalia, Senegal, Madagascar and the Comoro Islands. All of the boards have a vertical format.

The prevailing ornamental motifs on these boards are the following: geometric (cosmogonic ornaments with solar characteristics—rectangle and circle-centred compositions), vegetative (floral motifs), architectural and carpet motifs. Among the architectural motifs, ornaments of a stylized arch/prayer niche (*miḥrāb*) are present.

One of the most common decorative patterns on the Moroccan boards is the solar *palmette* and the reverse swastika. Some boards feature no ornaments (e.g., from Mauritania). A number of boards feature no text from the Holy Qur'an, for example those labeled 'mystical' or 'ritual' boards. In Nigeria and Sudan, the most popular ornaments belong to the carpet type (*allo zayyan/charafa*).

After carefully examining the collection from the Musée du quai Branly, I have arrived at the following regional classification: boards from Western Sudan and Nigeria; Moroccan boards; boards from Mauritania; boards from Ethiopia; boards from Somalia, and boards from Mali. Within each type, one can identify further subtypes. All the boards come from the period of the 1st half of the 14th century Hijri/late 19 th and early 20th century CE.

The specific characteristics of these groups are as follows:

- Moroccan boards are easily identifiable by the white couch and the vertical rectangular or almost square-like shape. These boards are richly decorated. They are mostly festive boards with ornamentation similar to the manuscripts, including the *'unwān*.

- There are different subtypes of Moroccan boards, based on ornamentation: (a) architectural boards, (b) vegetative-cosmogonic boards (featuring bright, local colours in oil paint), (c) boards resembling bookbinding and lacquers, and (d) boards decorated predominantly with text.
- Boards without a couch represent the traditions of Somalia, Guinea (Futa-Djalon), and Djibuti. These boards are also typical for the *Fulani* tradition.
- Narrow, vertically elongated boards come from Mali, Mauritania, Algeria, the Comoro Islands, Madagascar, and Guinea.
- Malian boards (from Timbuktu) can be distinguished by (1) shape: these are small-size boards with a single leg in the middle of the base; (2) decoration: the board drawings are in monochrome and consist of cosmogonic symbols (the sun, crescent, and stars) as well as vegetative elements—in addition to the Qur'anic text; on some boards ornaments are painted in colour; (3) the script is non-calligraphic.

Furthermore, non-decorated boards are always study boards rather than festive boards. Some boards feature a special 'plaiting' ornamental motif.

Board N° 74.1962.0.127 features a remarkable Temple Image (fig. 22) that deserves special discussion. The board comes from Fes in Morocco, and it combines all the ornamental types discussed earlier: architectural, geometric (cosmogonic symbols), vegetative, and calligraphic. While the composition on the board is pictorial in nature, it still follows the 'carpet' type. The board was made in the first half of the 14th century Hijri/early 20th century CE. The wood is covered with white couch. The text is written in brown ink. The illumination has been done in natural blue and red but the colours have faded. The board has a vertical format, and its size is 28.5 x 40.6 x 1.12 cm. There is no handle; on the top of the board there is a hole for hanging.

The board is decorated with a Temple Image of a common type: an arch between two columns marks off a complex composition representing an inner court of a mosque. The composition consists of three levels: (1) upper level: a minaret and roofs; (2) lower level: a pool for ritual washing in the centre of a court; (3) central level: connecting the previous two, shown in perspective. The two columns have ornaments and are painted in colour. The background within the composition is completely filled with text. The composition has a square-like rectangular frame filled with ornaments. More specifically, the frame is crowned by eight six-pointed stars, and above the frame there is a double *Basmala* inscription filling the empty space. On the reverse side of the board there are Qur'anic verses.

3.4 Exceptions from the Main Typology

3.4.1 The 'Brooklyn Board' (fig. 25a–e)

The collection of the Brooklyn Museum in New York City contains a rather unique Qur'anic board with zoomorphic and anthropomorphic images. The museum has listed the board's place of origin as Sudan, probably based on its place of purchase rather than the type of script, format, or decoration. As I have argued elsewhere,[21] the board belongs rather to the sacred caves in Marghi, Nigeria. I base my conclusion on a consideration of the text and ornaments, as well as the similarities between the images on the board and in rock graffiti in that particular area. As I have suggested, ornament can function as an *ex libris* of the sacred place where the object was manufactured and/or stored. I call this a *genizah* symbol (see above).

On the front of the board there are images of scorpions, one snake, and five tents. The tents may refer to a wedding camp, which would be further indicated by the flagpoles; on the other hand, the five triangles could be a schematic depiction of the 'leper hand' motif. The front of the board (fig. 25a) contains a fragment of the 97th *Sūra al-Qadr*, 'the Night of Power' (fig. 25c).[22] The textile ornament *zayyana* also occurs on the front of the board, framing the *Sūra*.

On the reverse side of the board (fig. 25b; fig. 25d), there are images of hunters as well as zoomorphic images including scorpions. Horsemen pursue an antelope and a giraffe (or it may be an ostrich). In West Africa, scorpions and snakes (fig. 25e) are frequently used as protective symbols against the evil eye.

The Brooklyn board exemplifies the phenomenon of the Islamic 'intrusion' into the field of indigenous ritual, and vice versa, thus forming a syncretic artistic language. Anthropomorphic and zoomorphic images circumvent the traditional Islamic ban on images within the domain of the sacred, including the Qur'anic text. This prohibition does not seem to apply to African Qur'anic boards, probably because they are not deemed to be exclusively Islamic objects.

---

21 See Grib 2009a, 22–34.

22 *Sūra al-Qadr* is commonly used for protection, which includes marriage and engagement ceremonies. This also points to the Marghi region, because of the *mba* wedding ceremony. Since students would never write this particular *Sūra* in a *zayyana*-like fashion, the board does not have an educational function. Furthermore, students are not allowed to put images on a training *allo* so as to distinguish them from a true master-healer. Apart from this, there are *'Ya Fattah, Ya 'Alim,'* the names of God (*'asmā' al-ḥusnā*) written on the top, which also speaks of a ritual purpose for this particular board.

Fig. 25: a–e. The 'Brooklyn Board', Brooklyn Museum Inv. N° 22.231 (a–b © Brooklyn Museum; c–e © Anastasia Grib).

The ornamentation on the board falls into three types:

1. the traditional 'carpet page' ornament, which is a version of the *zayyana* ornament;
2. the 'Temple Image' in the shape of a wedding camp (the second protective meaning here is 'leper hand' or a modified *khamsa*);
3. zoomorphic and anthropomorphic images like the ones on the rock graffiti: according to Le Quellec, their iconography dates back to the Neolithic period.

The Qur'anic text does not seem to belong in this particular decorative context. It was difficult to provide an accurate point of origin for the board, probably because the elements are painted chaotically and seem to be following their own internal logic. Yet one may argue, as I did, that the images from both sides of the board belong to the 'Nigerian graphic repertoire'. I have also noticed that these images are akin to those found among rock graffiti in the Marghi region of Nigeria, namely:

- the hunting scene,
- snakes and scorpions,
- the man with a spear (modified to a man with a flagpole).[23]

While the area with the graffiti covers over 400 sq. km, the Brooklyn board may have been manufactured near the villages of Uvu and Wondi, where there is a record of an *mba* ceremony from the year 1962.[24] As described by Le Quellec, there were a total of about ten caves with images belonging to the 'Nigerian graphic repertoire'; these images were directly involved in the *mba* ceremony, being drawn on the walls of the cave after the procession. Similar caves exist in Birnin Kudu. Two of the depictions from the caves of Rumfar Kurosha in Marghi have been interpreted by Le Quellec as boards from Qur'anic schools.[25]

3.4.2 Planchette Pyrogravée from the Musée du Quai Branly, Inv. N° 73.1963.0.268 (fig. 24a–b).

In the collection of the Musée du quai Branly there is only one board attributed to the Hausa region in Nigeria, yet the format and the decoration of the board are not typical for this particular region.

The board has a markedly ritual, non-Qur'anic character. The text is absent, but there is a series of nicely done images consisting of animals, scorpions, and human beings. All of these come as couples with their apparent gender features, both male and female.

23 Grib 2009a, 32.
24 Vaughan 1962, 49–52.
25 Le Quellec 2004, 77.

Fig. 26: Nigerian board from the collection of the Musée du quai Branly N° 71.1963.2.265 © Anastasia Grib (Courtesy: Musée du quai Branly).

One may see a connection with the 'Nigerian graphic repertoire', as one does on the board from the Brooklyn museum. I therefore believe this board to have originated in the Ceremony Mountains in the Marghi region.

The zoomorphic and anthropomorphic images are similar to the ones on the Brooklyn board, but the character of these images is different, including their silhouettes, proportions, technique (etched drawing on the board from the Musée du quai Branly vs. ink on the Brooklyn board), arrangement of figures (ordered procession with all the creatures presented as couples on the board from the Musée du quai Branly vs. chaotic composition and a Qur'anic text on the Brooklyn board). In other words, the depictions on the board from the Musée du quai Branly point to its ritual purpose and use.

The two boards reproduce the images that occur frequently in Nigerian (Yoruba and Hausa) wedding textiles, images of animals, insects, and human beings. On the other hand, wedding cloth can also feature images of the boards. This cross-use demonstrates the direct involvement of the board in wedding ceremonies.

To support my theory that both the Brooklyn board and the Planchette pyrogravée come from the Marghi region one can refer to yet another Nigerian board from the collection of the Musée du quai Branly (fig. 26). The board features the brownish Nigerian ink, a typical Nigerian *zayyana* decoration, and iconography similar to that of the Brooklyn board: a hunter, a snake and a tent.

4 Qur'anic Boards from West and North Africa: A Summarized Classification of Major Types

Table 1: Boards from Morocco

Local Type	Format	Material/ treatment	Handle	Decoration/illumination
boards from Morocco	vertical format, almost square in size	white couch	most have no handle, but there is a special hole for hanging in the upper section of the board	rich in decoration; decoration matches illumination of the Qur'anic manuscripts from the same region; use of *'unwān*

Subtypes *boards differ by the style of decoration and by ornamental motifs*				
(a) architectural				the reduced Temple Image
(b) vegetative-cosmogonic		bright, local colours, oil		
(c) similar to the binding cover		background with the ornamental filling		decoration resembles laquer bindings (with a diamond-shape in the centre of the ornamental field)
(d) training		monochrome colours prevail		less decorated than other subtypes; the main background is filled with the text

Table 2: Boards from Mali

Local type	**Format**	**Material/ treatment**	**Handle/base in the shape of supportive legs**	**Decoration/illumination**
boards from Mali	small format	no couch; in a number of boards ornaments are filled with coloured pigments	a comfortable long conic handle extending from the centre of the base; rounded top is frequent	monochrome cosmogonic symbols (solar symbols, crescent, and stars) and vegetative elements side by side with Qur'anic text; non-calligraphic script (used in training); motifs: 'plaiting' (frequently)
Subtypes *boards differ by format only*				
(a)	narrow, vertical, elongated			
(b)	small format			

Table 3: Boards from Mauritania

Local type	**Format**	**Material/ treatment**	**Handle/base in the shape of supportive legs**	**Decoration/ illumination**
boards from Mauritania	narrow, vertically elongated; the shape is not always symmetrical	no couch	smoothly rounded base; long thin handle with a hole for hanging; no additional festive decorations on the handle	no decoration

Table 4: Boards from Nigeria

Local type	**Format**	**Material/ treatment**	**Handle**	**Decoration/illumination**
boards from Nigeria (also identified as the *Western Sūdānī* type)	vertical symmetrical format, rectangular; proportions 1:1.8 (width:height)	additional decorative elements are attached to the handle; such elements are manufactured from a different material, e.g., leather, wood, or bone; the wood is polished	there is a handle; the base has legs; there is no hole for hanging	fine work
Subtypes *boards differ by decoration, shape and ornamental motifs*				
Hausa boards:				

(a) festive boards *allo zayyana*		bright, local colours, natural pigments	always topped by the *saukar fari* decoration (feast marking the completion of study at a Qur'anic school)	*zayyana* in the form of a carpet ornament; the text is the first *Sūra al-Fātiha* and/or an excerpt from the second *Sūra al-Baqara*
(b) draft, training boards *allo*			simple handle with no decor and no additional top	no decoration, only quotes from the Qur'an including drafts
Ritual boards (Marghi)	vertical and horizontal boards	natural colours, brown ink		zoomorphic and anthropomorphic images

Table 5: Boards from Somalia

Local type	**Format**	**Material/ treatment**	**Handle/base in the shape of supportive legs**	**Decoration/illumination**
boards from Somalia	narrow, vertically elongated	no couch	handle has a hole for hanging	study boards without ornamentation; the most basic type of QB
Subtypes *boards differ by the number of legs and handles*				
(a)			two legs and two handles	
(b)			single leg and single handle	

Table 6: Boards from Ethiopia

Local type	**Format**	**Material/ treatment**	**Handle/base in the shape of supportive legs**	**Decoration/illumination**
boards from Ethiopia	'traditional' shape (vertical and wide), proportions 1:1.8 (w:h)	no couch; darkened due to washing off the ink	handle with hole for hanging	study boards without ornamentation
Subtypes *boards differ by the quantity of legs*				
(a)			two legs	
(b)			single leg	

Conclusions

- In West and North Africa, one finds a semantic unity between two main types of objects belonging to the Material culture of the Qur'an: the Qur'anic manuscripts and the Qur'anic boards.
- The shape of QB is determined by its function.
- The QB is the material symbol of the Temple and the material image of the Text/ the Word.
- There are several types of ornamentation: architectural, geometric, carpet, figurative. An additional type of ornament is floral/vegetative.
- Motifs on both Qur'anic manuscripts and Qur'anic boards can be linked to local religious practices.
- The function of the cave is as a *genizah*, a repository of sacred objects including the QB. The place-tag is 'attached' to the manuscript and board by means of an ornament, which serves as the functional analogue of an *ex libris* in a book.
- Anthropomorphic and zoomorphic images on *Hausa* Qur'anic boards belong to the 'Nigerian graphic repertoire'.

Bibliography

Bakhtin, Mikhail M. (1984), *Rabelais and His World*, Bloomington.

Bivar, A. D. H. (1959), "Arabic Documents of Northern Nigeria", in: *Bulletin of the School of Oriental and African Studies, University of London* 22 (1/3), 324–349.

Blackmun Visona, Monika (2001), *A History of Art in Africa*, London.

Blair, Sheila S. (2008), "Arabic Calligraphy in West Africa", in: Shamil Jeppie and Souleymane Bachir Diagne (eds.), *The Meanings of Timbuktu*, 59–75.

Cifuentes, Frédérique (2008), "Sufi Sheikhs, Sheikhas, and Saints of the Sudan", in: *African Arts* 41 (2), 50–59.

Dobronravin, Nikolai (1999), *Arabograficheskaia pis'mennaia traditsiia Zapadnoi Afriki*, St. Petersburg.

Emerson, Caryl/Holquist, Michael (eds.) (1981), *The Dialogic Imagination. Four Essays by M.M. Bakhtin*, Austin.

Grib, Anastasia S. (2009a), "In-Between the 'Elite' and the 'Pagan'. Qur'anic Boards from West Africa", in: *Manuscripta Orientalia* 15 (1), 22–34.

Grib, Anastasia S. (2009b), *Problema lokalizatsii i geneologiia ornamentov koranicheskoi doski iz sobraniia Bruklinskogo muzeia (New York)* (St. Petersburg Museum of Anthropology and Ethnography, Russian Academy of Sciences), St. Petersburg https://www.academia.edu/6606378/The_Qur_anic_Board_from_the_Collection_of_the_Brooklyn_Museum_New_York_City_Localization_and_the_Genealogy_of_the_Ornaments (last accessed: 13.06.2016).

Hassan, Salah M. (1992), *Art and Islamic Literacy Among the Hausa of Northern Nigeria*, Lewiston.

Kirk-Greene, Anthony Hamilton Millard (1963), *Decorated Houses in a Northern City*, Kaduna.

Le Quellec, Jean-Loïc (2004), *Rock Art in Africa. Mythology and Legend*, Paris.

Moughtin, James Cliff (1985), *Hausa Architecture* (Ethnographic Arts and Culture Series 6), London.

Vaughan, James H. (1962), "Rock Paintings and Rock Gongs Among the Marghi of Nigeria", in: *Man. The Journal of the Royal Anthropological Institute of Great Britain and Ireland* 62 (4), 49–52.

Notes on Contributors

Sara Campanelli

(Dr.) Sara Campanelli graduated in Classics and specialized in Greek Epigraphy at 'Sapienza' – University of Rome, where she was also awarded her PhD in Ancient History. Since the Post-doctoral Fellowship at the Collaborative Research Center 933 – 'Material Text Cultures' at Heidelberg University, she has collaborated on EDR – Epigraphic Database Rome for the digital edition of the Jewish inscriptions in the city of Rome.

Her main field of research is Greek religion of Hellenistic and Roman periods. The epigraphic perspective allows her to focus on the material aspects of the religious phenomena, paying particular attention to ritual norms and practices, social dynamics, structure and organization of the cult groups, architectural layout and spatial arrangement of the cult places. For this reason, she has undertaken research on cult associations and cult foundations meant as structured religious expressions, which show the complex interaction of ritual, juridical, financial and social factors at the base of ancient religion.

Her geographic areas of interest are the Dodecanese (in particular the island of Cos) and Asia Minor, but she also deals with Greek inscriptions in the city of Rome.

Agnès Garcia-Ventura

(Dr.) Agnès Garcia-Ventura pursued a Degree in Humanities at the University Autònoma of Barcelona (Spain) with honours. After that, she specialized in Ancient Near Eastern Studies and Assyriology with two Master's Degrees at the University Pompeu Fabra and at the University of Barcelona (both in Barcelona, Spain), respectively. November 2012 she was awarded her PhD in History at the University Pompeu Fabra for her Dissertation entitled "Work and textile production during the Third Dynasty of Ur".

As Post-Doc she was awarded with a 'Material Text Cultures Research Fellow' at the Collaborative Research Center 933 – 'Material Text Cultures' at Heidelberg University, Germany (November 2012–May 2013). She also worked for one academic year as Assistant Professor at the University Autònoma of Barcelona, Spain (2013–2014). Since October 2014 she is Post-doctoral Researcher (with a Beatriu de Pinós fellowship, awarded by the Government of Catalonia, Spain) at the Sapienza, Università degli Studi di Roma, Italy.

Her main areas of interest are gender studies (especially their application to Assyriology), historiography of Ancient Near Eastern studies, ancient musical performance (in both Mesopotamia and the Phoenician and Punic contexts) and the organisation of work in Mesopotamia.

Anastasia Grib

(Dr.) Anastasia Grib is General Editor of A Guide to Islamic Calligraphy, a joint project with the Hermitage Museum. Her research career began in 2001 at the Metropolitan

Museum of Art in New York City where she analysed the collection of early Kufic manuscripts from the archive of Ernst Herzfeld. Shortly after, Anastasia entered the PhD program in Islamic Art at the State Hermitage Museum in St. Petersburg, Russia, under Prof. Dr. Mikhail Piotrovsky. In 2012, she had a viva voce of her dissertation at the Institute of Anthropology and Ethnography of the Russian Academy of Sciences (Kunstkamera); the title is "Local Traditions of the Material Culture of the Qur'an in the Islamic Centers of West and North Africa: Calligraphy and Ornaments of the Qur'anic Manuscripts and Boards, 19–20 c." She held a Post-doctoral Fellowship at the Sainsbury Research Unit of the University of East Anglia. Dr. Grib has produced a special issue of the Hermitage Magazine featuring the art and culture of Islam and, most recently, a catalogue on Qur'anic boards from the Galerie Frank Van Craen (Brussels). Her current project is a book about the Qur'anic board tradition in the larger Islamic world.

Flavia Manservigi
(Dr.) Flavia Manservigi graduated in Archaeology at the University of Bologna in 2008, and holds a PhD in 'Culture Letterarie, Filologiche, Storiche' from the same University; her thesis was about the 12th century Bolognese notary. She works for 'Genus Bononiae. Musei nella Città', where she coordinates the Secretary Office of the 'Festival della Scienza Medica', and where she coordinates the activity of the 'Centro Studi Sara Valesio'. She is an 'Expert' in Latin Palaeography, and she is an active member of the RAM (Ricerche e Analisi Manoscritti) Study Center at the University of Bologna. She has been a member of the Scientific Committee of the International Center of Sindonology of Turin since April 2016.

Melania Mezzetti
(Dr.) Melania Mezzetti got a degree in 'Arts and humanities', and holds a PhD in 'Romance philology and medieval culture' from the University of Bologna. She has been an 'Expert' in Latin Palaeography since 2010. She got a degree in 'Archival science, Palaeography and Diplomatics' at the State Archive of Bologna. She worked as an archivist for the project 'Una città per gli archivi' in two archives of Bologna. In 2013 she held a Post-doctoral Fellowship at the Collaborative Research Center 933 'Material Text Cultures' at Heidelberg University. She is currently a Tutor at the School of 'Arts, Humanities, and Cultural Heritage' of the University of Bologna.

Antonio J. Morales
(Dr.) Antonio J. Morales is University Lecturer in the 'Ägyptologisches Seminar' at Freie Universität Berlin and a Research Associate in the project A02 'Altägyptische Philologie' of the SFB 980 in the same institution. He is the director of the Freie Universität Berlin Mission to Deir el-Bahari, which aims at the documentation, study, and publication of Middle Kingdom tombs in Thebes (Luxor). His research focuses on religious literature and language in ancient Egypt, especially concerning the construction and

transmission of mortuary compositions such as the Pyramid Texts and Coffin Texts, and the history, society, and literature of the Middle Kingdom. Recent publications: The Pyramid Texts of Nut in the Old and Middle Kingdoms (Hamburg 2016); "Unraveling a thread: textual programs with Pyramid Texts in the Late Period tombs", in Studies in Ancient Egyptian Funerary Literature, ed. by L. Díaz-Iglesias and S. Bickel (Fribourg 2016); "Text-building and transmission of Pyramid Texts in the third millennium BCE: iteration, objectification, and change", Journal of Ancient Near Eastern Religions (2016); "Iteration, Innovation und Dekorum in Opferlisten des Alten Reichs. Zur Vorgeschichte der Pyramidentexte", Zeitschrift für Ägyptische Sprache und Altertumskunde 142.1 (2015); and Experiencing Power, Generating Authority: Cosmos, Politics and the Ideology of Kingship in Ancient Egypt and Mesopotamia, co-edited with J. Hill and Ph. Jones (Philadelphia 2013).

Nathan Morello

(Dr.) Nathan Morello graduated in Semitic Philology at the University of Turin (Italy). He has received his PhD in Assyriology and History of the Ancient Near East at the University of Udine, with a dissertation on the frontiers of the Assyrian Empire, focused on the role of fortified settlements.

After a Post-doctoral fellowship at the University of Leipzig and since the Post-doctoral Fellowship at the Collaborative Research Center 933 "Material Text Cultures" at Heidelberg University, he has been working at LMU University of Munich, at the *Reallexikon der Assyriologie und Vorderasiatischen Archäologie*, and as part of the research staff of the *Alexander von Humboldt-Professorship for the Ancient History of the Near and Middle East*, as member of the *Munich Open-Access Cuneiform Corpus Initiative* (MOCCI). Since 2013, he is part of the research team of the Archaeological Excavation Project of Ashdod-Yam (Israel), as architecture analyst.

His main field of research gravitates around the history of the Assyrian Empire, with particular interests for the dynamics of its frontiers, for the study of the Assyrian Royal Inscriptions corpus, and for the analysis of the Assyrian monuments and their perception.

Index

1 Subjects and Names

2 Places

3 Periods